NANOTECHNOLOGY SCIENCE AND TECHNOLOGY

NANOWIRES

PROPERTIES, SYNTHESIS AND APPLICATIONS

NANOTECHNOLOGY SCIENCE AND TECHNOLOGY

Additional books in this series can be found on Nova's website under the Series tab.

Additional E-books in this series can be found on Nova's website under the E-book tab.

NANOTECHNOLOGY SCIENCE AND TECHNOLOGY

NANOWIRES

PROPERTIES, SYNTHESIS AND APPLICATIONS

VINCENT LEFÉVRE
EDITOR

Nova Science Publishers, Inc.
New York

Library of Congress Cataloging-in-Publication Data

Nanowires : properties, synthesis, and applications / editor, Vincent Lefhvre.
p. ; cm.
Includes bibliographical references and index.
ISBN 978-1-61470-129-3 (hardcover)
1. Nanowires. I. Lefhvre, Vincent, 1964-
[DNLM: 1. Nanowires--chemistry. 2. Biosensing Techniques. 3. Nanostructures--chemistry. QT 36.5]
TK7874.85.N365 2011
620'.5--dc23

2011020779

Published by Nova Science Publishers, Inc. † New York

CONTENTS

PREFACE

Nanowires are nanostructures that have a thickness or diameter constrained to tens of nanometers or less and an unconstrained length. In addition, many different types of nanowires exist, including metallic, semiconducting and insulating. This book presents current research in the study of the properties, synthesis and application of nanowires. Topics discussed include semiconductor nanowires and heterostructure based gas sensors; transport properties of nanostructured materials; nanowire array electrodes in biosensor applications and analogies between metallic nanowires and carbon nanotubes.

Chapter 1 - This chapter encompasses a review of the design, synthesis, fabrication and applications of nanowires derived using the bottom up approach as well as top-down lithography approaches. The fabrication of nanowires based on self-assembly processes of oligonucleotides, peptides, lipids and micro-organisms such as viruses are discussed in detail. Further, fabrication methods based on non-lithographic techniques as well as patterning methods have been compared. Although methods like molecular beam epitaxy, chemical vapor deposition and laser ablation have been used for several years for the growth of nanowires, the use of biological nanowires as templates combining chemical tuning and biological properties allows for the development of building blocks of unique specificity and desired properties due to the presence of inherent functionalities within the biological moieties. Further, biological nanowires can often be prepared by environmentally friendly methods at room temperature, without the use of harsh chemical conditions. For the development of nanodevices for sensors, nanoelectronics or optoelectronics applications, frequently the nanowires are conjugated with metal or semi-conductor nanocrystals, which are grown directly on the nanowire surfaces. In some cases, the growth of the specific nanocrystals on the nanowires is promoted by the recognition sequences of the oligonucleotides or the peptides used. Furthermore the morphologies of the nanocrystals formed can be fine-tuned depending upon the diameters of the nanowires. In recent times, nanotubes, nanofibers and nanowires have also been found to be useful for biomedical applications for the development of tissue engineering materials, or as biomarkers for diseases and as biosensors. Here in, some examples where nanowires and nanotubes have been used for inducing the growth of bone cells, regeneration of nerve fibers or for detecting neuronal activity adjacent to blood vessels, and a variety of other biological and chemical monitoring processes are discussed. A brief overview of and the significance of preparing highly biocompatible nanowires is also presented. Overall, a combination of nanoelectronic, optical and magnetic methods allow for the fabrication of highly sensitive and functional devices and opens new doors for enabling a wide range of applications of bio-based nanowires.

Chapter 2 - In recent years, significant interest has emerged in the synthesis of nanoscale materials owing to their superior and enhanced functional properties. The most attractive class of materials for functional nanodevices are based on semiconductors, in particular metal oxides based nanostructures. For application as gas sensor, nanostructures offer several advantages including high surface area-to-volume ratio, dimensions comparable to the extension of the surface charge region, relatively simple preparation methods allowing large-scale production and sensors that are convenient to use. Of the various nanostructures, nanowires (NWs) are particularly useful for gas sensing application as they offer various advantages. These include pathway for electron transfer (length of NWs), enhanced and tunable surface reactivity implying possible room temperature operation, faster response and recovery time and ease of fabrication and manipulation. The smaller dimension further enhances the possibility of high integration density thereby leading to smaller size of the actual sensor device and low power consumption. All the above mentioned features definitely make NWs a potential candidate for the development and realization of next generation sensing devices. This chapter deals with the progress made towards the effective use of semiconducting NWs for achieving superior sensing performance has been critically addressed. In particular, different sensor configurations like single-NW based, multiple-NW based, NW films and as grown films have been investigated in detail. Besides, the result obtained using the investigations of doping element, incorporation into the polymer matrix and use of heterostructures on improvement in the sensing characteristics has been elaborated with examples. Steps taken towards commercialization of ultimate sensor device and the major obstacles involved are also discussed.

Chapter 3 - This review focuses on the transport properties of nanostructured materials. The complete theoretical understanding of transport properties such as electrical resistivity, thermal conductivity and thermoelectric power is presented in this article. Resistivity in metallic phase of Zn composite nanowires is analyzed within the framework of Bloch-Gruneisen (BG) model of resistivity as well the effect of electron-electron scattering is evaluated. The resistivity in Semiconducting phase of Zn composite nanowires is discussed with small polaron conduction (SPC) model. Mott expression is used to generate the electron diffusive thermoelectric power ($S_c^{diff.}$) and phonon drag thermoelectric power (S_{ph}^{drag}) is calculated within the relaxation time approximation when thermoelectric power is limited by the scattering of phonons with impurities, grain boundaries, charge careers and phonons in the nanowires. The thermal conductivity (κ) and S_{ph}^{drag} shows anomalous temperature dependent behaviour, which is an artifact of various operating scattering mechanisms. The anomalies are well accounted in terms of interaction among the phonons-impurity, phonon-grain boundaries, phonon-electron and the umklapp scattering. Furthermore, the effect of embedding nanoparticles on thermal conductivity of crystalline semiconductors is presented under phonon scattering mechanism. Numerical results obtained from the theoretical analysis are also compared with experimental results.

Chapter 4 - This chapter investigates the area of nanowire based biosensors for different analyte detection. The electrochemical properties of two different platforms made of gold nanowire array (NAE) and Pt nanoparticle modified gold nanowire array (PtNP/NAE) are examined in details for the detection of hydrogen peroxide. Third-generation H_2O_2 biosensor is prepared by covalent immobilization of horseradish peroxidase (HRP) on the self-assembled monolayer modified NAEs. Also PtNP/NAEs are used to fabricate oxidase enzyme based biosensor. A comparative study between these two systems has been performed and a state of the art comparison with these two systems is reported. Moreover, a detailed electrochemical and physical characterization has been performed using cyclic voltammetry, amperometry, scanning electron microscopy (SEM) and transmission electron microscopy (TEM).

Chapter 5 - In relation to quantum transport, the more remarkable analogies between metallic nanowires and carbon nanotubes are discussed from both qualitative and quantitative standpoints. In fact, the authors investigate in what aspects of electron conductance metallic nanowires are similar to metallic multiwalled carbon nanotubes. Within this framework, the authors establish a general mathematical relationship for the electrical conductance of both perfect and imperfect metallic nanowires and multiwalled carbon nanotubes.

Chapter 6 - Carbon Nanotubes (CNTs), especially Single Walled Carbon Nanotubes (SWCNTs) are good candidates for the devices of nano-electromechanical systems (NEMS). Intuitively, the elasticity of SWCNT and electromechanical coupling of SWCNT should be chirality dependent. By carefully studying the structure and symmetry of SWCNT, an analytical approach to obtain SWCNT's chirality-curvature dependent anharmonic anisotropic elastic constants is developed, the harmonic elastic constants, $c_{11}, c_{12}, \ldots,$ c_{66} and anharmonic elastic constants $c_{111}, c_{112}, \ldots, c_{666}$ are expressed analytically as series expansion of curvature parameter (r_0/R), i.e., the ratio of carboncarbon bond length to the radius of tube. The authors found constants reflecting the coupling between axial strain and circumferential strain, such as c_{16} and c_{116} have terms proportional to $(r_0/R)^2 \sin(6\theta_c)$, which imply the asymmetric axial-strain induced torsion (a-SIT) phenomenon. Base on the authors' analytical method, the authors reproduced recently reported Molecular Dynamic (MD) simulations on asymmetric a-SIT accurately. Present method can be used to study various chirality-curvature dependent electromechanical coupling phenomena of SWCNTs.

Chapter 7 - Growth of multicomponent nanostructures (or "nanoscale heterostructures") is critical for the development of complex nanodevices and multifunctional platforms. However, to fully understand the field of nanoscale heterostructures, it is important for nanotechnologists and nanoscientists to fundamentally understand single-component nanostructures. This review article provides an outline on critical concepts in single-component nanowires that could be further utilized to fabricate multicomponent axially heterostructured nanowires. This article comprises of an introduction focused on nanostructures and their applications, followed by a classification of nanostructures with a particular desire to develop and study nanowires, growth and characterization studies of single-component nanowires (semiconducting, metallic, etc.), using this knowledge to design and fabricate multicomponent axially heterostructured nanowires, characterization advances for understanding such nanowires, and applications of axially heterostructured nanowires. This article attempts to introduce a complete roadmap for the development of multicomponent axially heterostructured nanowires from single-component nanowires and approaches to integrate the former in device architectures. Such nanowire systems hold immense promise for applications in chemical and biological sensors, imaging tools, nanomagnetics, optoelectronics, and nanoelectronics.

In: Nanowires: Properties, Synthesis and Applications
Editor: Vincent Lefevre
ISBN: 978-1-61470-129-3

Chapter 1

FROM SELF-ASSEMBLY TO NANOELECTRONICS, SENSORS AND MEDICINE - A BIOLOGICAL APPROACH FOR THE GROWTH AND APPLICATIONS OF NANOWIRES – A REVIEW

Stacey N. Barnaby, Nako Nakatsuka and Ipsita A. Banerjee [*]
Department of Chemistry, Fordham University, Bronx NY, U. S.

ABSTRACT

This chapter encompasses a review of the design, synthesis, fabrication and applications of nanowires derived using the bottom up approach as well as top-down lithography approaches. The fabrication of nanowires based on self-assembly processes of oligonucleotides, peptides, lipids and micro-organisms such as viruses are discussed in detail. Further, fabrication methods based on non-lithographic techniques as well as patterning methods have been compared. Although methods like molecular beam epitaxy, chemical vapor deposition and laser ablation have been used for several years for the growth of nanowires, the use of biological nanowires as templates combining chemical tuning and biological properties allows for the development of building blocks of unique specificity and desired properties due to the presence of inherent functionalities within the biological moieties. Further, biological nanowires can often be prepared by environmentally friendly methods at room temperature, without the use of harsh chemical conditions. For the development of nanodevices for sensors, nanoelectronics or optoelectronics applications, frequently the nanowires are conjugated with metal or semi-conductor nanocrystals, which are grown directly on the nanowire surfaces. In some cases, the growth of the specific nanocrystals on the nanowires is promoted by the recognition sequences of the oligonucleotides or the peptides used. Furthermore the morphologies of the nanocrystals formed can be fine-tuned depending upon the diameters of the nanowires. In recent times, nanotubes, nanofibers and nanowires have also been

[*] E-mail address: banerjee@fordham.edu

found to be useful for biomedical applications for the development of tissue engineering materials, or as biomarkers for diseases and as biosensors. Here in, some examples where nanowires and nanotubes have been used for inducing the growth of bone cells, regeneration of nerve fibers or for detecting neuronal activity adjacent to blood vessels, and a variety of other biological and chemical monitoring processes are discussed. A brief overview of and the significance of preparing highly biocompatible nanowires is also presented. Overall, a combination of nanoelectronic, optical and magnetic methods allow for the fabrication of highly sensitive and functional devices and opens new doors for enabling a wide range of applications of bio-based nanowires.

INTRODUCTION

The convergence of basic sciences such as chemistry, physics, biology as well as engineering and materials science has triggered an exhilarating research focus in the field of nanobiotechnology [1]. In particular, by utilizing the bottom-up approach, specific nanomaterials of a range of shapes and sizes can be tuned for various device fabrications by exerting control of conditions at the molecular level [2]. The bottom up approach for manipulation of individual atoms and molecules was proposed by Richard Feynman in 1959 in his well-known statement: “There’s plenty of room at the bottom” [3]. Since then, the self-assembly approach has been utilized as an extremely versatile method for the preparation of building blocks for nanodevices using biomolecules such as peptides [4], viruses [5], oligonucleotides [6], DNA, [7] and lipids [8]. The multitude of advances in the design, synthesis and characterization of materials at the nanoscale allows scientists to recognize and manipulate interactions for the formation of nanomaterials such as nanofibers [9], nanocrystals [10], nanowires [11], nanoparticles [12], and nanobelts [13].

Of the various types of nanomaterials, nanowires have attracted considerable attention. They are projected to play important roles as inter-connects and practical components in the fabrication of nanoscale electronic, optoelectronic, electrochemical and electromechanical devices with potential applications as memory storage devices [14-15], biosensors [16], nanofluidic circuits [17], biomaterials, biomedical devices, electronics [18], barcodes for biological multiplexing [19], and photothermal sensors [20]. Fascinating properties arise from the extreme size quantization effects in nanowires. For example, it has recently been reported that Zhu and co-workers have developed coiled silicon nanowires that have the potential to be stretched beyond their original length, leading to the possibility of incorporating stretchable electronic devices into clothing and implantable health monitoring devices [21]. Carbon nanotube based nanowires [22] and oxidized silicon nanowires [23] may potentially function as ballistic conductors where electrons can travel through the conductor without collisions due to negligible electrical resistivity caused by scattering [24]. In other studies, using piezoelectric materials, scientists have succeeded in creating nanowires that generate electricity from kinetic energy [25].

Historically, nanowires have been prepared by a variety of methods. One common approach for the synthesis of nanowires is templating [26], where physical templates such as mesoporous materials, porous alumina [27], polycarbonate membranes [28], or biomolecules [29-31] have been utilized. The template can then be selectively removed upon creation of the nanowires. In particular, the use of biomolecules such as DNA, and peptides as templates is advantageous because their surface chemistry facilitates biological recognition and three-dimensional structures are often times generated by molecular self-assembly.

For mimicking the properties of proteins for device fabrications, in general, once the candidate self-assembling building blocks are astutely selected from a library of sequences, the constituents are mapped onto the nanostructure shape [32] and the stability of the device is reliant on the favorable association between the building blocks [33]. As early as 1981, Eric Drexler declared that protein design could be used to fabricate devices by the "bottom-up" approach in which proteins can be used as monomeric building blocks for the manufacture of higher order structures via self-assembly [34]. The benefit of employing peptide self-assembly to construct nanowires is the ability to tailor morphologies and functions of the resultant material, which is dictated by the individual building blocks comprising the assembly [35]. Currently, most *de novo* design endeavors of proteinaceous structures have been extended toward the fabrication of larger nanoscale devices and materials. There is a dual challenge to this progression: first, to judiciously devise building blocks amenable to self-assembly that form nanostructured materials and second, to confer the desired functionality [36].

In addition to proteins, over the past decade, DNA has also emerged as an attractive template for the formation of nanowires because it can be modified via different functional groups [37]. Pioneering work by Mirkin and Alivisatos has resulted in the development of DNA directed assembly for the preparation of various types of nanowires. DNA has been used as a template to prepare nanowries via conjugation with materials such as metals [38], conducting polymers [39], nanoparticles (such as Au [16b, 40], calcium phosphate [41], CdS [42], Ni, and Co [43]), fluorophores [44], and single-walled carbon nanotubes [45]. Specifically, DNA is the template of choice for metallization because of its ability to self-assemble into a variety of intricate structures [46] due to the fact that DNA is negatively charged and allows for binding with positively charged metal cations, which can consequently be reduced to form metal nanoparticles [47]. Further, the double helix of DNA provides mechanical strength [48]. A variety of metal nanowires have been synthesized from DNA such as Ag [49], Au [50], Pt [51], Cu [52], CdSe [53], Pd [54], Co [55], and Fe_2O_3 [16d, 56], where the use of the DNA has assisted in driving the assembly of ordered nanostructures. Single-stranded DNA (ssDNA) has also been used in the formation of metal nanowires, such as Au [57] and CdSe [58]. ssDNA has the ability to hybridize with complementary strands, thus allowing for further modifications. For example, Mallouk and co-workers hybridized ssDNA to Ag nanowires at specific sites. They were able to remove the metal coatings from the tips, and soak them in another strand of DNA, thus creating hybrid wires, where DNA sequences were only present on the tips and the rest of the wire was barren. Furthermore, Wang and co-workers showed that when circular DNA is self-assembled in the presence of cocaine, DNA nanowires could be formed [59]. Furthermore it was found that upon functionalization of glucose oxidase (GOx) and horseradish peroxidase (HRP) to the DNA template, it has potential in the area of biocatalysis.

Another common method is the vapor-liquid deposition method, which first became popular in the 1960s as a means to generate Si wires [60], and since then has been used in the controlled synthesis of nanowires and nanorods, despite the fact that it lacks the ability to create materials with uniform crystallinity. In recent times, Mirkin and co-workers have developed several lithographic techniques such as electron beam lithography, dip-pen nanolithography (DPN), focused ion-beam lithography, and nanoimprint lithography [61]. An improved lithographic method termed as on-wire lithography (OWL) has been used to create gapped nanowires and single or multi-component nanowire structures in a controlled manner, with dimensions ranging from 2 nm to many micrometers [62]. This method has many advantages over previous methods because the gap size can be controlled and it combines template-directed synthesis with electrochemical deposition and wet-chemical etching.

This review provides an in depth look at the various methods utilized in the preparation of nanowires, with emphasis on biomolecule generated nanowires and their applications and future perspectives.

DNA Based Nanowires

Over the past decade, DNA self-assembly into hierarchical structures has been extensively explored [32, 64-65]. A chief aspect of DNA nanotechnology involves the use of self-assembled DNA lattices to scaffold assembly of other molecular components [66]. In general, DNA has been used in the creation of periodically patterned structures [67], nanomechanical devices [65a-b], and computing systems [68]. Self-assembling DNA nanostructures defining intricate curved surfaces in three-dimensional (3D) space using the DNA origami folding technique have also been constructed, resulting in a series of DNA nanostructures with high curvature such as 2D arrangements of concentric rings, 3D spherical shells, ellipsoidal shells and nanoflasks [69]. DNA has also been functionalized for suitable attachments to direct the assembly of other functional materials [70]. Binding metal nanoparticles to strands of DNA, or allowing the DNA strands to direct the formation of metal nanoparticles has led to the generation of DNA-based nanowires for a multitude of applications. In general, there are two basic methods for generating nanowires on DNA templates—solution suspended DNA method or substrate immobilized DNA method [71]. The first method involves attaching specific metal cations to DNA templates followed by subsequent reduction to generate nanowires [57b]. Because the nanowires have mobility in solution, they often form coils and cross-links [48b], which makes them difficult to align in an ordered manner, and consequently are relatively complex for applications in nanodevices and nanosensors. The second method involves using a substrate to first align the DNA followed by the addition of a metal ion solution to make the nanowires [15d]. This method however, results in the growth of parasitic nanoparticles adjacent to the nanowires [39b, 52a]. In this section we will examine the aforementioned approaches, as well as shed light on additional methods, which address and improve upon the above limitations.

Fabrication of Metal Nanowires (Au, Ag and Pt)

The recent trend toward utilization of bottom-up fabrication has led to the synthesis of metallic and semiconductor nanowires, which are predominately useful in areas such as photonics [72], electronics [73], nanoelectromechanical systems (NEMS) [74], and life sciences [75]. In particular, the construction of gold nanowires is advantageous because of the many applications of gold nanoparticles such as surface-enhanced Raman scattering (SERS)-based sensors [76], electrical and optical nanodevices [77-78], colorimetric biosensors [79], drug delivery [80], cancer imaging, therapeutics [81-83], biological markers [84], sensors [85], and molecular recognition systems [86]. Nakao and co-workers synthesized highly ordered Au nanoparticle assemblies via a novel surface-functionalization method through the reduction of $HAuCl_4$ using aniline (AN) to link the nanoparticles to the surface of DNA [41a]. Electrophoresis analysis and zeta-potential measurements indicated positive charges on the surface, thus suggesting that the AN-Au nanoparticles had an aniline monolayer on their surface, which helped in the attachment of the nanoparticles to the DNA molecules. The nanoparticles were strongly bound to DNA via electrostatic interactions between the negatively charged phosphate backbone and the positively charged aniline shell. Two methods were then explored to modify the formed gold nanowires, as detailed in Figure 1a. In Method I, (Figure 1b) DNA molecules were pre-stretched and fixed on surfaces, and it was found that DNA attachment of AN-AuNPs was limited to one side. However, DNA attachment of AN-AuNPs in the solution phase (Method II) was found to occur from multiple directions. Thus, in Method II binding of multiple particles surrounding DNA at close positions was observed [87].

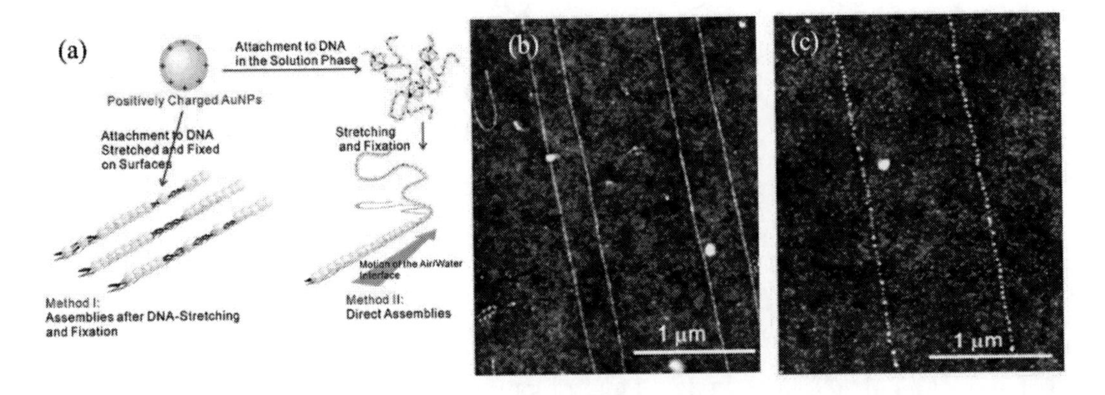

Figure 1. (a) Scheme for preparation of Au nanoparticles on DNA using methods I and II; (b) AFM image of ordered nanoparticle assemblies obtained using Method I; (c) AFM image of ordered nanoparticle assemblies obtained using Method II. (Ref. [41a] *Copyright*, Reproduced with permission from the *American Chemical Society* 2003).

In another study, DNA-Au nanoparticle wires were generated by the incorporation of an intercalator (psoralen) functionalized Au nanoparticle into double-stranded DNA, followed by the photochemical cross-linking of the intercalator to the DNA matrix in the presence of UV-light [88]. Mbindyo and co-workers reported the DNA-directed assembly of Au nanowires up to 6 μm in length. Oligonucleotides were adsorbed as monolayer coatings on the wires through Au-thiol linkages. It was observed that duplexes formed between strands on the nanowires. Further, the nanowires were modified with ssDNA exclusively at the tips, with the rest of the wire covered by passivating monolayers, thus allowing for site-specific DNA directed assembly [89]. In a recent study, Chen and co-workers created nanoscale ssDNA patterns on PDMS substrates by a master-replica transfer process from LPNE (lithographically patterned nanowire electrodeposition) for sensing applications [58]. Arrays of ssDNA "nanolines" were produced on streptavidin-coated polymer (PDMS) surfaces by transferring biotinylated ssDNA from a master pattern of gold nanowires which were attached to a glass substrate. The gold nanowires were lithographically patterned by electrodeposition (LPNE), and then “inked” with biotinylated ssDNA to a thiol-modified ssDNA monolayer attached to the gold nanowires. The ssDNA nanolines were capable of hybridizing with ssDNA from solution and formed double-stranded DNA (dsDNA) patterns. This procedure is summarized in Figure 2. Optical diffraction patterns were used to detect the hybridization adsorption of unlabeled ssDNA target molecules. The results were indicative of potential applications of these arrays in bioaffinity sensing.

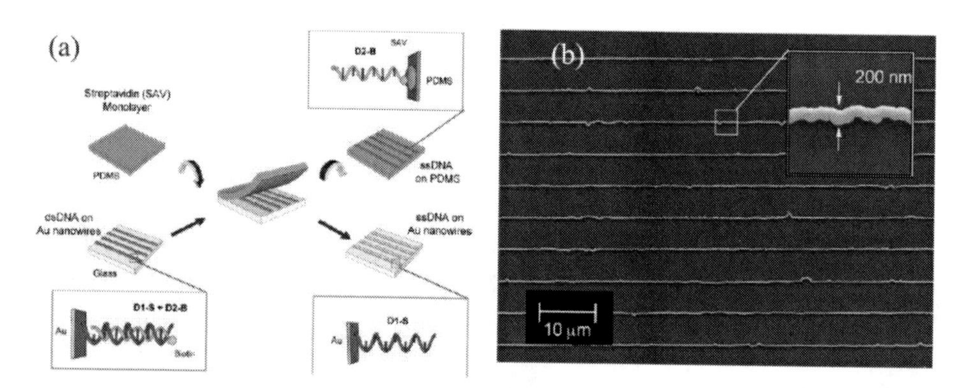

Figure 2. (a) Scheme for the master-replica transfer process for preparation of nanoscale ssDNA patterns on PDMS substrates. A thiol-modified ssDNA (D1-S) array master was created then hybridized with complementary biotinylated ssDNA (D2-S) to create a dsDNA monolayer on the nanowire surfaces. A streptavidin-modified PDMS substrate was then placed in contact with the gold nanowire array. When peeled apart, D2-B was bound to PDMS replica by biotin-streptavidin interactions, leaving D1-S on the nanowire master. The D1S sequence used was 5’-HS$(CH_2)_6$-TTT TTT TTT TTT TTT TTT TTT TTT TTT TTT TTT TTT TTT TTT TTT-3’; D2-B sequence utilized was 5’-Biotin$(CH_2)_6$ AAA AAA AAA AAA AAA AAA AAA AAA AAA AAA-3’ (b) SEM image showing a portion of the gold nanowire array. The inset is a high magnification image of a single gold nanowire. (Ref [58] *Copyright*, Reproduced with permission from the American Chemical Society 2010).

In addition to gold nanowires, DNA also lends itself as a template for designing silver nanowires [90]. Silver ions become embedded within the double helix, thus forming stable complexes [91]. Wei and co-workers showed a simple approach for the construction of Ag nanowires using DNA templates. Networks of DNA were immersed into a solution of silver nitrate, which allowed for the absorption of silver ions onto the DNA strands, and a concomitant immersion into sodium borohydride solution allowed for the silver reduction. In order to modify the dimension and morphology of the silver nanostructures, the DNA concentration, as well as the reduction time was modified. The DNA networks served both as a substrate for the absorption of silver ions as well as a template for controlled growth [92]. Braun and co-workers demonstrated the construction of conductive silver nanowires using oligonucleotide templates [39b]. Briefly, 12-base oligonucleotides, were derivatized with disulfide groups at their 3’ ends, and attached to the gold electrodes. Each electrode was functionalized with specific oligonucleotide sequences and a connection was then made by hybridizing two distant surface-bound oligonucleotides with fluorescently labeled λ-DNA containing two 12-base sticky ends, where each of the ends was complementary to one of the two different sequences attached to the gold electrodes. Hybridization on both ends was facilitated by covering the electrodes with a solution containing the λ-DNA and inducing a flow perpendicular to the electrodes, thereby stretching the λ-DNA molecules in the flow direction. The flow was stopped when a DNA bridge was observed by fluorescence microscopy. To make the DNA bridge conductive, silver was then deposited on it by a three-step chemical deposition process [93]. To form silver nanowires, the silver ion-exchanged DNA was then reduced to form silver aggregates on the DNA skeleton, which acted as a catalyst to form silver nanowires using an acidic solution of hydroquinone and silver ions under low light conditions [94]. The wires were found to be highly conductive and had low resistance.

In a separate study, Yan and co-workers efficiently templated conductive nanowires on self-assembled DNA nanostructures by metalizing 4 x 4 nanoribbon patterns with silver [32]. DNA structures of four 4-arm junctions were oriented to form a square, and the 4x4 tiles were subject to programmable self-assembly, whereby both uniform-width nanoribbons as well as 2D-nanogrids were formed. It is predicted that the 4 x 4 DNA tiles can be efficiently programmed by varying sticky ends to build unique arrays for applications in the construction of logical molecular devices such as quantum-dot cellular automata arrays.

In another study, Keren and co-workers created patterned DNA nanowires using silver ions via localization of the reducing agent glutaraldehyde [95]. This use of this method for silver nanowire formation lends itself to application in wiring of molecular-scale electronic circuits, as well as patterning the metallization of branched DNA structures. Another templating method involving DNA also generated silver nanowires, by utilizing a three-helix bundle (3HB) [32]. A 3HB consisting of three double helices of DNA that were not coplanar was utilized. Through programming of the sticky-ends, both 1D filaments and 2D lattices were generated. The 1D filaments were then utilized for electroless chemical deposition of silver, which led to silver nanowires with an average diameter of 25-35 nm. The silver nanowires were then deposited with chromium-gold double layer electrodes by electron-beam lithography. It was found that the silver nanowires were more resistive than polycrystalline silver and more conductive than other dsDNA templated silver nanowires [39b].

Eichen and co-workers also utilized molecular self-assembly of DNA to create nano-sized electrical circuits using silver nanowires [96] by first assembling DNA molecules to form a network. Photolithographic techniques were utilized to place gold electrodes on a glass substrate. Each of the electrodes were then draped with a unique monolayer of single-stranded oligonucleotides. Different oligonucleotides were utilized for each electrode to selectively respond to a specific complementary sequence. The device was then placed into an aqueous solution of DNA with specific sequences and sticky ends. The interactions between the complementary DNA sequences allowed for hybridization with oligonucleotides attached to electrodes. Finally to achieve conductivity, the metal cluster-DNA binding systems were bound to specific sequences found in the DNA network. The methods for forming conductive silver nanowires are similar to those previously described [39b, 93-94, 97].

Recently, DNA-templated photo-induced silver nanowires were prepared by a simple, low cost method wherein it was found that upon exposure to UV-light, Ag ions bound to DNA template networks were reduced to Ag nanoparticles [98]. It was seen that the reduction only occurred when the system was exposed to UV light at 254 nm. By controlling both the concentration of DNA and the irradiation time, the mesh size of the Ag-DNA network and the diameter of the resulting nanowires were controlled. The Ag-DNA network was then plated onto comb-like gold electrodes on a silicon substrate in order to test conductivity. Further, in order to test the ability of the network to act as a humidity sensor, water vapor was chemisorbed onto the surface of the nanowires, which allowed for the incorporation of hydroxyl groups at the surface [99]. It was found that the water molecules absorbed on the surface directly correlated to the electrical response.

A majority of research related to metal nanowires has focused on gold and silver nanoparticels because of their well characterized properties, but DNA templating approaches have also been utilized for the formation of nanowires consisting of other metal nanoparticles. For example, Seidel and co-workers synthesized chains of platinum clusters on templates of DNA using a selectively heterogenous, template-controlled mechanism [100]. It was reported that specific incubation times were necessary for forming Pt (II) complexes with double stranded DNA in order to achieve template-directed formation of uniform cluster chains after chemical reduction of the DNA/salt solution. It was interesting to note that the incubation time played a vital role in the formation of the nanowires, because under shorter incubation periods, DNA acted as a non-specific capping agent for the formed clusters, and the formation of random cluster aggregates was observed.

Another means for generating heterogenous Pt nanowires involved the use of Pt(II) complexes, which were covalently bound to the DNA bases prior to the reduction [101]. The group conducted first-principle molecular dynamics (FPMD) simulations at 300 K prior to experimental work. The simulations showed that in order to obtain strong-bond dimer geometry, nucleotide ligands were needed, thus elucidating the crucial role of DNA as a catalyst in the formation of nanowires. Further, it was shown that the metal dimers formed on the DNA were the preferred sites for further reduction, and the continued formation of the dimers on the surface of DNA allowed for an increased uptake of Pt atoms from solution. Furthermore, Mertig and co-workers found that Pt(II)•DNA adducts were the preferred sites for nucleation, and therefore began the process of in-situ metallization by increasing the rate of heterogenous metal cluster formation.

DNA Templated Transition Metal (Nickel and Copper) Nanowires

Formation of nanowires from nucleic acids, by incorporation of transition metal ions [102-103], followed by in situ reduction is a common method for the formation of tunable nanowires for a wide range of applications. For example, DNA-Ni^{2+} complexes were developed in order to prepare DNA templated Ni nanowires. It is believed that such nanowires have the potential to be applied as templated catalyst lines, nanoscale magnets, or as directed protein localizers [104]. As shown by Woolley and co-workers, allowing surface-aligned DNA substrates to mix with an aqueous solution of nickel ions resulted in the formation of DNA-Ni^{2+} complexes. In order to form the nickel nanoparticles, the DNA-templated nickel complexes were reduced with $NaBH_4$. In addition, DNA-nickel-protein nanoconjugates were also formed from the association of the DNA-nickel complexes with histidine-tagged phosducin-like protein (His-PhLP). This technique is particularly advantageous because intricate patterning was formed without the need for lithography.

Monson and co-workers designed a methodology that resulted in the unique formation of DNA templated Cu nanowires [52a]. In order to do so, DNA was first aligned on a silicon substrate before it was treated with an aqueous solution of $Cu(NO_3)_2$. The Cu(II) ions bound to DNA due to electrostatic interactions, and upon reduction with ascorbic acid, the DNA was coated by a sheet of metallic copper. Through repetition of the reduction step, the Cu nanoparticles were coated completely on the DNA strands. Recently Bu and co-workers designed Cu-mediated GC and AT base pairs to theoretically investigate appropriate building blocks for DNA-based molecular wires. A three-copper-mediated guanine-cytosine ($G_{3Cu}C$) and a two-copper-mediated adenine-thymine ($A_{2Cu}T$) base pair were designed. In order to examine the effect of H-by-Cu substitution on conductivity, three-layer-stacked $G_{3Cu}C$ and $A_{2Cu}T$ of repeat and cross sequences were studied. Their studies indicated that the multi-Cu-mediated $G_{3Cu}C$ and $A_{2Cu}T$ pairs could be promising units for building blocks of the Cu based DNA nanowires [105].

DNA-Templated Pd Nanowires

One of the earlier studies involving the growth of Pd nanoparticles on DNA structures was investigated by Richter and co-workers in the year 2000 [48a], which proved to be the first of a series of reports on the subject by this group. Previously, DNA-Pd studies indicated that upon binding to DNA, the Pd-DNA complex formed caused the DNA planes to bend during activation [106]. The technique of cluster deposition was utilized to metallize λ-DNA [48a]. Initially, DNA was treated with Pd ions, followed by reduction. A shorter reduction time (<1 minute) led to the formation of numerous distinct Pd nanoclusters 3-5 nm in diameter, whereas a longer reduction time led to a more uniform coating of Pd nanoparticles, resulting in grain-like metallic structures on DNA. Attempts were made to align DNA by immobilizing it on a substrate by evaporating a DNA solution onto a glass coverslip, where the meniscus moved the strands. Parallel orientation of the strands occurred when the DNA strands were adsorbed to the surface as they were pulled by the receding meniscus. In their next report, Richter and co-workers [15c] synthesized conductive Pd nanowires. The DNA metallization was carried out in two steps, where the first step involved activating the template, and the second step involved reduction of the Pd clusters using lactic acid and dimethylamine borane. In order to improve the measurements, the wires were fixed to gold pads by imposing electron-beam-induced carbon lines on the system. The resistance of each wire was then determined by breaking a wire using a micromanipulation device integrated with an optical microscope, so that the resistance before and after cutting could be measured as shown in Figure 3. This method allowed for a large quantity of measurements (>100 wires), which lent itself to a more statistically accurate determination of the conductivity. It was seen that there was a linear relationship between the resistance and the distance between Au electrodes, thus indicating that the length of the wire was related to its resistance. It was found that ohmic behavior occurred at a resistance of 743 ohms, and when the nanowires were cut, they became insulating samples. Finally, it was seen that the nanowires were free of both nonconducting regions and Coulomb blocking behavior [107].

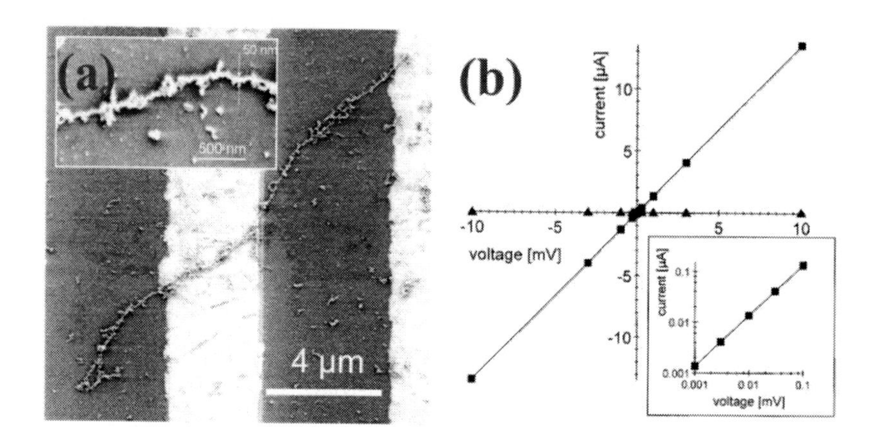

Figure 3. (a) SEM image of a single palladium metallized DNA strand, wherein the right-hand side of the strand connects two gold electrodes over a SiO_2 substrate. The inset shows a magnification of the middle part with a diameter of 50 nm; (b) Two-terminal current–voltage curves of the single, pinned nanowire before (□) and after (Δ) cutting. The inset shows the ohmic $I-V$ characteristic of the nanowire down to 1 mV. After cutting, the sample was insulating. (Ref. [15c] *Copyright*, Reproduced with permission from *American Institute of Physics* 2001).

In another third study [108], Richer and co-workers examined the formation of Pd nanowires under low-temperature. Earlier, low temperature studies of disordered metallic structures have indicated that because of the weak localization and/or electron-electron interactions, these systems exhibited quantum behavior [109-110]. Because the systems have a low dimension, the resistance increased as the temperature decreased. Similar Pd wires were utilized as described in the previous study [15c], and they were reduced until aggregates of homogenous clusters were no longer observed, which led to metallic wires that followed the geometry of the DNA. The resistance ratio as a function of temperature was determined graphically, which confirmed that DNA-templated nanowires possessed a similar value as that of the Pd films [111]. Because quantum behavior was seen in the wires due to the disorder present, the wires were subsequently annealed at 200°C in order to reduce any defects seen. This allowed for an improvement in the crystal structure, thus overcoming any disorder in the system. Because the Pd nanowires assembled on DNA templates exhibited the same inverse relationship of temperature and resistance of disordered metal systems and the properties studied yielded the same results as films of Pd, it was concluded that using DNA as a template does not change any known metallic properties.

Kundu and co-workers also investigated the formation of conductive Pd nanowires using DNA as a template [112]. A rapid photochemical method was utilized to form thin conductive Pd wires, where DNA was utilized as both a reducing agent and a capping agent. A solution of DNA/Pd salt was exposed to UV light (260 nm) for 4h. No additional reducing agent or additive was needed for the nanowire formation. Other studies using UV irradiation techniques indicated that it is much more efficient than conventional thermal convection in generating nanowires [47a, 90]. Current-voltage measurements showed that the nanowires were governed by Ohmic behavior with low contact resistance, thus lending the resulting nanowires as potential interconnections in nanoscale integrated circuitry, functional nanodevices, and optoelectronics.

DNA-Based Photonic Nanowires

DNA labeled quantum dots allows for a system of many applications such as fluorescent biomarkers [113], catalysts [114], Raman spectroscopy [115], and nanoscale electronics. Sarangi and co-workers reported that monocrystalline cubic CdSe nanowires approximately 4.0 nm in width with string-like morphology were formed when synthesized in the presence of both poly Gss DNA (containing 30 guanine bases) and its conjugate, Css DNA (containing 30 cytosine bases) [54]. Such a method is advantageous because previous methods of forming quantum dots on the surface of DNA required a thiol modification [116]. This method required allowing multivalent cations to electrostatically interact with ssDNA, which caused a charge condensation and gave a complex of positively charged cations and ssDNA in solution. After stabilization, CdSe-Gss DNA complexes were synthesized electrochemically [117].

Hannestad and co-workers utilized DNA as a scaffold for a chromophore with overlapping absorption and emission bands enabling fluorescence resonance energy transfer (FRET) between pairs of chromophores, leading to sequential transfer of the excitation energy along DNA nanowires, thus leading to the formation of self-assembled photonic wires. Further, this method allowed for a large span in wire lengths. It is interesting to note that the intercalating agent (yttrium oxide, YO) was chosen for its homotransfer capability, enabling effective diffusive energy migration along the wire without loss in energy in contrast to heterotransfer FRET. By utilizing injector and detector chromophores at opposite ends of the wires, directionality of the wires were achieved. Thus, by utilizing self-assembly, the group created two component DNA-based photonic wires capable of long-range energy transfer [118]. It was seen that energy transfer depended on both the length of the wire and the density of the chromophore. Overall, efficient energy transfer occurred in wires up to $\approx$20 nm in length. Figure 4 shows the schematic for the multichromophoric DNA wire that was constructed.

Hybrid multi-fluorophore DNA-photonic wires have also been generated by self-assembly around semiconductor CdSe-ZnS core-shell QD's by Medintz and co-workers [7b]. The QDs functioned as both central nano-scaffolds as well as UV energy harvesting donors that allow FRET through the DNA wires with emissions close to the near-infrared region. In order to assemble the wires, DNA fragments labeled with a series of increasingly red-shifted acceptor-dyes were hybridized in a prearranged linear manner to complementary DNA templates that was chemo-selectively modified with a hexahistidine peptide. The peptide was necessary to provide metal-ion coordination with multiple hybridized DNA-dye structures to a central QD. Efficiencies determined by steady-state and time-resolved spectroscopy revealed that acceptor dye quantum yields were the predominant limiting factor. Such DNA-based photonic structures with QDs can lead to the development of a new generation of biophotonic wire assemblies.

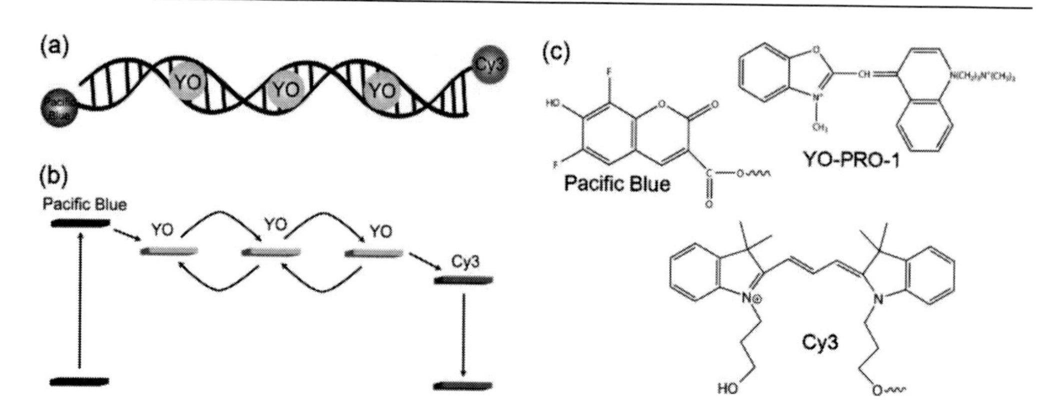

Figure 4. Scheme showing the multichromophoric DNA wire with attached Pacific Blue (injector) and Cy3 (detector) and intercalated YO. (b) Corresponding energy level diagram, showing the diffusive energy migration in the YO section of the wire. (c) Structures of Pacific Blue, YO-PRO-1, and Cy3, the three chromophores used in the construction of the photonic wire. (Ref. [118] *Copyright*, Reproduced with permission from *American Chemical Society* 2008).

In a separate study, DNA strands were used as a template for the growth of 1D chainlike assemblies of CdS nanoparticles as well as uniform wires [119]. The DNA strands were either surface-immobilized or in solution, allowing them to act as templates to control the growth of the CdS nanoparticles. Based on the reaction conditions and the state of DNA, the growth was controlled in order to generate quantum-confined CdS as either 1D chainlike assemblies of particles or as uniform nanowires. For surface immobilization, two different surfaces were explored—mica and alkyl monolayers on single-crystal Si(111). The mica proved to be the optimum surface because the DNA molecules anchored to it, in large part due to interactions between the metal ions, the surface oxygen functionalities, and the phosphate groups. In the case of the Si(111) surface, the DNA was unreactive towards it, and thus combing was needed in order to align the DNA [15d, 120]. Therefore, marked differences were seen in the QD's formed based on the surface utilized. In the case of the mica, beads of homogenous CdS nanoparticles were formed on the DNA chains. In comparison, when the Si(111) surface was utilized, the reaction yielded randomly coiled strands, thus indicating the lack of adherence of the DNA strands to the surface. The DNA was covered with random aggregates of particles, which resulted in a lack of homogeneity in terms of particle size. In the solution method, λ-DNA was allowed to react with cadmium nitrate and sodium sulfide and incubated overnight (4°C) before Cd^{2+} and Na_2S were added for another 24h, prior to allowing the samples to incubate at room temperature for 48h. In this case, uniform coatings of CdS on the DNA strands were observed, leading to the formation of nanowires of uniform length. The reasoning behind the generation of wires as opposed to mere particles was the longer reaction time and the additional idle time of the reaction before observation. When the reaction was left undisturbed for only 24h instead of 48h, more particles of CdS were observed. The blue shift in the PL spectrum also confirmed the role of DNA in controlling the growth of the QD's [121]. The electrical potential of the nanomaterials was also measured, as they were integrated into a two-terminal electrical device. A single CdS nanowire was placed between two electrodes, and a two-terminal *I-V* measurement confirmed charge transport in the CdS nanowires [122], as well as a reproducible nonlinear response at room temperature.

CdS nanowires have also been generated on DNA templates by employing PDMS transfer method [71]. In general, CdS nanoparticles were selectively deposited and confined on DNA strings which were aligned on a PDMS sheet to form CdS nanowires. The nanowires were then transferred to the substrate. The formation of CdS nanowires was believed to occur via a three-step mechanism. The first step involved the activation, leading to cadmium ion-DNA complexes (first 24h). The second step was the addition of TAA (thioacetamide) under UV irradiation, which allowed for S^{2-} to be slowly released [47a]. The final step was the sequential and ordered growth of the CdS nuclei along the DNA during incubation, which allowed for the formation of homogenous CdS nanowires. The width and length of the nanowires could be controlled by adjusting the incubation time on the PDMS sheet. Further, the nanowires could stretch over 10 μm after 96 h of incubation. The scheme for the formation of CdS nanowires as well as the AFM and SEM images of the thick and uniform nanowires obtained using this method is shown in Figure 5. This approach may be utilized for photoelectronic nanodevices and nanosensors. By merging the alignment of DNA on a PDMS sheet and the formation of CdS nanowires in solution, well-aligned nanowires can be transferred to other substrates. An advantage of this method is that no additional substrate surface modification is required, and one can control the size and morphology of the nanowires by changing the incubation times.

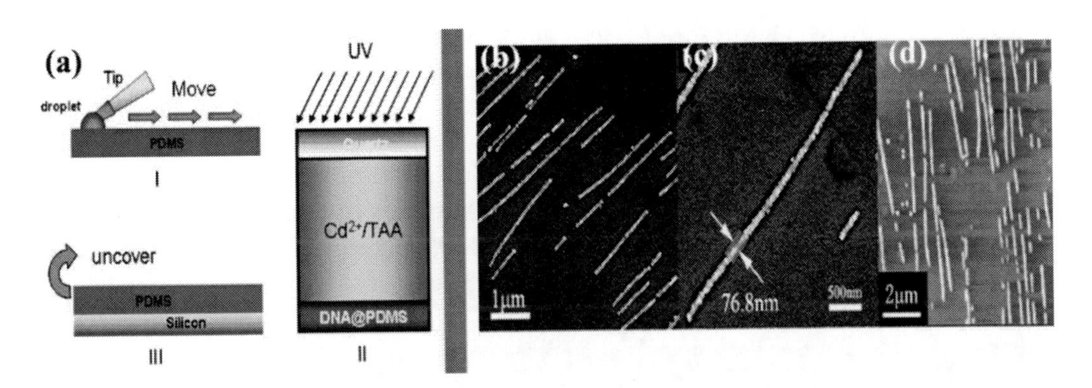

Figure 5. (a) Scheme for fabrication of CdS nanowires on fixed DNA scaffolds by using a PDMS transfer method. (I) DNA is first aligned and fixed on a PDMS surface. (II) A PDMS sheet is submerged into a solution containing cadmium ions and thioacetamide (TAA) under UV irradiation at room temperature. (III) CdS nanowires on PDMS are transferred to other substrates. (a) FESEM image of CdS nanowires on silica, transferred from PDMS after 72hr synthesis under UV irradiation. (b) FESEM image showing an individual CdS nanowire. (c) AFM image of CdS nanowires on silica, transferred from PDMS. (Ref. [71] *Copyright*, Reproduced with permission from *American Chemical Society* 2009).

DNA has also been employed for the growth of highly stable and efficient infra-red emitting PbS quantum dots [123]. In order to synthesize the PbS nanoparticles onto DNA templates, different sites on DNA including the phosphate backbone and the two bases were examined so that optimum chemical interactions could be achieved. The formation of DNA-supported PbS quantum dots depended on the lead and sulfur precursor concentrations as well as the ratio of DNA to the precursors and temperature. Furthermore, the nanocrystals obtained were aged in blood plasma and luminescence quantum efficiency indicated a half-life of one week.

DNA Templated Polymeric Nanowires

Polyaniline, polypyrrole, polythiophene, and various other π –conjugated polymers have been utilized for their specific electronic, optical, and catalytic properties [124-125]. Polymer microstructures have been generated using numerous methods such as photolithography [126], microcontact printing [127], membrane-template synthesis [128], and electrochemical dip-pen lithography [129]. Taking advantage of the well known recognition properties of biomolecules, DNA has also found a variety of applications as a template to bind π-conjugated polymers in order to generate nanowires [40, 130]. Various π-conjugated polymer-functionalized DNA nanowires have been reported by Nakao and co-workers [124]. For example, polyphenazasiline containing alkylammonium salts (PPhenaz-TMA) were synthesized and directly attached to DNA. The DNA template was further stretched and aligned to form the evenly coated polymeric nanowires. The amount of stretching determined the quantity of gaps seen in the wires. The strong binding of the polymer to the DNA was indicated by absorbance spectroscopy, where a characteristic peak for aromatic amines was seen at 340 nm, which indicated π- π^* absorption. As a control, additional DNA was added, which decreased the peak at 340 nm. This hypochromism has been seen in other systems of DNA and intercalator dyes [134], which confirmed that the aromatic units of PPhenaz-TMA were intercalated into stacks of DNA. If metallic nanowires were desired, the PPhenaz-TMA/DNA nanowires were incubated with metal salts such as $HAuCl_4$. The metal deposition occured by a redox reaction between the PPhenaz-TMA nanowires and the metal ions. Thus, the reaction was found to be selective to the system and was confirmed by NSOM and AFM. The DNA nanowires obtained after attachment to PPhenaz-TMA followed by metallization with gold are shown in Figure 6.

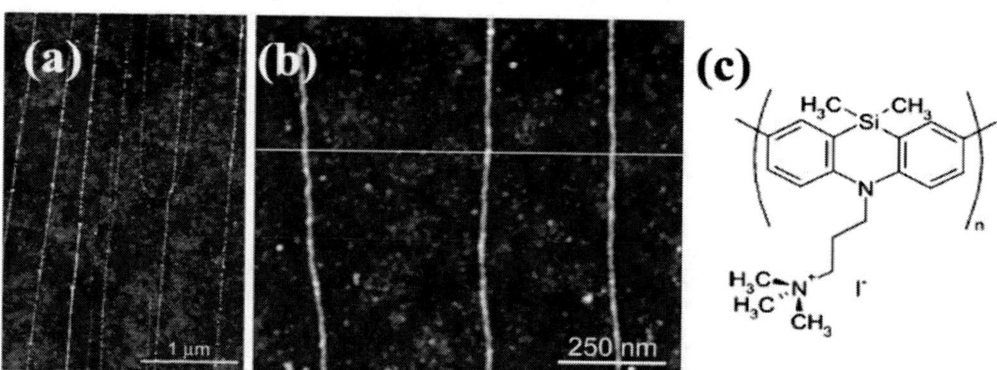

Figure 6. AFM image of stretched and aligned DNA molecules after DNA-attaching of PPhenaz-TMA in solution. (b) AFM image of PPhenaz-TMA/DNA nanowires after immersion in an Au^{3+} solution for 30 min. (c) Chemical structure of PPhenaz-TMA. (Ref. [124] *Copyright*, Reproduced with permission from *American Chemical Society* 2005).

Houlton and co-workers recently demonstrated the synthesis of supramolecular conducting nanowires by utilizing the self-assembly approach to promote interactions between DNA and pyrroles [132]. They reported that oxidation of pyrrole with $FeCl_3$ in DNA-containing solutions yielded both the cationic polypyrrole (PPy) and the anionic DNA-polymers. Close interactions between the two polymer chains in the self-assembled nanowires were observed. Individual, conformationally flexible nanowires which then aligned by molecular combing were obtained. The DNA-PPy system provides a convenient method for fabricating electrical devices by stretching the nanowires across the electrode gaps. I–V curves obtained proved that the nanowires were conductive. It is interesting to note however that when polymerization of pyrrole on a surface immobilized DNA-template was attempted, continuous coverage was not observed on the wires, instead, beads-on-a-string like structures were observed, suggesting that immobilization hinders the assembly process. The supramolecular formation of the PPy-DNA nanowires as well as the AFM image showing well aligned nanowires is shown in Figure 7.

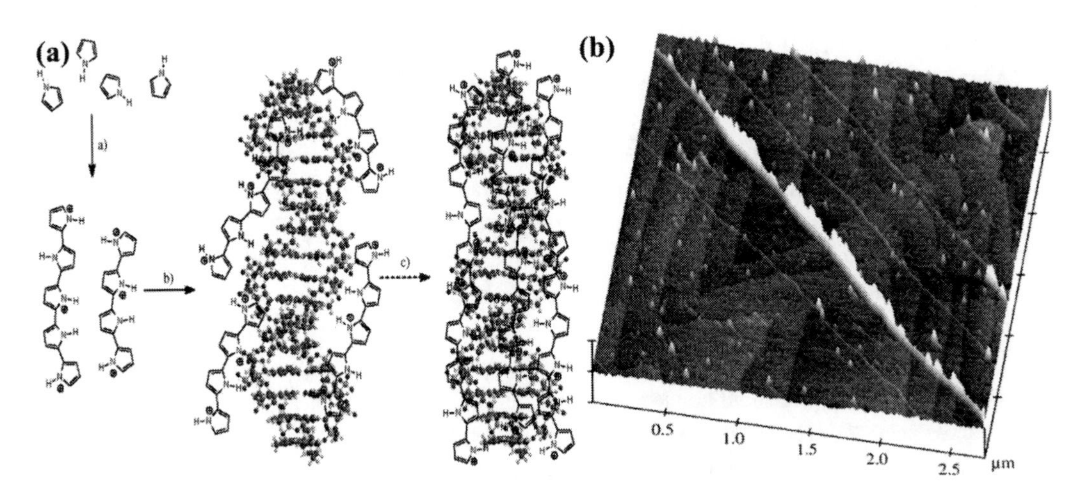

Figure 7. (a) Scheme 1. Proposed mechanism for self-assembly of DNA/PPy nanowires. a) Oxidation of pyrrole monomer with $FeCl_3$, b) Interaction between PPy oligomers on DNA, c) polymer growth on DNA template. (b) DNA/PPy stretched on alkylated silicon (111) surfaces. a) Surface plot AFM image showing the height variation DNA/PPy nanowires. The materials isolated from the reaction contained bare DNA-scaffold molecules and thicker DNA/PPy nanowires. (ref [132] *Copyright*, Reproduced with permission from *Wiley-VCH* 2007).

Poly-indoles are another class of polymers with numerous applications in sensing and electronics because of their well characterized photo-luminescent, conductive, and electrical properties [133-134]. Hassanien and co-workers prepared nanowires by utilizing λ-DNA as a template in the presence of the indole monomer, which was then oxidized using $FeCl_3$. The nanowires were formed by non-covalent interactions between the DNA and the polymer chains. It was shown that the reaction time affected the height and width of the nanowires, as more reaction time led to a thicker coating and a larger diameter (5-20 nm). It is interesting to note that the thicker strands were seen to consist of several smaller strands of nanowires, which began to appear as the ropes unraveled. Conductive atomic force microscopy (c-AFM) allowed for resistance to be measured as a function of distance along the nanowire. The group found that there was a linear relationship between the resistance and the distance. In order to measure the conductivity and thermal stability of the nanowire, temperature-dependent two-terminal current-voltage measurements at Au electrodes were performed. It was found that factors such as polymer structure, degree of crystallinity, and dimensionality played a role. The conductance was found to be 40 S cm^{-1}, which was higher than typical bulk PIn prepared by a similar method ($10^{-2} - 10^{-1}$ S cm^{-1}) [134c].

In the quest toward preparation of smart devices, fabrication of conducting polyaniline nanowires on thermally oxidized Si surfaces by use of DNA as templates was recently reported. The use of stretched and immobilized DNA strands as templates resulted in significantly lesser agglomeration of DNA due to shielding of charges on DNA when polyaniline/DNA complexes formed in solution. Further, the oriented DNA strands immobilized on the Si surface predetermined the position and the orientation of the nanowires. This method was an innovative step toward amalgamating programmable-assembly of DNA motifs with the distinct electronic properties of conducting polymers for the development of high-density functional nanodevices. In order to generate the nanowires, DNA was stretched along the Si substrate to provide the template for the HRP (horse radish peroxidase) enzymatic reaction for the polymerization of polyaniline. The aniline monomers aligned and formed a complex, promoting extended conjugation of the resulting polyaniline chains [135]. The smooth coating of the conducting polymer polyaniline allowed for controlled electrical conductivity along the length of the DNA, which was an improvement upon DNA metallic nanowire structures, which lacked the ability to control the conductance. The smooth coating was achieved by pH control. It was found that a pH value of 4 was the optimum condition to produce uniform smooth polyaniline coating. Because of the extreme sensitivity of the nanowires to the process of proton doping-undoping, the nanowires were particularly useful as chemical sensors. It was seen that exposing the nanowires to HCl vapor allowed for a large increase in the current generated, and exposing the wires to NH_3 had no effect. When HCl-doped nanowires were exposed to NH_3 gas, the conductance dropped significantly because of the deprotonation of polyaniline. Thus, such nanowires have potential applications in gas sensing [136].

An interesting aspect of polymer-based nanowires is that in addition these materials being grown on DNA templates, some of these materials can form nanowires on their own. Further, many of these nanowires are not only conductive but also serve as a template for metallic nanowires in order to enhance their conductivity [137]. Polymeric nanowires have proven useful in a variety of applications such as chemical and biological sensors [138], field-effect transistors [139], interconnects in electronic circuitry [140], and tools for observing 1D charge transport in materials [141]. In addition, luminescent conductive polymer nanowires are currently under investigation for their potential as components in polymer LEDS [142]. Both templating approaches as well as lithographic techniques have been used to align the nanowires. Different lithographic techniques include dip-pen lithography [142c], soft lithography [138a], edge lithography [143], and electron beam lithography [144] will be discussed later in the chapter. Conjugated polymeric nanowires have also been employed in light-harvesting action spectroscopy [145]. To this end, single molecular nanowires, and perlyne dye-endcapped multichromatic conjugated polymers were studied based on their exciton migration as a function of excitation energy, which allowed for the determination of the molecules and chromophores specifically contributing to absorption at a given wavelength. In general, the study was geared toward uncovering the role of intramolecular energy-dependent light-harvesting properties of the conjugated polymer system. Interestingly enough, specific molecules that absorb at long wavelengths were also shown to absorb at short wavelengths, which indicated a very broad chain absorption spectrum. Overall, it was found that the polymer-endcap donor-acceptor coupling was controlled by the coupling strength of the final chromophore in the chain to the endcap. Therefore this study demonstrated that polymers and endcaps were not isolates.

Currently, there exists a need to produce polymer based nanowires with a high aspect ratio [146] for enhancing applications. Wang and co-workers devised a method in which *in situ* polymerization of *o*-anisidine led to conducting polymer nanowires [147]. Intramolecular hydrogen bonding between substituents on the monomer and the polymer backbone promoted the alignment of the nanowires due to π-π stacking. Interestingly enough, reaction time and polymerization additives changed the morphology of the system from a random nanofiber arrangement to that of nanowire bundles. A mixture of nanowire bundles and loose nanofibers were formed, and therefore by drop-casting the mixture onto a solid substrate, the nanofibers were peeled off by cellophane tapes, which allowed for them to be imaged directly, as shown in Figure 8. Highly ordered nanowires were formed because poly-*o*-anisidine was utilized with its additional methoxy moiety, when compared to polyaniline, which allowed for a higher propensity of intermolecular non-covalent interactions. To prove the role of the methoxy group, two different polymers were synthesized (*N*-methyl-1,2-diamine and 2-methylthioaniline), where the oxygen in *o*-anisidine was replaced by nitrogen and sulfur. It was found that a somewhat aligned intertwined nanowire assembly resulted for the substitution of *N*-methyl-1,2-diamine, whereas a random arrangement of short nanofibers resulted with the use of 2-methylthioaniline. Although N-H•••N hydrogen bonding still occurred in the first case, it was not as extensive as the N-H•••O hydrogen bonding, which explained the lesser organization. Because sulfur did not form extensive hydrogen bonding networks, little organization was seen. As a control, *o*-anisidine was polymerized without any initiators, which resulted in nanowires with substantially less organization, as aggregates were the dominant structure formed. Various dopant acids were experimented with, but only HCl allowed for orderly nanowire formation. Because of the high order present in the nanowire alignment, these conductive nanowires could serve as efficient chemical sensors [136b, 138b], molecular memory devices [148], and capacitors [149].

In a separate study, conductive nanowires arrays of polypyrrole (PPY) and polyaniline (PANI) were formed using electrodeposition between electrodes in channels created on semiconducting and insulating surfaces [150]. The polymer deposition occurred via a three-electrode system, where deoxygenated pyrrole or aniline was placed in an electrolyte channel between the two electodes. The electro-polymerization occurred in the galvanostatic mode, where a current was applied between the two gold electrodes. A drop in the potential signified that polymer nanowires were formed and the reaction was terminated. The PANI nanowires grown between the electrodes were found to be continuous throughout the diameter between the electrodes, which was an improvement over earlier work, where the nanowires were heterogenous [144]. Depending on deposition time, uniform nanowires of PANI and PPY were made to vary in length between 0.5-13μm and 100 nm- 1 μm in diameter. In order to determine whether the polymeric nanowires could serve as sensors, a change of resistance as a function of environmental pH was examined for the PANI nanowires. The addition of HCl reduced the resistance. This was also seen in PPY nanowires, but the correlation was not as good. Therefore, these nanowires can potentially be embedded with bioreceptors, because of their facile and benign synthesis, and may potentially be utilized as nanobiosensors, which would be a big improvement over current multi-step processes needed to fabricate nanowires or carbon nanotubes and modify their surfaces.

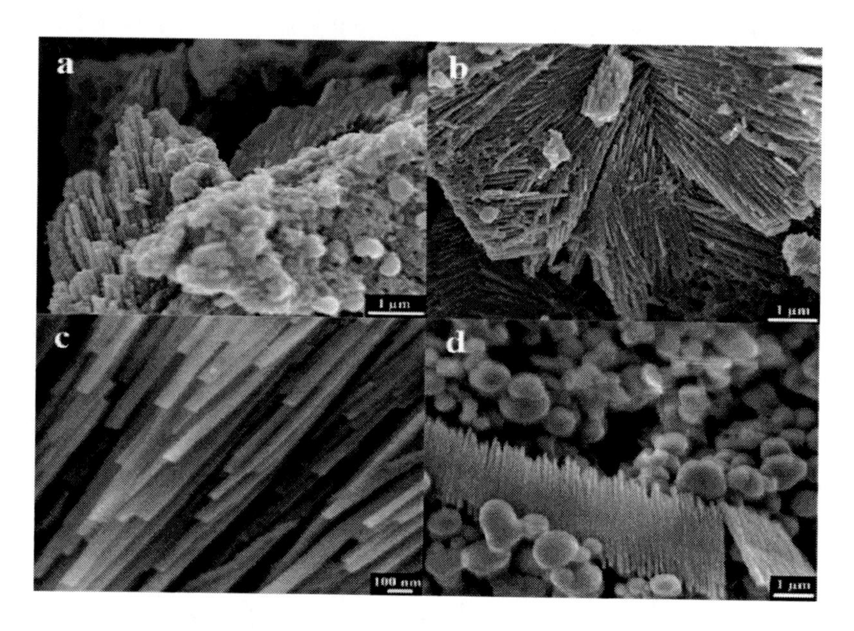

Figure 8. SEM images of poly-*o*-anisidine nanowires. (a) A drop-cast sample composed of 2 layers, entangled, nonoriented nanofibers on top of a bottom layer of highly aligned nanowires. (b) A view of the bottom layer after peeling off the top layer with a piece of cellophane tape. (c) High magnification image of aligned array of nanowires. (d) Poly-*o*-anisidine synthesized in the absence of initiators. (Ref. [147] *Copyright*, Reproduced with permission from *American Chemical Society* 2009).

In order to increase the conductivity, polymeric nanowires are often times functionalized with metal nanoparticles. For example, silver nanowires were generated on a polymer template by Metwalli and co-workers, using a flow-stream technique [137b]. This technique relied on the deposition of gold nanoparticles onto a microphase-separated cylinder, forming a polystyrene-*block*-polyethyleneoxide P(S-b-EO) diblock copolymer film. Because the copolymer film formed parallel cylinders, the attachment of the film to a fluidic cell allowed a gold nanoparticle solution to be pushed into the flow channel. The nanoparticles that interacted with the two copolymers began to assemble, which allowed for a hybrid metal/polymer film. Increased x-ray intensity was used to measure gold deposition, which occurred from a steady and continuous flow of gold nanoparticles. It was seen that during the flow-stream method, because the particles were mobile in the aqueous solution and are under hydrodynamic flow, particle interaction was dispersive as opposed to attractive, which resulted in less aggregates being formed. Therefore, once the gold nanoparticles anchored on the polymer (PS), there was an immediate attraction between the two systems. In addition, the PEO surface was hydrophilic, which also prevented particle aggregation. Thus, the polymer chains most likely stretched around and encompassed the particles, which resulted in larger particle sizes [151], favoring the formation of 1D structures, which presented a novel method in which gold nanowires were synthesized from a polymer template.

Virus-Assisted Nanowire Formation

The pursuit for new, environmentally friendly, and benign methods for the formation of nanowires has led to significant research into biomimetic methods to prepare nanowires with excellent size and shape control. One of the areas of research includes examining the potential for viruses to act as biological templates for the formation of nanowires [152]. It is interesting to note that viruses, with the ability to form crystalline semiconductor, metallic, oxide, and magnetic nanowires, prove to be an attractive template. One of the common viruses studied is M13, which is a bacteriophage with a high potential because it is inexpensive, easily produced, readily available, and only infects *E. coli* bacteria. By altering the phage-packed DNA, the virus can be manipulated so that it expresses specific peptides on its surface [153]. This protein may be filled with desired analyte receptors, which readily bind to small molecules [154], proteins [155] and DNA [156]. In a recent study, [157] Arter and co-workers examined a new strategy in which the properties of conducting polymer-based electrochemical sensors and virus-based molecular recognition were merged in order to build devices for biosensing. Briefly, M13 bacteriophage was grafted into an array of poly (3,4-ethylenedioxythiophene) (PEDOT) nanowires, leading to the formation of hybrids of conducting polymers and viruses. The virus was incorporated into the polymeric backbone of PEDOT during electropolymerization via lithographically patterned nanowire electrodeposition. The resultant arrays of virus-PEDOT nanowires allowed for real-time, reagent-free electrochemical biosensing of analytes. The scheme for the synthesis of the hybrids and the nanowire device formed is shown in Figure 10. This method is highly significant because it provides real-time, reagent-free biosensing method, that could potentially be utilized for early disease detection and diagnosis and provides a valuable route for electrical resistance-based sensing at room temperature.

Magnetic nanowires of CoPt and FePt and semiconductor nanowires of ZnS and CdS have also been synthesized through the incorporation of specific peptides into the M13 virus coat, which provides a template for the growth of the nanowires [158]. Various peptides were incorporated into the self-assembled virus capsid, which provided the linear template. This was accomplished by generating 1D nanostructures via virus directed growth and assembly of crystalline nanoparticles, followed by an annealing process, which resulted in high aspect ratio crystalline nanowires via aggregation-based crystal growth [159]. In addition to M13, tobacco mosaic virus (TMV) has also been utilized in the synthesis of CoPt and $FePt_3$ nanowires. It was found that TMV had the capability to biomineralize Co-Pt alloys [160]. Further, it was found that approximately one-third of the TMVs showed the formation of nanowires in the central channels, although the inner channels were not completely occupied and the nanowires were found to display ferromagnetic behavior. Such materials have applications in data storage applications and in spintronics.

Pioneering work by Belcher and co-workers recently demonstrated that cobalt oxide nanowires synthesized using viruses can be useful as lithium ion battery electrodes [161] by integrating specific gold binding peptides into the viral filament coat, thus resulting in the formation of hybrid gold-cobalt oxide nanowires. Specifically, the sequence tetraglutamate (E4-) was fused to the N terminal of each copy of the major coat p8 protein. E4 was utilized because of its known ability to bind metal ions through the interaction of the carboxylic acid side chain participating in ion exchange [162]. Furthermore, nanowires of Co_3O_4 were produced by first incubating E4 with cobalt chloride, followed by reduction with $NaBH_4$, which resulted in crystalline nanowire formation. In order to obtain hybrid nanowires, a peptide sequence that expressed both Au and Co_3O_4 was utilized. Positive electrodes were then utilized in order to determine the electrochemical potential of the Co_3O_4 nanowires. Reversible capacity was observed for the nanowires at a value twice that of carbon-based negative electrodes. Control experiments indicated a lack of any electrochemical actions, which further confirmed that the capacitance is a complete result of the nanowires. Cyclic voltammetry was utilized to determine the electrochemical potential of the hybrid nanowires.

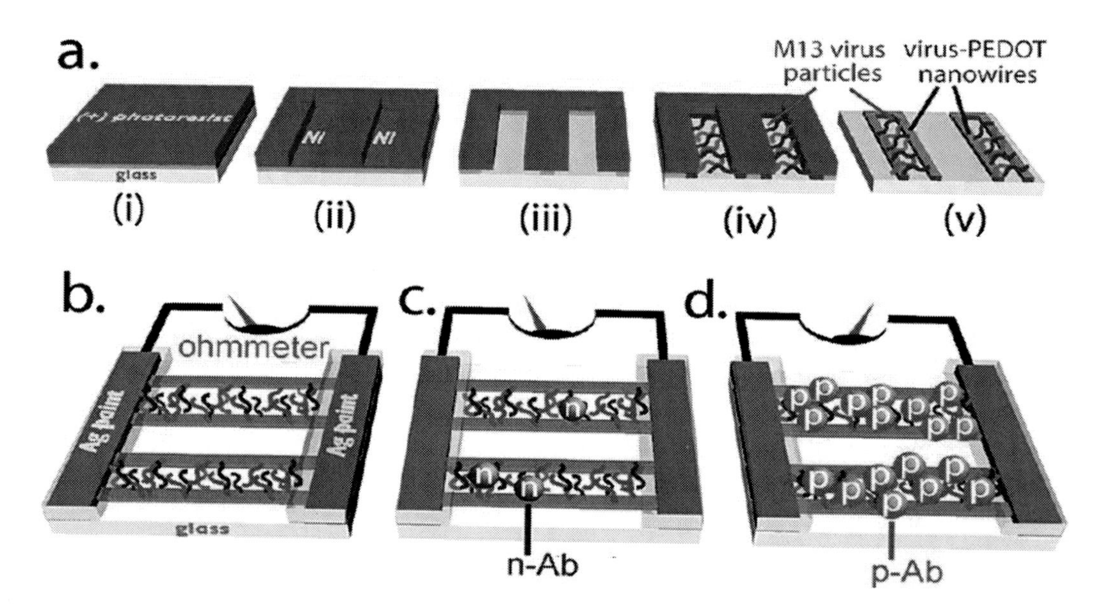

Figure 9. (a) Scheme showing the LPNE process for the synthesis of virus-PEDOT nanowires. (b) The fabricated virus-nanowire device utilized for resistance-based measurements. (c) Scheme showing minimal change in resistance after treatment with a non-binding negative control antibody (n-Ab, blue circles). (d) Ample change in resistance in the presence of virus-binding positive antibody (p-Ab, red circles). (Ref. [157] *Copyright*, Reproduced with permission from *American Chemical Society* 2010)

It was determined that the hybrid nanowires had a higher lithium storage capacity than that of the Co_3O_4 nanowires, which was a result of the electrochemical activity of Au [163]. Further, the hybrid nanowires were found to possess a 30% greater capacitance than Co_3O_4 nanowires. Therefore, the hybrid nanowires may be a new direction for producing flexible Li batteries at the nanoscale. Belcher's group has also investigated the spontaneous reduction of silver nanoparticles by introducing hexamer peptide sequences into yeast and M13 viruses to facilitate the growth of templated silver nanocrystals [164]. The main focus was on the function of carboxylic acid containing peptides expressed on yeast surfaces since the -COOH group is recognized to coordinate metal ions, which may possibly act as a nucleation site for nanoparticle formation [165]. Furthermore, glutamic and aspartic acid-rich peptides were found to play a central role in biomineralization as evident in their high population in the protein sequences. The genetically engineered yeasts were found not only to mediate the reduction of silver ions through expressed peptides, but also serve as templates for highly controlled spatial growth of nanoparticles. The principle of peptide-mediated reduction revealed by the genetically engineered yeast was further extended to a filamentous M13 virus scaffold for constructing crystalline silver nanowires.

Peptide Based Building Blocks

Natural proteins and oligopeptides adopt distinct secondary structures through interplay of hydrophobic interactions, hydrogen bonding, electrostatic, and van der Waal forces [166]. By mimicking these interactive forces, artificial peptides and peptide amphiphiles [167] can be utilized to form novel nanostructures. An affluence of stable nanostructures have been constructed from simple peptide building blocks [168] and the wide range of morphologies and macroscopic arrangements arising from the organization of peptide nanomaterials into macroscopic materials endeavor to match the requirements for specific applications [169]. The ability to predictably manipulate the self-assembly of autonomous units derived from biological macromolecules such as proteins is expected to lead to significant advances in diagnostics, bioimaging [170], and biomedical applications. Amino acids present a natural unique material for the design of intricate architecture as their combinations in oligopeptides and sizeable proteins correspond to an enormous chemical diversity [171]. In order to effectively engineer nanostructures using naturally occurring constituents, two conditions must be satisfied. First, the selected building block should have a high population time in the desired conformation, which requires relatively stable configurations such as a structural repeat taken from commonly occurring repeat protein architecture [172]. Secondly, the connection between the building blocks should be constructive, with an energy gap between the desired self-assembly and all other prospective associations. Based on concepts of hierarchical protein folding, the protein building block is defined by means of compactness, degree of isolation, and hydrophobicity of candidate constituents [173]. Interestingly, Nussinov and co-workers presented an algorithm in order to locate building blocks for a given protein tertiary structure. The algorithm allowed multiple dissections at each iterative level, creating a descending array of fragments. Each node in the descending anatomy tree was considered a single-segment building block, while the complete native structure of the protein was considered the starting root-node of the anatomy tree [174]. This model is considered a "practical" model for protein folding since the process is depicted as hierarchical [175]. The basic unit from which a fold is constructed is the outcome of a combinatorial assembly of a set of building blocks, a highly populated fragment in a given protein structure [176]. Thus when the building block is cut from the protein chain, the most highly populated conformation of the resulting peptide is believed to be similar to that of the building block when in the native protein. The ultimate objective was to be able to formulate a constructive shape, and create it by putting together building block parts. From a technological perspective, the principal advantage is the anticipation that they will be simpler to synthesize and handle. Furthermore, smaller sizes offer a greater number of shapes, chemical properties, and tuning, as well as relative shorter synthetic routes, which in turn presents greater opportunities such as introducing mutations with variants of native residues. For large nanostructure designs which are required to be rigid, the mutual stabilization between building blocks is imperative, which elucidates the significance of self-assembly, where stabilization increases with the number of units in the assembly.

Zheng and co-workers utilized molecular dynamics simulations to construct stable nanotubes by stacking monomeric, naturally occurring building blocks from left-handed β-helices [177], as shown in Figure 10. They predicted the ability to manipulate the peptide by introducing a conformationally constrained synthetic residue, 1-aminocyclopropanecarbo-xylic acid, and studied its capacity to enhance the stability of nanotubular structures. In fact, one of the early self-assembling fibrous systems based on designed linear peptides that adopted α-helical conformations were described in 1997 by Kojima and co-workers where a single-peptide system (the α3-peptide) that assembled to form fibers several microns long and 5-10 nm thick was examined [178].

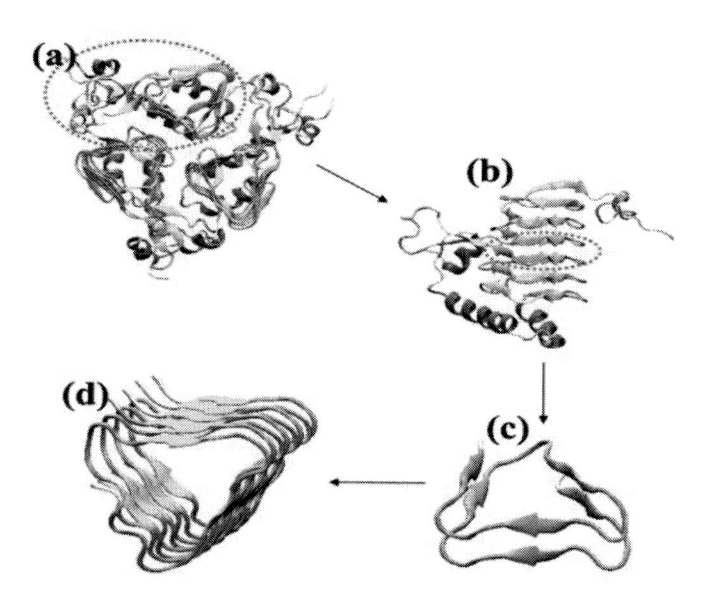

Figure 10. Scheme showing the construction of a nanotube using the naturally occurring protein building block from a *β*-helix (taken from galactoside acetyltransferase, PDB code 1KRR). (a) The trimeric crystal structure of galactoside acetyltransferase (GAT) from *E. coli*, with three left-handed parallel *β*-helix domains. (b) The monomeric structure of GAT (circled) taken from the trimeric GAT structure. (c) A single building motif (circled) taken from the monomeric 1KRR structure with selected residues 131-165. (d) A nanotubular structure obtained by stacking four repetitive building motifs on top of each other. (Ref. [177] *Copyright*, Reproduced with permission from *American Chemical Society* 2007).

In 2000, Woolfson and Ryadnov introduced a self-assembling fiber (SAF) system comprised of two short *de novo* leucine-zipper peptides intended to assemble in a sticky ended manner to provide building blocks for elongated coiled coil-based fibrils which in turn assembled to form fibers [179]. More recently, scientists have turned to testing the sensitivity of 'smart materials' that have systems which respond to solvent conditions (pH, salt), temperature, electronic or photonic materials, metals, and other complex ligands present [180]. Finding specific means to incorporate responsiveness or 'smartness' using peptides by sequence manipulations gives rise to an astounding number of different structures, and this is nurturing the expansion of highly efficient and hybrid materials. The prospective applications of smart materials based on peptide based building blocks ranges from the construction of scaffolds for adhesion, growth and differentiation of cells, to templating inorganic materials [181] and sensing. Technologically, metallization of peptide nanostructures [182] leads to conducting nanocircuits, (similar to those mentioned earlier for DNA), nanosensors and ordered monolayers that can serve as templates for biomineralization and directed photonic crystal growth [183].

Biomedical Applications of Peptide Based Nanotubes and Nanowires

The capacity to manipulate protein structure not only enables an enhanced comprehension of the principles of protein folding, but also advances the development of novel therapeutics and drug-delivery vehicles [184]. However, the fabrication of structures that resemble natural biomaterials, even at the lowest level of hierarchical organization, is complicated because of the involvement of two dissimilar organic and inorganic nanophases that have specific spatial relations with respect to one another [185]. In order to overcome this challenge in an artificial system, an organic nanophase designed to exert control over crystal nucleation and growth of the inorganic component must be arranged [186]. Stupp and co-workers have reported the use of self-assembly and mineralization to prepare peptide nanotubular composite materials that could potentially reconstruct the structural orientation between collagen and hydroxyapatite observed in bone [187]. The nanotubular composite was assembled by the self-assembly, covalent capture, and mineralization of a peptide-amphiphile (PA). Such biomaterials are anticipated to present a provisional three-dimensional scaffold to interact biocompatibly with cells to control their function, guiding the multifaceted multi-cellular progression of tissue formation and regeneration [188]. Due to the inability to treat many diseases solely by small-molecule drugs, cell-based therapy is rapidly emerging as an alternative approach [189]. Consequently, attempts have been made to formulate new scaffolds to aid in controlled drug release, tissue repair, and tissue engineering [190]. The ideal biologically compatible scaffold for sustaining cell attachment and growth should have no cytotoxicity, possess features that promote cell-substrate interactions, and incite minimal immune responses and inflammation [191]. When investigating tissue dynamic regeneration after damage, the role of the extracellular matrix (ECM) and its interactions with cells must be accentuated to tackle mimicry and manipulation of those interactions for tissue engineering [192]. The intricate, fibrillar architecture of natural ECM components has inspired researchers to synthesize materials with similar structures such as nanofibers [193]. Significant progress has been made using supramolecular self-assembly to form nanofibrillar matrices *in situ* where self-assembly of oligomeric peptide and amphiphilic building blocks formed higher order structures via noncovalent interactions [194]. While many of these systems entail self-assembly under environments intolerable to cells, several can gel at close to-physiological conditions.

For example, Zhang and co-workers developed a class of nanofibrillar gels with very high water content, cross-linked by the spontaneous self-assembling of ionic self-complementary oligopeptides [195]. The components of the hydrogel scaffold were amphiphilic oligopeptides with alternatively repeating constituents of positively charged lysine or arginine and negatively charged aspartate and glutamate; amino acid sequences specifically designated to facilitate the formation of a hydrogel scaffold through spontaneous molecular self-assembly [196]. The periodic repeats of alternating ionic hydrophilic and hydrophobic amino acids in the sapeptides were first found as a segment in the yeast protein, zuotin [197]. Several type I sapeptide scaffolds are known to support cell attachment of a range of mammalian and avian primary and tissue culture cells [198]. Furthermore, under appropriate culture conditions, these matrices have been demonstrated to retain the functions of differentiated neural cells and chondrocytes [199] and promote the differentiation of liver progenitor cells [200]. Thus, despite not being outfitted with any specific biofunctional ligands, these gels are scaffolds that biomechanically coordinate cells in a 3D manner. Deming and co-workers prepared fibrillar hydrogels from diblock copolypeptide amphiphiles [201], which self-assembled at low solid content and mild gelation conditions and support cell encapsulation [202]. Stupp and co-workers have shown the advancement in such supramolecular gels by synthesizing self-assembling oligomeric-amphiphiles that permit integration of specific biomolecular signals, which highlights the prospective of incorporating both biomechanical and biomolecular cues within scaffolds [203]. Innovative medical advances may arise from combining new scaffolds for cell-based therapies with stem cell technology. The encapsulation of stem cells in the sapeptide scaffold may allow them to differentiate into desired cell types with specific growth factors and cytokines, which will then be followed by the appliance of cell-scaffold systems into desired tissues. Thus these biocompatible and biodegradable sapeptide scaffolds have an expansive range of applications for tissue repair and engineering [204].

Longer designed peptide surfactants with a hydrophilic head of one or two charged amino acids and a hydrophobic tail of four or more consecutive hydrophobic amino acids are recognized to possess the ability to form a complex of cationic or anionic open-ended nanotubes upon dissolution in water [205]. Specific amino acids play a vital role in determining the final structure of the nanoconstruct: alanine and valine produce homogenous and stable nanotubes, while for cationic peptides, lysine or histidine are preferred. Due to steric effects [206], aspartic acid (D) has sparked interest since it is believed to have been present in the prebiotic environment of the early earth [207]. Envisaged applications consist of carriers for the encapsulation and distribution of small, water-insoluble molecules and large biological molecules, including negatively charged nucleic acids inside the cell, as well as cosmetic practices.

Ghadiri's group discovered a class of organic nanotubes based on judiciously designed cyclic polypeptides with an alternating number of D- and L- amino acids which interact through hydrogen bonding into an array of self-assembled nanotubes [208]. These cyclic peptides stacked through extensive intermolecular hydrogen bonding to form long cylindrical structures with an anti-parallel β-sheet structure [209] as shown in Figure 11. The outer surface of the nanotubes is defined by all the amino acid side chains and thus can be controlled by peptide design or by covalent incorporation of polymers, producing polymer shells surrounding the nanotubes thus leading to the development of new controlled-release drug delivery carriers [210].

Synthetic peptides of alternating hydrophilic and hydrophobic amino acid residues have a tendency to adopt a β-sheet structure [211]. Rich and co-workers have probed a 16-residue peptide, EAK 16, originally found in zuotin, a yeast protein, due to their distinctive ability to form insoluble macroscopic membranes. Owing to the molecule's hydrophobic alanine side chains on one side and self-complementary pairs of positively charged lysine- and negatively charged glutamic acid side chains on the other surface, spontaneous association to form a macroscopic membrane was rendered achievable. The efficacy of the membrane in the field of drug delivery arises from its resistance to digestion with several proteases including trypsin, α-chymotrypsin, papain, protease K, and pronase, in addition to remaining stable when heated to 90°C at various pH values. Thus the combination of the EAK16 membrane's extreme stability in serum, high resistance to proteolytic digestion, simple composition, apparent lack of cytotoxicity, and easy synthesis in large quantities makes it a perfect candidate for slow-diffusion drug-delivery systems. Furthermore, the likeness of EAK16 fibers to the insoluble proteins found in numerous pathological diseases suggest that it may prove to be a constructive model scheme for investigating the characteristics of structures and sequences that fabricate such remarkable properties [212]. Drugs that inhibit the self-assembly of the peptides may be valuable for the treatment of these diseases [213].

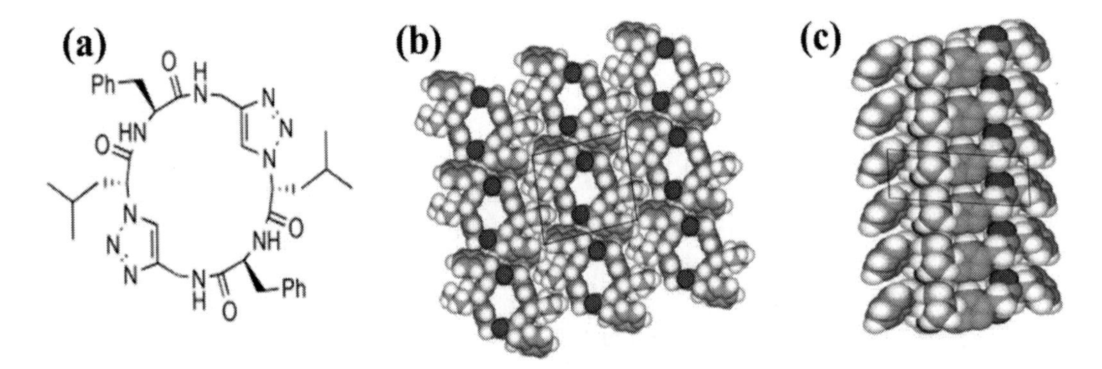

Figure 11. (a) Chemical structure of a single macrocyclic heteropeptide (b) crystal packing viewed along the tube axis (solvent omitted), (c) one tube viewed from the side (solvent omitted). (Ref. [209b] *Copyright* , Reproduced with permission from *American Chemical Society* 2003).

The selective delivery of newly synthesized therapeutic agents to target cells or organs is a central challenge in every clinical procedure, where peptide nanoconstructs are touted to have numerous applications [214]. Another area where there is intensified attention for tissue engineering is the ideal artificial arterial substitute [215]. Collagen gel scaffolds were first used in the construction of arterial grafts by Weinberg and Bell. Their arterial graft consisted of an adventitia of bovine fibroblasts implanted in a collagen gel. However, this three-layered arterial structure could not be used for implantation since after a period of static culture, it had a burst strength of 90 mm Hg which is significantly lower than the normal systolic blood pressure, and lacked elastin, an essential element to the compliance of a vascular prosthesis [216]. Consequently, investigators have focused their attention on improving the strength of these constructs and their approach has been to surround them with a supporting sleeve of collagen which resulted in artery constructs with a burst pressure of up to 650 mm Hg, significantly higher than without the support sleeve [217]. These collagen-based tissue engineered vessels have been implanted into dogs as vena cava bypass grafts with a 64% patency at 24 weeks [218]. Since the shortcoming of this model was found to be the presence of a non-degradable support, an innovative method involving the amalgamation of the collagen-gel scaffold with endothelial seeding and a biodegradable support sleeve is now considered a more practical option [219]. Furthermore, L'Heureux and co-workers developed a model involving smooth muscle cells and fibroblasts grown in cell sheets to create a blood vessel substitute derived from their work with skin substitutes [220]. The cell self-assembly model was made with intact layers of human vascular cells grown to over confluence in order to form viable sheets of cells and extracellular matrix. In order to assemble a construct, the smooth muscle cell sheet was rolled over a permeable mandrel to adopt a nanotubular configuration forming the medial layer [221]. Similarly, a fibroblast sheet was rolled over the top of the smooth muscle sheet to form an adventitial layer. The nanotubular structure was cultured to maturity and during this progression, the fibroblasts produced and organized matrix proteins that enhanced the structural characteristics of the tissue analog [222]. Human endothelial cells were then seeded onto the lumen of the matured vessel to form a confluent monolayer and thus L'Heureux and co-workers were able to create constructs that could withstand more than 2000 mm Hg pressure before bursting [223]. In this way, numerous cross-linking techniques have been investigated in the struggle to find the ideal procedure to stabilize the collagen-based structure of the tissue while maintaining its mechanical reliability and natural conformity [224].

For prosthetic and robotic applications, the development of a robust and low-power artificial skin that is responsive to the environment through touch is of profound interest [225]. Currently, artificial electronic skin, known as "e-skin" has been developed, which requires large-scale incorporation of pressure-sensitive mechanisms with an active-matrix backplane on a thin plastic support substrate. Previously, organic transistors that are considered malleable such as polymers and rubbers had been used as this backplane, suggesting the achievability of the notion [226]. However, the disadvantage of using such organic molecules, although they offer capable material systems for adaptable electronics, is their comparatively low carrier mobility which often calls for large operating voltages. Fabrication of sensitive skin consists of thousands of pressure sensors which cannot be realized with present-day silicon-based electronics [227]. Thus, inorganic crystalline semiconductors nanowires present an evident advantage since they demonstrate high carrier mobilities with exceptional mechanical flexibility arising from their miniaturized dimensions [228]. Their diameter of approximately 30 nm makes them exceedingly flexible with superb mechanical reliability and robustness and even from the cost perspective, the capacity to synthesize and print single-crystalline structures without the use of crystalline wafers and/or complex epitaxial growth processes is highly attractive [229] as shown in Figure 12. In order to exploit their distinctive physical properties, a uniform assembly of ordered nanowire arrays is essential. Javey and co-workers utilized the contact-printing method to assemble Ge/Si core/shell nanowire arrays on a polyimide substrate followed by device fabrication processing [230]. The fully fabricated e-skin can effortlessly be bent or rolled to a small radius curvature, which validates the outstanding mechanical flexibility of the substrate and its integrated electronic components.

A high degree of alignment and uniformity for the contact-printed nanowire arrays is critical to achieve high-performance and large-scale electronics that are configured as functional systems at such a high degree of complexity, thus a laminated pressure-sensitive rubber (PSR) is used as the sensing element [231]. A longer-term goal would be to use the e-skin to restore the sense of touch to patients with prosthetic limbs, which would involve substantial progression in the integration of electronic nanowires as sensors with the human nervous system.

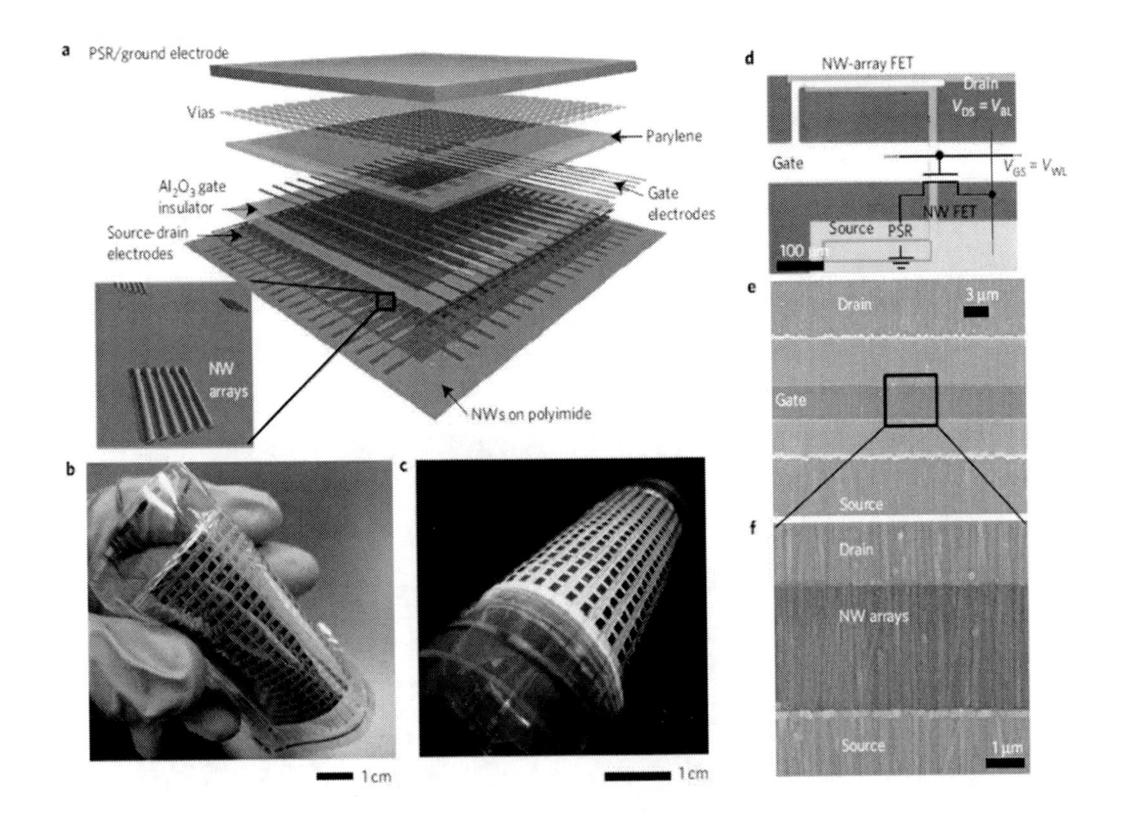

Figure 12. (a) Scheme of the passive active layers of nanowire e-skin; (b, c) Optical photographs of a fully fabricated e-skin device (7 x 7 cm^2 with 19 x 18 pixel array) under bending (b) and rolling (c) conditions; (d) Optical microscope image of a single sensor pixel in the array depicting a Ge/Si nanowire-array FET (channel length ~ 3μm, channel width ~ 250 μm) integrated with a PSR. The circuit structure for the pixel is shown in e, f, Scanning electron micrographs of a NW-array FET, showing the high degree of nanowire alignment and uniformity achieved by contact printing with a density of ~ 5NWs μm^{-1}. (Ref. [229] *Copyright,* Reproduced with permission from *Nature Publishing Group 2010*).

Neural interfaces are anticipated to play a central part for diagnosis and therapy of medical conditions such as pain control, facilitation of motor control after stroke, trauma, or neuro-degenerative disorders [232]. In order to stimulate and trace nerve signals while being biocompatible, these interfaces need to be recurrently implanted into the central nervous system (CNS) [233]. Recent experiments have revealed that nanostructured surfaces potentially improve electrical properties of electrodes and elicit reduced tissue responses [234]. Upon fibers that are tens of microns in diameter, cells appear to respond as though to a 2-D substrate, obtaining an abnormal flat shape, leading to a nonphysiological, asymmetrical occupation of adhesion receptors; such matrices have previously revealed outstanding accomplishments in tissue engineering applications such as the reconstruction of a dog urinary bladder [235]. This model has served as scaffolds for neural stem cells to facilitate regeneration after brain injury in a mouse stroke model led by Snyder and co-workers. Neural stem cells (NSCs) defined as self-renewing, primordial cells with the capacity to give rise to differentiated progeny within all neural lineages in all regions of the neuraxis, are assumed to exit in embryonic and fetal germinal zones, where they participate in CNS organogenesis [236]. Next, "biobridges" composed of NSCs seeded upon polyglycolic acid (PGA)-based scaffolds were implanted in an attempt to heal an extensively injured brain [237]. Other examples of exploring artificial stem cell microenvironments involve Mahoney and Saltzman's design of a synthetic microenvironment useful as a transplantation medium based on polylysine-coated poly (lactide-co-glycolide) microparticles loaded with nerve growth factor-beta [238]. The integration of a cell-adhesive matrix and a controlled-release scheme for a morphogenetic factor allowed them to control fetal brain cell survival and differentiation in a rat model [239].

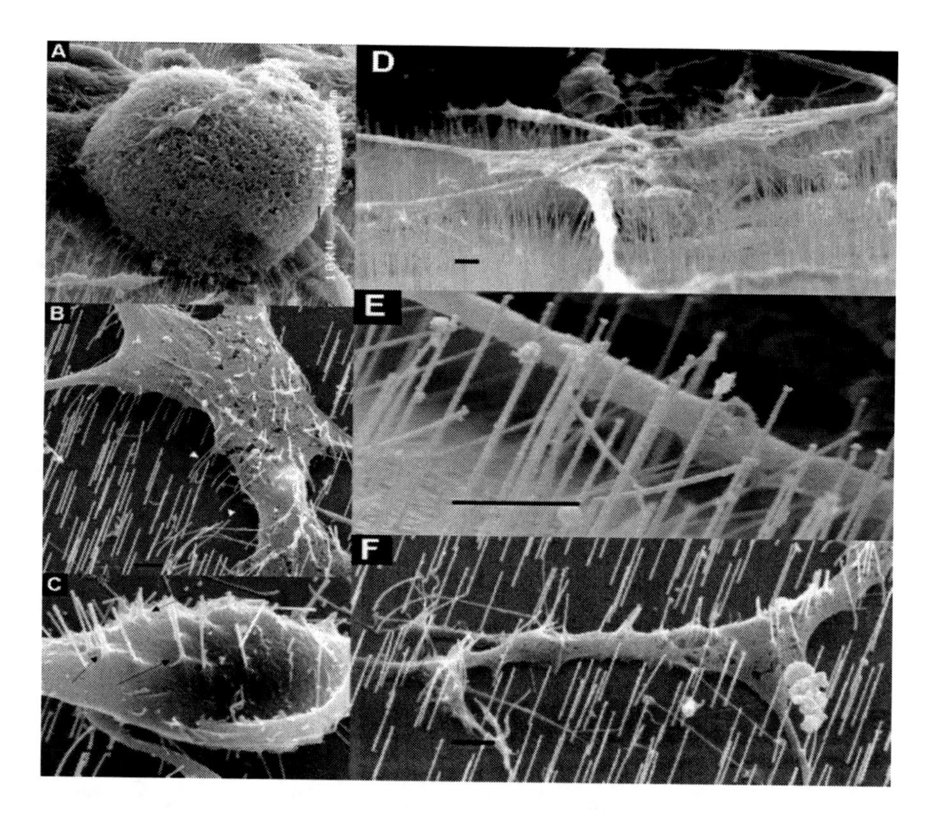

Figure 13. SEM-images showing interaction between GaP nanowires and neuronal cells. (A) Neural cell body on nanowires surfaces. (B) Well-spread non-neuronal cell penetrated by numerous nanowires. The bending of the nanowires is indicated by white arrow heads. (C) Underside of cell body (flipped over at rinsing) penetrated by nanowires and membrane adhesion to the nanowires shown by arrows. D-F demonstrate three different types of process growth. (D) Processes growing on top of the wires, attached to their tips. (E) Axon growing in the space between the substrate and the wire tips, adhering to the sides of the wires. (F) Process spreading over the bulk substrate, apparently engulfing nanowires encounterd along its path. Scale bars 1 μm. (Ref. [242] *Copyright*, Reproduced with permission from *The American Chemical Society* 2007).

Martin and co-workers investigated neuronal probes using the synthetic peptide DCDPGYIGSR which was co-deposited alongside a conductive polymer polypyrrole (PPy) onto the electrode surface by electrochemical polymerization [240]. It was found that the peptide was entrapped in the PPy film and did not diffuse away within weeks of soaking in aqueous solution. Thus the coated probes were implanted in a guinea pig brain and significantly more neurofilament positive staining was found on the coated electrode which indicated that the coatings had established strong connections with the neuronal structure in vivo [241]. Danielsen and co-workers have shown that neurons are capable of thriving on gallium phosphide (GaP) nanowire substrates [242], suggesting that GaP nanowires are biocompatible. By having a diminutive cell-electrode distance, recording from individual neurons is rendered possible and currently, the recording and stimulation of single neurons using nanowire transistor arrays is feasible [243]. In order to investigate the tissue response after nanowire injection into the brain, they tackled the biocompatibility of the nanowire as a nanorod, and to eradicate potential concerns arising from the chemical nature of the nanowires, they were coated with chemically biocompatible materials [244]. SiOx surfaces are known to have biocompatible properties with the only potential hazard for the body being their size and shape [245] whereas GaP is known to partially dissolve and accumulate in the kidney and liver when implanted into the rat soft tissue and thus cannot be considered a chemically biocompatible material [246]. The nanowires were suspended in a physiological buffer solution and implanted in the rat striatum and the tissue response was probed with respect to quantification of essential components such as the microglia that detect and engulf any infectious substance as well as dead cell debris, astrocytes involved in brain-tissue repair by forming a protective scar around the injury to limit the spread of inflammatory response to the injured area, neurons, and the total number of cell nuclei [247]. It was found that nanowire implantation elicited a considerably large astrocyte reaction as well as a microglial response, both of which declined over time. The SEM images showing the interactions between neuronal cells and GaP nanowires is shown in Figure 13. Additional data has also suggested that some nanowires were able to pass the blood-brain barrier and leave the brain. Although there was no evidence of chronic nanowire toxicity, there was instability in the numbers, suggesting short- and long-term effects of the nanowires and thus necessitating precaution and additional investigation.

Biosensing

Biological sensors based on nanowire/nanotube field effect transistors (FETs) are one of the most capable applications of bionanotechnology and have been used to detect a large variety of biological molecules such as proteins [248], as well as monitor enzymatic activities [249] and observe cellular signaling/responses [250] with sensitivity and response time comparatively better than conventional techniques [251]. Nanostructure-based sensors have served as a dominant detection platform for a broad range of applications including biological sensors, electrochemical sensors, gas sensors, optical sensors, pH sensors, and orientation sensors [252]. These nanosensor devices have an array of critical features that sets them apart from other sensor technologies available today including high sensitivity, exquisite selectivity, rapid response and recovery, and the potential for integration of addressable arrays on a massive scale [253]. Hahm and Lieber demonstrated that p-type silicon nanowires

(SiNWs) functionalized with peptide nucleic acid (PNA) receptors can serve as ultrasensitive and selective real-time DNA sensors [254]. Furthermore, they examined a highly sensitive detection scheme for identifying small molecule inhibitors that did not require labeling of the protein that could be carried out using SiNWs FET devices [255]. They connected the Abl tyrosine kinase to the surface of SiNWs FET within microfluidic channels to create active electrical devices and investigated the binding of ATP and competitive inhibition of ATP binding with organic molecules [256].

Until recently, in order to achieve high sensitivity in gas phase sensors, physical separation techniques such as chromatography or spectroscopic fingerprinting were exploited since protein bimolecular sensors, although capable of utilizing the "lock-and-key" interactions to achieve selectivity, are limited [257]. However, the associated instrumentation is limited in portability, thus ruling out the possibility of implantable or wearable sensors, and requires skilled human operators for its manufacture [258]. Consequently, "Electronic noses," a form of vapor analyzer designed to mimic the olfactory system via the integration of sensor arrays, offer a promising alternative with the potential for continuous real-time monitoring and discrimination of large families of gases [259]. The sensors are designed in a combinatorial fashion to yield varying responses to different analytes and for the case of conventional e-noses, selectivity is augmented by utilizing larger sensor libraries [260]. In recent times, electronic noses based on arrays of semiconducting nanowires, labeled "nano-noses" have been implemented [261]. These novel structures possess ppb sensitivities, an outcome of the nanostructure diameters being similar to the width of the surface space charge region [262]. In some cases, these nanosensors have shown some selectivity to certain molecules aided by chemoselective polymer coatings [263], surface chemistry functionalizations [264], and varying metal oxide material compositions [265].

Heath and co-workers established that covalently coupling oligopeptides to the surfaces of silicon nanowires (SiNWs) provide an appealing model system for selective sensors by displaying high degrees of selectivity to small molecules such as acetic acid (AcOH) and ammonia (NH_3) so they combined theoretical modeling with experimental data in an attempt to comprehend the selectivity mechanism of the peptides on their target analytes [266]. It was found that the three-element nanowire/peptide sensor arrays, fabricated to evaluate the sensitivity of the NW-Peptide sensors, had the capacity to separately detect AcOH and NH_3 from gas mixtures even in the presence of background gases. The SiNWs were fabricated using the superlattice nanowire pattern transfer (SNAP) method [267] and their surfaces terminated in intrinsic silica, which permits facile nanowire surface modification without strongly affecting the semiconducting core [268]. The nanowires were found to perform as excellent field-effect transistors on both solid and flexible plastic substrates with mobilities comparable to that of bulk silicon [269]. These outcomes provide a model platform in the use of sensors for targeted applications such as non-invasive breath monitoring for molecular disease indicators, as well as food spoilage or chemical threat detectors [270]. The advancement in this field is evident by observing the work of those such as Zhou and co-workers, who have tackled the calibration techniques of indium oxide nanowire biosensors which give radically suppressed device-to-device variation in sensing response [271]. Their mathematical method will be useful for other nanowire FET-based sensor arrays, making multiplexed sensor arrays a functional solution for measuring and monitor multiple analytes (biomarkers) in tandem, and thus is a substantial step toward the broader application of nanowire biosensors.

LIPID NANOWIRES AND NANOTUBES

The query that scientists are endeavoring to answer is whether there is an efficient way to coalesce artificial devices to biological mechanisms to attain superior functionality, given that such integration could potentially lead to paradigm-shifting advances in biotechnology and medicine [272]. Lipids are the basic building block of biological membranes and can self-assemble in liquid media into diverse aggregate morphologies, depending on the molecular shape and solution condition such as the lipid concentration, electrolyte concentration, pH value, and temperature [273]. A limited number of lipid molecules are also capable of self-assembling into open-ended, hollow cylindrical structures, which are composed of rolled-up bilayer membrane walls [274]. The consequential lipid nanotubes (LNTs) have unique properties that promote template-synthesized one-dimensional nanostructures such as nanowires, which no other individual templates possess [275]. Firstly, in sharp contrast to carbon nanotubes, the tubular structures are of interesting hydrophilic internal and external membrane surfaces and can provide asymmetrical surfaces as well as identical ones [276]. These advantages make them ideal candidates for controlled release, chemical reaction and selective filling of nanomaterials [277]. Secondly, the capacity to precisely control the inner and outer diameters, length, and wall thickness in an extensive range from several nanometers to micrometers permits the direct determination of their aptness for diverse technological applications [278]. For example, Spector and co-workers proposed that reducing the LNT membrane thickness may cut the lipid material cost by at least 80% in practical applications [279]. Thirdly, the incorporation of desired functional groups into lipid monomers allows the outer and inner surfaces of the LNT to be functionalized and modified [280]. Fourthly, the LNTs can be easily manipulated, positioned, and aligned on diverse substrates with various techniques including microextrusion, microfluidic network, magnetic field, and biorecognition [281]. Furthermore, templating the LNTs enables the production of diverse 1-D nanostructures, given that the neat hollow cylinders of the LNT are capable of modulating the nucleation, growth, and deposition of inorganic substances on not only their external and internal surfaces, but in the hollow cylinder and in their bilayer membrane wall [282]. This potential architecture is deemed one of the largest self-organized nonliving structures yet observed, and thus templating of the LNTs is a promising route to making diverse 1-D nanostructures. In the near future, both the exploitation of the molecular recognition ability of LNTs and the recent progress in the development of a 3-D manipulation technique will facilitate the association of templated nanowires to external nanodevices in a meticulous and convenient way [283].

Supramolecular nanotube architectures for self-assembled lipid nanotubes have attracted much awareness and there are current investigations focused on precise dimension control [284], alignment on solid substrates [285], and rational functionalization of their surfaces [286]. It has been found that hydrophilic 10-100 nm scaled hollow cylinders of the lipid nanotubes can supply suitable nanospace for biomacromolecules at least 10 times larger in dimension compared to the molecules used for conventional host-guest chemistry [287]. Consequently, they have the capacity to serve as containers for nanoscale reactions [288], channels for nanoscale separation [289], and are potential carriers for drug delivery [290]. Lipid nanotubes with different inner-and outermost lipid membrane surfaces are fascinating nanoarchitectures applicable to the particular adaptation of each surface [291], selective filling of nanomaterials into the hollow cylinders [292], controlled release [293], and formation of templates for the manufacture of inorganic nanomaterials [294].

Shimizu and co-workers covalently linked fluorescence-donor dye as an optical sensing moiety to the amino groups on the inner surfaces of the lipid nanotubes [295]. They found that the self-assembly of a series of unsymmetrical bolaamphiphiles, *N*-(2-amino-ethyl)-*N*′-(β-D-glucopyranosyl)-alkanediamide in alkaline aqueous solutions can selectively give nanotubes by optimizing the initial molecular packing and the length of the oligomethylene spacer of the bolaamphiphiles [296]. The obtained nanotubes, named "W-nanotubes" were found to have distinctive inner and outer surfaces covered with amino- and glucose head-groups respectively, and had the capacity to encapsulate negatively charged nanoparticles as well as the spherical protein ferritin into the hollow cylinder. Combining time-lapse fluorescence microscopy with the fluorescence resonance energy transfer (FRET) system enabled the visualization of dynamic encapsulation and nanofluidic behavior of the spherical protein ferritin, gold nanoparticles, and a label reagent in the nanochannels shaped by the nanotube hollow cylinders [297].

The functionalization of nanowires with lipid bilayers represents a vigorous tactic for using membrane proteins in bionanoelectric devices. The impermeability of lipid bilayers to ions and large molecules makes them an instinctive choice to avert non-specific adsorption as well as providing electrical insulation [298]. The importance of lipid molecule design is evident from the fact that common phospholipids are not capable of spontaneously forming 1D tubular structures in aqueous solutions unless they are forced into a metastable configuration by an external force [299]. In order to successfully form a lipid bilayer on a highly curved support surface of a nanowire is to ensure that the energy of the interactions of that surface with the lipid bilayer counterbalances the elastic energy of the bilayer deformation [300]. In addition, bilayer fluidity, the ability of the lipid molecules to diffuse freely into the bilayer plane, is the key to stability, defect healing, and membrane protein incorporation [301]. Noy and co-workers revealed that lipid vesicle fusion onto carbon nanotubes (CNTs) customized with several hydrophilic polymer layers generates a continuous lipid bilayer shell around the nanotube [302] and consequently showed that an analogous lipid bilayer shell could assemble around a hydrophilic polysilicon nanowire without any polymer cushion layer [303], as shown in Figure 14. They also discovered that these bilayer shells are capable of assembling on larger nanowire structures such as that of SnO_2 nanowire waveguides [304]. Additionally, McEuen and co-workers who studied the diffusion of lipid molecules over a CNT covered by a planar lipid layer established the astounding aptitude of lipid bilayers to adapt to the nanoscale curvature of 1D substrates, and their experiments revealed that small single-walled nanotubes did not present a barrier for lipid molecule diffusion [305].

Martinez and co-workers have integrated their versatile, self-assembled nanoplatform with a field-effect transistor (FET) that involves membrane-bound proteins that arbitrate the majority of signaling in living systems [306]. The nanoplatform is made up of a 1D lipid-bilayer membrane covering a nanowire template [307], which effectively mimics a cellular membrane composed of phospholipids, molecules with a hydrophobic head and two hydrophobic tails that arrange themselves into a bilayer. By mimicking the outer layer of a membrane, the bilayer functions as an impermeable barrier around the nanowire while still allowing for the incorporation of membrane proteins that act as channels for specific signaling ions to pass through. In their experiment, the small semiconducting SiNW was fixed to a pair of metallic sources and drain electrodes passivated with a layer of silicon nitride, which protects and insulates the electrodes from the buffer solution used for the experimentation involving biological molecules [308], as shown in Figure 15. The crucial step to create a FET that permitted current to conduct between sources and drain electrodes when a suitable voltage was applied, was the formation of the lipid bilayer on the surface of the SiNW using the native silicon oxide as a template. In the self-assembly process, lipid vesicles (small spheres of lipid membrane), were fused on the nanowire surface forming a continuous coating, which was proven via cyclic voltammetry to provide an effective shield that insulates the nanowire surface from the surrounding solution [309].

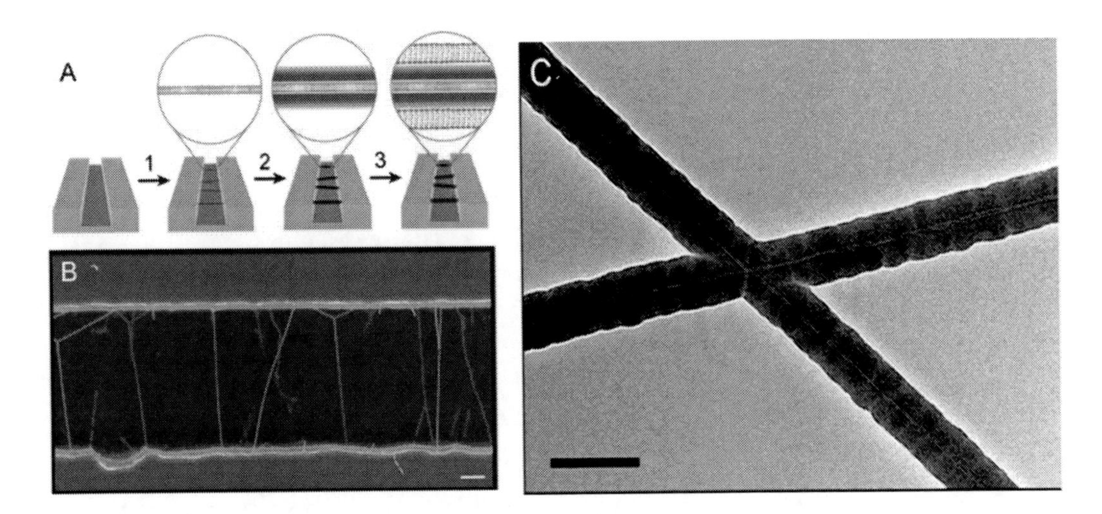

Figure 14. Silicon-coated carbon nanotubes as substrates for studying 1-D lipid bilayers. (A) Scheme showing the assembly of lipid bilayer membranes on silicon nanowires. Step 1: CVD growth of suspended carbon nanotubes; step 2: deposition of Ti adhesion layer and amorphous silicon on carbon nanotubes; step 3: formation of supported bilayer by vesicle fusion. (B) SEM image of silicon-coated carbon nanotubes suspended over a 5 μm wide channel. (C) TEM image of two intersecting carbon nanotubes coated with a layer of amorphous silicon. Scale bars: 1 μm (B), 100 nm (C). (Ref. [300] *Copyright,* Reproduced with permission from *American Chemical Society* 2007).

Further experiments demonstrated the capability of the device platform to not only use biological components as functional parts, but to control the actual functionality of the biological materials. For example, when a voltage-gated peptide Alamethicin (ALM) was incorporated into the nanowire device, the electric field of the device could be manipulated to control the opening and closing of the peptide pore with little response to pH changes in the environment [310]. Evidently, the lipid bilayers provided a matrix for a virtually unlimited number of membrane proteins with diverse functions and these biomimetic devices have the potential to power many applications in biosensing and bioelectronics. This was fascinating because the group succeeded in manipulating matter to a degree corresponding to the extent of biological objects. Even primordial steps where nanowire electronics were interfaced with membrane channels and pumps, neurons, and heart tissue cells, insinuate the power of this integrated device model.

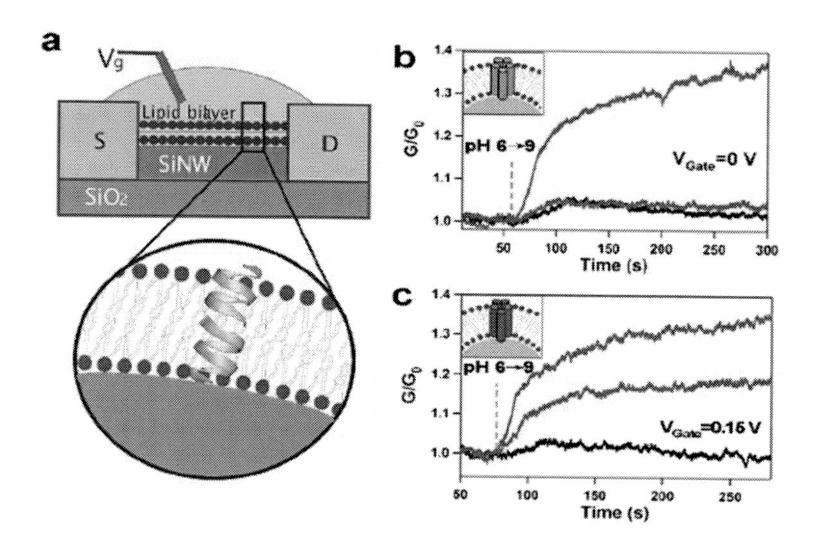

Figure 15. Scheme for devices showing an ion channel embedded in lipid bilayer wrapping the silicon nanowire (SiNW). S and D indicate source and drain electrodes and Vg is the applied gate voltage. (b) SiNW normalized conductance, G/Go , recorded as the solution's pH was changed from 6 to 9 for the uncoated (blue) and coated NW (black), and the coated-NW device incorporating Alamethicin pores (red). Time traces of similar experiments recorded at a gate bias of 0.15 V. (Ref. [308] *Copyright,* Reproduced with permission from *National Academy of Science* 2009).

Lithographic and Templating Methods for Nanowire Formation and Alignment

Lithographic methods, by which nanowires are templated in a highly controlled and reproducible manner, have emerged as a promising "top-down" alternative approach to conventional synthesis techniques [311]. In the final sections of this review, some of the various top-down methods for fabrication of nanomaterials, specifically nanowires, are discussed. Fabrication techniques such as imprint lithography, edge-lithography, dip-pen nanolithography, and microfluidic stamping for the preparation of tailored architectures are discussed. In general, lithographic methods have significantly expanded the design of colloidal nanomaterials, allowing enhanced applications in self-assembled materials for sensors and electronics.

In recent work by the Li group, magnetic molecularly imprinted nanowires were synthesized using a nanoporous alumina template [312]. Historically, molecular imprinted synthetic polymers have proven useful in generating artificial macromolecular receptors [313]. In order for them to be utilized for biological macromolecules, the imprinted sites would need to be accessible and homogenous so that they have the same binding affinity. Current methods render the imprinted sites inaccessible because they are completely encased within the polymer. In order to allow for the binding site to be exposed, the polymer must be ground up, which in many cases results in obvious deformation, which irreversibly changes specificity of the sites, consequently changing their selectivity. One aspect Li and co-workers wished to explore was whether a monomer with vinyl or acrylic groups could be utilized in the process, as well as whether or not superparamagnetic nanocrystallittes could be entrapped within the nanowires, which were derived from theophylline, which was used for the templating and held stationary on the porous walls of the alumina membrane [312]. The nanopores were then filled with a pre-polymerization mixture, which contained superparamagnetic $MnFe_2O_4$ nanocrystallites. The alumina membrane was removed by chemical dissolution following the completion of the polymerization, which subsequently yielded magnetic polymer nanowires of a controlled size, which contained theophylline binding sites with a saturated magnetization. The nanowires revealed superparamagnetic behavior at 300 K, which signified the incorporation of the $MnFe_2O_4$ nanocrystallites. Significantly, the imprinted magnetic polymeric nanowires were able to bind theophylline more efficiently than nanowires that were not imprinted.

Yun and co-workers used a combination of electrochemical and lithographic methods in their report on the growth of Pd and polypyrrole nanowires using electrochemical growth through e-beam patterned electrolyte channels, which were applied as pH sensors [144]. In general, E-beam lithography (EBL) has been utilized to fabricate nanostructures smaller than 100 nm [15-16, 56, 144]. The benefits of EBL include superior spatial resolution, extended field of focus, and reasonable beam requirements (ie synchroton source not needed). The first step in the process involved spin-coating an electron sensitive polymer (typically polymethyl methacrylate (PMMA)) onto a substrate, upon which patterns were designed into the deposited polymer by a highly collimated electron beam, followed by selective dissolution of the polymer regions damaged due to the electron exposure. Thermal evaporation then allowed for the sample to be a coated with a thin layer of Au metal. The last step involved removing the polymer via dissolution, which allowed for only the metal nanoparticles to be left in their pre-determined arrangement on the substrate. A single conducting polypyrrole nanowire was synthesized with a 200 nm or 500 nm diameter and up to 3 μm in length. This polymeric nanowire was found to be suitable for applications in pH sensing. [314].

Soft Lithography using Polymeric Nanowires

Soft lithography has been utilized to produce both narrow polyaniline lines [315] and broad PEDOT-PSS electrodes [316], whereas PEDOT-PSS has been utilized to improve performance in electronic devices, as a modified anode on top of indium tin oxide (ITO) in photodiodes [317], and for enhanced rectification ratio in polymer diodes [315]. Soft lithographic techniques were employed to synthesize conducting polymer poly (3,4-ethylenedioxythiophene) based nanowires doped with poly(4-styrenesulfonate) (PEDOT-PSS) which were patterned by micro-molding in capillaries (MIMIC) in order to form nanowires on a solid substrate of either glass or Si [138a]. Capillary action was utilized to transform an aqueous dispersion of PEDOT-PSS into nanowires in MIMIC structures. The set-up was left idle until the solution entered the capillaries and dried out, so that the stamp was peeled off, which gave the polymeric nanowires a grating from the stamp. If it was desired to change the height of the resulting nanowires, a force was to be applied during MIMIC, which resulted in a spiderlike web, as not all the material is deposited. The resistance was measured by evaporating gold line parallel electrodes on the substrate, followed by molding the polymer pattern on top. It was determined that during the MIMIC process, the polymer attained anisoptropic conductivity, as some defects in the wire were observed.

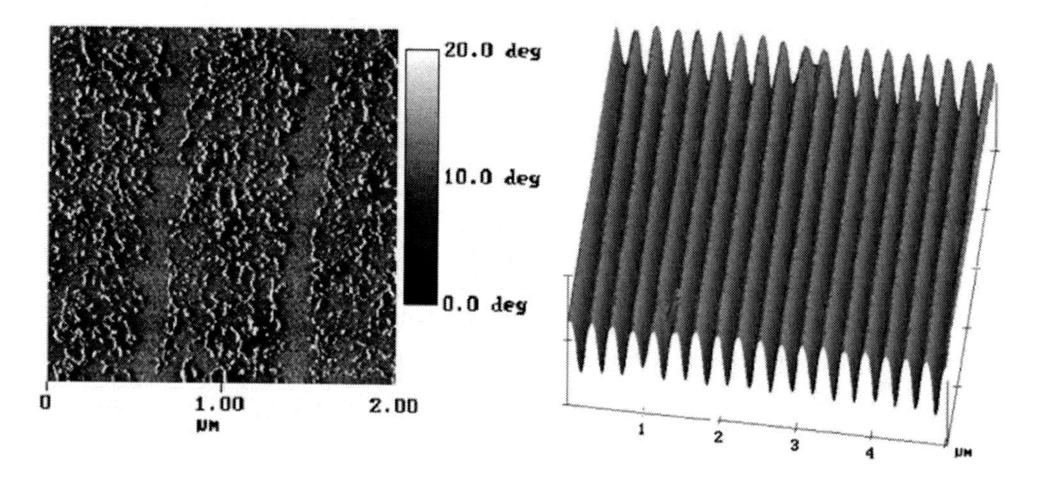

Figure 16. Left: SFM image of the conducting polymer PEDOT-PSS molded on a Si wafer with a compressed stamp. The phase image show that the nanowires are separate from each other. SFM image of nanowires of semiconducting polymer POMeOPT by soft-embossing and the molecular structure of POMeOP T(right) (Ref. [138a] *Copyright*, Reprinted with permission from the *American Chemical Society* 2002).

Some problems seen during stamping is that a film bridge sometimes formed in areas that should be only be filled with polymer, which is disadvantageous if the polymeric nanowires were to be utilized in an electronic system because the individual wires would be unable to connect to the device. Therefore, in order to synthesize smaller nanowires, the use of soft embossing was explored for the semiconducting and fusible polymer POMeOPT. This technique first involved spin-coating the polymer from a solution on a clean film, followed by heating the film with an elastomer stamp on top until the polymer began to flow. Pressure was applied atop the stamp to increase the length of the depths. Because of their tunable properties, these polymeric nanowires may be used to design electrodes, photonic structures, and pixel elements for microelectronic devices and biochips. Soft embossing was also explored, where polymers deposited from dispersion were patterned in order to form nanowires and 2-D nanodots on semiconducting polymer (poly(3-(2'-methoxy-5'-octyphenyl) thiopene)) POMeOPT. In addition, poly(3-[(*S*)-5-amino-5-carboxyl-3-oxapentyl]-2,5-thiophenylene hydrochloride) POWT [318] was also patterned using the same lithographic techniques of MIMIC and liquid printing. Comparison of SFM image of nanowires obtained by stamping and soft embossing are shown in Figure 16. It was determined that soft embossing resulted in smoother nanowires when compared to the previously synthesized PEDOT-PSS nanowires. Therefore soft lithography can be utilized to pattern both semiconducting polymers. There were limitations to this system, in that the template used in this experiment could only handle periods over 100 nm and could only use the customary PDMS materials. During MIMIC compression, the diameter of the channels decreased, which reduced the area for the deposition.

Edge Lithography

In recent work, George Whitesides and his group developed the technique of Edge Lithography to synthesize conjugated polymeric nanowires [143]. This is a three-step process, whereby a composite film of alternating layers of a conjugated polymer and a sacrificial material is spin coated. The two polymers utilized were MEH-PPV (poly(2-methoxy-5-(2′-ethylhexyloxy)-1,4-phenylenevinylene) and BBL (poly(benzimidazobenzophenanthroline ladder). BBL was the sacrificial material for MEH-PPV. The process is designed so that every other layer served as a sacrificial material which was subsequently removed from the desired final structure. The film was then embedded in an epoxy matrix and sectioned with an ultramicrotome (nanoskiving). Essentially, nanoskiving allowed for the nanostructures to be synthesized using a microtome to section patterned or stacked thin films of inorganic materials [319]. This was considered a type of edge lithography because the ultramicrotome exposed an area of an embedded thin film in order to form the dimension of the nanowires [62b]. Dissolution of the BBL with methanesulfonic acid resulted in uniaxially aligned MEH-PPV nanowires, which had rectangular cross sections. Further, the MEH-PPV nanowires were etched with an oxygen plasma to give the BBL nanowires. Electronic studies indicated the conductivity (4 S cm^{-1}) of the MEH-PPV nanowires changed rapidly and reversibly in the presence of I_2 vapor, which had been shown to cause a reversible increase in the conductivity of conjugated polymers [320]. These nanowires may potentially be utilized in sensing because they allowed for an analyte to be rapidly diffused in and out of the wire (or adsorption/desorption from its surface) [138b, 321], which led to an improved electrical response and recovery rate when compared to thin films and fibrous networks. This facile method has proven the ability to fabricate conductive nanowires in a simpler manner when compared to electron-beam lithography, and may also prove useful when nanowires are desired to have a small length (< 100 mm).

Dip-Pen Nanolithography

Dip-Pen nanolithography (DPN) is a technique developed by the Mirkin group that has been utilized to design patterns with specific chemical functionality [142, 322], dendrimers [323], raised gold wires [324], modified oligonucleotide on metals, insulators [325], and protein nanoarrays [326]. Noy and co-workers combined dip-pen nanolithography with single molecule optical detection in order to design nanoscale patterns down to the single-molecule deposition level using luminescent materials. This technique was further utilized to synthesize light-emitting polymer nanowires of a desired size, which could be directly alligned to optoelectronic devices and thus allowed for an improvement over template-directed synthesis [327] and stretching of polymer bridges [141]. In their work, rhodamine 6G (R6G) was utilized as the writing “ink” for the luminescent nanoscale patterns because of its strong affinity for negatively charged glass surfaces, which eliminated the need of an additional substance to encourage transfer of the ink. After a drop of R6G was dropped on a thin paper membrane, it was brought into contact with the AFM tip for less than a minute (“inking”), after which lithography was performed. Simultaneous lithography and imaging of the patterns was made possible using an instrument that was a hybrid of atomic force microscopy and scanning confocal microscopy [142c], as shown in Figure 17. In order to design the luminescent polymer nanowires, polymer lines of poly [2-methoxy-5-2'-ethylhexyl)oxy-1,4-phenylenevinylene] (MEH- PPV) were written on a glass surface using chloroform as ink. The authors found three limitations that prohibited the growth of nanowires of long length—a) the available range of the force microscope scanner, b) the amount of material on the tip of the probe, combined with c) the ability of the tip to deliver ink continuously. Different tip speeds were utilized in order to create nanowires of different thicknesses, where faster tip speeds yielded a lesser wire thickness. This was significant as dip-pen lithography could now be utilized for very small amounts of material, thus creating structures based on a single molecule.

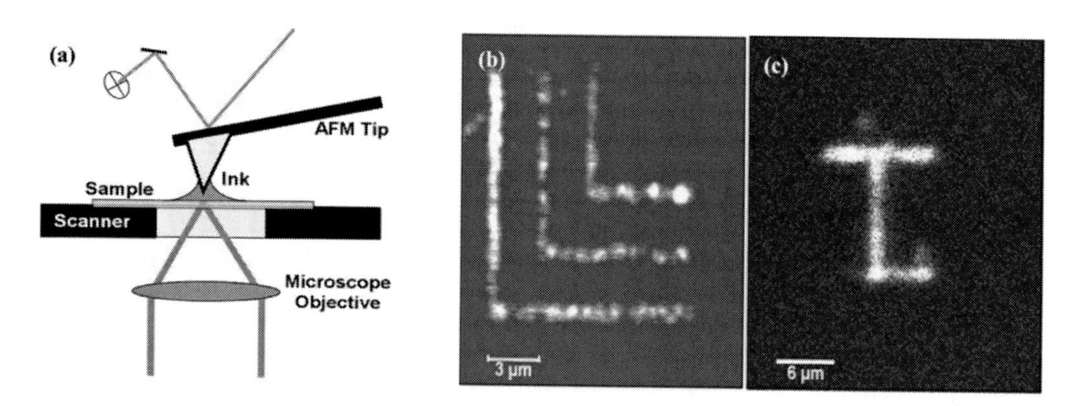

Figure 17. (a) Scheme for the experimental setup used for dip-pen nanolithography combining atomic force microscopy and scanning confocal microscopy functionalities. A Bioscope AFM head was mounted on a Nikon Eclipse 300 inverted microscope equipped with a custom built stage that incorporated a closed-loop piezo scanner. The scanning stage was used in both the lithography procedure and confocal imaging. The fluorescence excited by a 514 nm line of Ar-Ion laser was collected with 100 x 1.4 NA lens, passed through a 50mm pinhole to reject stray light and then detected by an avalanche photodiode (Perkin-Elmer Optoelectronics, Santa Clara, CA). Excitation light was rejected by a long-pass filter (Omega Optical, VT). (b) Scanning confocal microscopy image of a Rhodamine 6G patterned in a series of lines on glass surface. (c) Scanning confocal microscopy image of a HCG antibody pattern fabricated on glass. Each line in the pattern was overwritten 10 times to maximize the protein surface density. (Ref. [142c] *Copyright*, Reproduced with permission from the *American Chemical Society* 2002).

Conclusions and Future Perspectives

The versatility of nanowires arises from the diverse methods from which they can be designed, fabricated, and fine-tuned for widespread of applications. Overall, the field of bionanotechnology has made significant progress in advancing the design of nanowires with large surface area and tunable surfaces. Nanowires derived from DNA can be coated with a variety of nanoparticles and other biological materials and the addition of metal nanoparticles has been shown to substantially increase the conductivity and optoelectronic properties. Specific sequences of oligopeptides and nucleotides allow for the formation of tunable nanowires, because different binding sequences could bind to different entities, and allows the formation of highly controlled assemblies. Work in the area of peptide based nanowires, nanotubes, and nanofibers has resulted in the development of not only biocompatible and highly stable nanomaterials, but in some cases, those materials have also have shown stability even in harsh or proteolytic conditions. Nanowires have the potential to revolutionize the biomedical field by enabling disease detection, serving as powerful devices for electronic and chemical recording from cells, as scaffolds for tissues and organs, as well as forming the bases of potential implants for highly functional and powerful prosthetics. Additionally, the potential architectures of lipid based nanotubes are deemed to be the largest self-assembling non-living structures yet observed and thus there is much anticipation over the potential diversity of such nanostructures and their applications. The fine control over the resulting nanostructures is particularly advantageous in applications such as smarter device fabrications in applications such as electronics, spectroscopy, and sensing. Developing new synthetic strategies and a more in depth fundamental understanding of the dynamics of the various processes is a necessity for fine-tuning nanowires for enhanced applications. Utilization of natural biocompatible materials has the prospect of opening limitless applications that could transfigure everyday life. The future for bio-based nanowires holds much promise for transformative technologies.

REFERENCES

[1] Leung, KC; Wang, YJ. Mn-Fe Nanowires Towards Cell Labeling and Magnetic Resonance Imaging. In Lupu, N (Ed.). Nanowires Science and Technology. Croatia: Intech; 2010.

[2] Yang, P. The Chemistry of Nanostructured Materials (1st ed.). Singapore, Singapore: World Scientific Publishing Co. Pte. Ltd; 2003.

[3] Feynman, R. There's plenty of room at the bottom. *Engineering and Science*, 1960, 99, 22-36.

[4] [a] Martos, A; Alfonso, C; Lopez-Navajas, P; Ahijado-Guzm, R; Mingorance, J; Minton, AP; Rivas, G. Characterization of Self-Association and Heteroassociation of Bacterial Cell Division Proteins FtsZ and ZipA in Solution by Composition Gradient-Static Light Scattering. *Biochemistry* 2010, 49, 10780-10787; [b] Liu, G; Prabhakar, A; Aucoin, D; Simon, M; Sparks, S; Robbins, KJ; Sheen, A; Petty, SA; Lazo, ND. Mechanistic Studies of Peptide Self-Assembly: Transient alpha-Helices to Stabilible Beta-Sheets. *J. Am. Chem. Soc.* 2010, 132, 18223-18232; [c] Dhathathreyan, A; Nair, BU. Influence of Sequence on the Self-Assembly of Peptide Nanoribbons on Silicon Substrates. *J. Phys. Chem. B.* 2010, 115, 16650-16654.

[5] [a] Doni, G; Kostiainen, MA; Danani, A; Paven, GM. Generation-Dependent Molecular Recognition Controls Self-Assembly in Supramolecular Dendron-Virus Complexes. *Nano Lett.* 2010 in press; [b] Lin., Y; Su, Z; Xiao, G; Balizan, E; Kaur, G; Niu, Z; Wang, Q. Self-Assembly of Virus Particles on Flat Surfaces via Controlled Evaporation. *Langmuir* 2010, in press.

[6] Patwa, AN; Gonnade, RG; Kumar, VA; Bhadbhade, MM; Ganesh, KN. Ferrocene-Bis(thymine/uracil) Conjugates: Base Pairing Directed, Spacer Dependent Self-Assembly and Supramolecular Packing. *J. Org. Chem.* 2010, 75, 8705-8708.

[7] [a] Xu, Y; Wu, Q; Sun, Y; Bai, H; Shi, G. Three-Dimensional Self-Assembly of Graphene Oxide and DNA into Multifunctional Hydrogels. *ACS Nano* 2010, in press; [b] Boeneman, K; Prasuhn, DE; Blanco-Canosa, JB; Dawson, PE; Leninger, JS; Ancona, M; Steward, MH; Susumu, K; Juston, A; Medintz, IL. Self-Assembled Quantum-Dot-Sensitized Multivalent DNA Photonic Wires. *J. Am. Chem. Soc.* 2010, 132, 18177-18190.

[8] Kameta, N; Masuda, M; Minamikawa, H; Mishima, Y; Yamashita, I; Shimizu, T. Functionalizable Organic Nanochannels Based on Lipid Nanotubes: Encapsulation and Nanofluidic Behavior of Biomacromolecules. *Chem. Mater.* 2007, 19, 3553-3560.

[9] Guy, MM; Tremblay, M; Voyer, N; Gauthier, SF; Pouliot, Y. Formation and Stability of Nanofibers from a Milk-Derived Peptide. *J. Agric. Food Chem.* 2010, in press.

[10] Forbes, LM; Goodwin, AP; Cha, JN. Tunable Size and Shape Control of Platinum Nanocrystals from a Single Peptide Sequence. *Chem. Mater.* 2010, 22, 6524-6528.

[11] Berchmans, S; Nirmal. RG; Madhu, PS; Yegnaraman, V. Templated synthesis of silver nanowires based on the layer-by-layer assembly of silver with dithiodipropionic acid molecules as spacers. *J. Coll. Int. Sci.* 2006, 303, 604-610.

[12] Pokorski, JK, Steinmetz, NF. The Art of Engineering Viral Nanoparticles. *Mol. Pharmaceutics* 2011, 8(1), 29-43.

[13] Zhou, L; Yang, L; Yuan, P; Zhou, J.; Wu, Y; Yu, C. Alpha-MoO3 Nanobelts: A High Performance Cathode Material for Lithium Ion Batteries. *J. Phys. Chem. C.* 2010, 114, 21868-21872.

[14] [a] Zhai, TY; Fang, XS; Zeng, HB; Xu, XJ; Bando, Y; Golberg, D. A Comprehensive Review of One-Dimensional Metal-Oxide Nanostructure Photodetectors. *Sensors* 2009, 9(8), 6504-6529; [b] Ross, CA. Patterned magnetic recording media. *Ann. Rev. Mater. Res.* 2001, 31, 203-235; [c] Gu, Q; Cheng, CD; Haynie, DT. *Nanotechnology* 2005. 16, 1358-1363.

[15] [a] Jagannathan, H; Ivanisevic, A. Circular dichroism study of enzymatic manipulation on magnetic and metallic DNA template nanowires. *Coll. Surf. B.* 2008, 67, 279-283; [b] Gu, Q; Cheng, C; Gonela, R; Suryanarayanan, S; Anabathula, S; Dai, K; Haynie, DT. DNA nanowire fabrication. *Nanotechnology* 2006, 17, R14-R25; [c] Richter, J; Mertig, M; Pompe, W; Monch, I; Schackert, H. Construction of highly conductive nanowires on a DNA templates, *Appl. Phys. Lett.* 2001, 78, 536-538; [d] Deng, Z; Mao, C. *Nano Lett.* 2003, 3, 1545-1548.

[16] [a] Yogeswaran, U; Chen, S. A review on the electrochemical sensors and biosensors composed of nanowires as sensing material. *Sensors* 2008, 8, 290-313; [b] Njamjav, D; Kinsella, JM; Ivanisevic, A. *Appl. Phys. Lett.* 2005, 86, 093107; [c] Njamjav, D; Ivanisevic, A. *Biomaterials* 2005, 26, 2749-2757; [d] Kinsella, JM; Ivanisevic, A. Enzymatic Clipping of DNA Wires Coated with Magnetic Nanoparticles. *J. Am. Chem. Soc.* 2005, 127(10), 3276-3277.

[17] Reches, M; Gazit, E. Casting Metal Nanowires Within Discrete Self-Assembled Peptide Nanotubes. *Science* 2003, 300, 625-627.

[18] [a] de Waele, R; Koenderink, AF; Polman, A. Tunable Nanoscale Localization of Energy on Plasmon Particle Arrays. *Nano Lett.* 2007, 7(7), 2004-2008; [b] Graeter, SV; Huang, J; Perschmann, N; Lopez-Garcia, M; Kessler, H; Ding, J; Spatz, JP. *Nano Lett.* 2007, 7(5), 1413-1418; [c] Garcia, EJ; Hart, AJ; Wardle, BL; Slocum, AH. Fabrication of composite microstructures by capillarity-driven wetting of aligned carbon nanotubes

with polymers. *Nanotechnology*. 2007, 18, 165602/1-165602/11; [d] Neeta, LL; Ramakrishnan, R; Li, B; Subramanian, S; Barhate, RS; Liu, Y; Ramakrishnan, S. Fabrication of nanofibers with antimicrobial functionality used as filters: protection against bacterial contaminants. *Biotechnol. Bioeng*. 2007, 97, 1357-1365.

[19] Keating, CD; Natan, MJ. Striped metal nanowires as building blocks and optical tags. *Adv. Mater.* 2003, 15, 451-454.

[20] Gunawidjaja, R; Jiang, C; Peleshanko, S; Ornatska, M; Singamaneni, S; Tsukruk, VV. Flexible and Robust 2D Arrays of Silver Nanowires Encapsulated within Freestanding Layer-by-Layer Films. *Adv. Funct. Mater.* 2006, 16, 2024-2034.

[21] Xu, F; Lu, Wei; Zhu, Y. Controlled 3D Buckling of Silicon Nanowires for Stretchable Electronics. *ACS Nano*, 2011, 5, 672–667

[22] White, CT; Todorov, TN. Carbon nanotubes as long ballistic conductors. *Nature*, 1998, 393, 240-242.

[23] Greer, JC; Fagas, G. Ballistic Conductance in Oxidized Si Nanowires. *Nano Lett.* 2009, 9, 1856-1860.

[24] Takayanagi, K; Oshima, Y; Kurui, Y. Conductance quantization of gold nanowires as a ballistic conductor. *Interconnect Technology Conference, 2009. IITC 2009. IEEE International*. 2009, 47-50.

[25] Wang, ZL. The new field of nanopiezotronics. *Materials Today*, 2007, 10, 20-28.

[26] Yang, P. Chemistry of Nanostructured Materials. London: World Scientific Publishing Co. Pte. Ltd; 2003.

[27] Sander, MS; Prieto, AL; Gronsky, R; Sands, T; Stacy AM. Fabrication of high-density, high aspect ratio large area-bismuth telluride nanowire arrays by electrodeposition into porous anodic alumina templates. *Adv. Mater.* 2002, 14, 665-667

[28] Schonenberger, C; van der Zande, BMI; Fokkink, Henny, M; Schmid, C; Kruger, M; Bachtold, A; Huber, R; Birk, H; Staufer, U. Template Synthesis of Nanowires in Porous Polycarbonate Membranes: Electrochemistry and Morphology. *J. Phys. Chem. B*, 1997, 101, 5497–5505.

[29] Kinsella, JM, Ivanisevic, A. DNA-Templated Magnetic Nanowires with Different Composition: Fabrication and Analysis. *Langmuir,* 2007, 23, 3886-3890.

[30] [a] McMillan, AR; Paavola, CD; Howard, J; Chan, SL; Zaluzec, NJ; Trent, JD. Conformation ordered nanoparticle arrays formed on engineered chaperonin protein templates. *Nat. Mater.* 2002, 1, 247-252; [b] Patolsky, F; Weizmann, Y; Willner, I. Actin-based metallic nanowires as bio-nanotransporters. *Nat. Mater.* 2004, 3, 692-695.

[31] Huang, Y; Chiang, CY; Lee, SK; Gao, Y; Hu, EL; De Yoreo, J; Belcher, AM. Programmable assembly of nanoarchitectures using genetically engineered viruses. *Nano Lett.* 2005, 5, 1429-1434.

[32] Yan, H; Park, SH, Finkelstein, G; Reif, JH, LaBean, TH. DNA-Templated Self-Assembly of Protein Arrays and Highly Conductive Nanowires. *Science,* 2003, 301, 1882-1884.

[33] Tsai, CJ; Zheng, J; Nussinov, R. Designing a nanotube using naturally occurring protein building blocks. PLoS Comput. Biol. 2006, 2, 311-319.

[34] Tsai, CJ, Zheng, J; Zanuy, D; Haspel, N; Wolfson, H; Alema, C; Nussinov, R. Principles of Nanostructure Design With Protein Building Blocks. *Proteins: Structure, Function, and Bioinformatics,* 2007, 68, 1–12.

[35] Drexler, KE. Molecular engineering: an approach to the development of general capabilities for molecular manipulation. *Proc. Natl. Acad. Sci.* 1981, 78, 5275-5278.

[36] Lu, K; Jacob, J; Thiyagarajan P, Conticello VP, Lynn DG. Exploiting amyloid fibril lamination for nanotube self-assembly. *J. Am. Chem. Soc.* 2003, 125, 6391 6393.

[37] Rajagopal, K; Schneider, JP. Self-assembling peptides and proteins for nanotechnological applications. *Curr. Opin. Struct. Biol.* 2005, 14, 480–486.

[38] Mbindyo, JK; Reiss, BD; Martin, BR; Keating, CD; Natan, MJ; Mallouk, TE. DNA-Directed Assembly of Gold Nanowires on Complementary Surfaces. *Adv. Mater.* 2001, 13, 249- 254.

[39] [a] Guan, J; Yu, B; Lee, JL. Forming Highly Ordered Arrays of Functionalized Polymer Nanowires by Dewetting on Micropillars. *Adv. Mater.* 2007, 19, 1212-1217; [b] Braun, E; Eichen, Y; Sivan, U; Ben-Yoseph, G. DNA Templated Self-Assembly of a Conductive Wire Connecting Two Electrodes. *Nature,* 1998, 391, 775-778; [c] Hopkins, DS; Pekker, D; Goldbart, PM; Bezryadin, A. Quantum Interference Device Made by DNA Templating of Superconducting Nanowires. *Science*, 2005, 308, 1762-1765.

[40] Ma, Y; Zhang, J; Zhang, G; He, H. Polyaniline Nanowires on Si Surfaces Fabricated with DNA Templates. *J. Am. Chem. Soc.* 2004, 126, 7097-7101.

[41] [a] Nakao, H; Shiigi, H; Yamamoto, Y; Tokonami, S; Nagaoka, T; Sugiyama, S; Ohtani, T. Highly Ordered Assemblies of Au Nanoparticles Organized on DNA, *Nano Lett.* 2003, 3, 1391-1394; [b] Stsiapura, V; Sukhanova, A; Baranov, A; Artemyev, M; Kulakovich, O; Oleinikov, V; Pluot, M; Cohen, JHM; Nabiev, I. DNA-assisted formation of quasi-nanowires from fluorescent CdSe/ZnS nanocrystals. *Nanotechnology* 2006, 17, 581-587.

[42] Sokolova, V; Prymak, O; Meyer-Zaika, W; Cölfen, H; Rehage, H; Shukla, A; Epple, M. Synthesis and characterization of DNA-functionalized calcium phosphate nanoparticles, *Mat.-wiss. u. Werkstofftech* 2006, 37, 441-445.

[43] Willner, I; Patolsky, F; Wasserman, J, Photoelectrochemistry with Controlled DNA-Cross-Linked CdS Nanoparticle Arrays. *Angew. Chem.* 2001, 113, 1913-1916.

[44] Samson, J; Varotto, A; Nahirney, PC; Toschi, A; Piscopo, I; Drain, CM. Fabrication of Metal Nanoparticles Using Toroidal Plasmid DNA as a Sacrificial Mold, *ACS Nano* 2009, 3(2), 339-344.

[45] Heilemann, M; Tinnefeld, P; Sanchez Mosteiro, G; Garcia Parajo, M; Van Hulst, NF; Sauer, M. Multistep Energy Transfer in Single Molecule Photonic Wires, *J. Am. Chem. Soc.* 2004, 126, 6514-6515.

[46] Keren, K; Berman, RS; Buchstab, E; Sivan, U; Braun, E, DNA-Templated Carbon Nanotube Field-Effect Transistor, *Science* 2003, 302, 1380-1382.

[47] [a] Kundu, S & Liang H. Photochemical Synthesis of Electrically Conductive CdS Nanowires on DNA Scaffolds, *Adv. Mater.* 2008, 20, 826-831; [b] Le, JD, Pinto, Y; Seeman, NC; Musier-Forsyth, K; Taton, TA; Kiehl, RA, DNA-Templated Self-Assembly of Metallic Nanocomponent Arrays on a Surface, *Nano Lett.* 2004, 4, 2343-2347; [c] Seeman, N, The design and engineering of nucleic acid nanoscale assemblies, *Current Opin. Struct. Biol.* 1996, 6, 519-526; [d] Zhang, J; Liu, Y; Ke, Y; Yan, H, Periodic Square-Like Gold Nanoparticle Arrays Templated by Self-Assembled 2D DNA Nanogrids on a Surface, *Nano Lett.* 2006, 6. 248-251; [e] Martin, MCS; Gruss, C;

Carazo, JM, Six molecules of SV40 large T antigen assemble in a propeller-shaped particle around a channel, *J. Mol. Biol.* 1997, 268(1), 15-20.

[48] [a] Braun, E; Eichen, Y; Sivan, U; Ben-Yoseph, G. *Nature* 1998, 391, 775; [b] Richter, J; Seidel, R; Kirsch, R; Mertig, M; Pomper, W; Plaschke, J; Schackert, HK, Nanoscale palladium metallization of DNA, *Adv. Mater.* 12, 507-510 **(2000)**; [c] Mirkin, CA; Lestinger, RL; Mucic, RC; Storhoff, JJ. A DNA-based method for rationally assembling nanoparticles into macroscopic materials, *Nature* 1996, 382(6592), 607-609; [d] Warner, MG & Huchison, JE, Linera Assemblies of nanoparticles electrostatically organized on DNA scaffolds, *Nat. Mater.* 2003, 2(4), 272-277.

[49] Wirtz, D, Direct measurement of the transport of a single DNA molecule, *Phys. Rev. Lett.* 1995, 75(12), 2436-2439.

[50] Kundu, S & Liang, H. Microwave Synthesis of Electrically Conductive Gold Nanowires on DNA Scaffolds, *Lanmuir* 2008, 24, 9668-9674.

[51] [a] Keren, K; Krueger, M; Gilad, R; Ben-Yoseph, G; Suvan, U; Brauen, E, Sequence-specific molecular lithography on single DNA molecules, *Science* 2002, 297(5578), 72-75; [b] Deng, Z; Tian, Y; Lee, S-H; Ribbe, AE; Mao, C, DNA-encoded self-assembly of gold nanoparticles into one-dimensional arrays, *Angew. Chem. Int. Ed.* 2005, 44(23), 3582-3585; [c] Braun, G; Inagki, K; Estabrook, RA; Wood, DK; Levy, E; Cleland, AN; Strouse, GF; Reich, NO, Gold nanoparticle decoration of DNA on silicon, *Langmuir*, 2005, 21, 10699-10701.

[52] [a] C. F. Monson, A. T. Woolley, DNA-Templated Construction of Copper Nanowires, *Nano Lett.* 2003, 3(2), 359-363; [b] Ford, WE; Harnack, O; Yasuda, A; Wessels, JM, Platinated DNA as precursors to templated chains of metal nanoparticles, *Adv. Mater.* 2001, 13(23), 1793- 1797.

[53] Seidel, R; Ciacchi, LC; Weigel, M; Pompe, W; Mertig, M, Synthesis of Platinum Cluster Chains on DNA Templates: Conditions for a Template-Controlled Cluster Growth, *J. Phys. Chem. B.* 2004, 108(30), 10801-10811.

[54] Sarangi, SN; Goswami, K; Sahu, SN. Biomolecular recognition in DNA tagged CdSe nanowires, *Bio sens. Bio electron.* 2007; 22, 3086-3091.

[55] Gu, Q. & Haynie, D. T. Palladium nanoparticle-controlled growth of magnetic cobalt nanowires on DNA templates, *Mater. Lett.* 2008, 62(17-18), 3047-3050.

[56] [a] Ross, CA. Patterned magnetic recording media. *Ann. Rev. Mater. Res.* 2001, 31, 203-235; [b] Gu, Q; Cheng, C; Haynie, DT, Cobalt metallization of DNA: toward magnetic nanowires, *Nanotechnology* 2005, 16(8), 1358-1363.

[57] [a] Yogeswaran, U & Chen, S, A review on the electrochemical sensors and biosensors composed of nanowires as sensing material, *Sensors* 2008, 8, 290-313; [b] Nyamjav, D; Kinsella, JM; Ivanisevic, A, Magnetic wires with DNA cores: A magnetic force microscopy study, *Appl. Phys. Lett.* 2005, 86, 093107-1 – 093107-3; [c] Nyamjav, D & Ivanisevic, A, Templates for DNA-templated Fe_3O_4 nanoparticles, *Biomaterials* 2005, 26(15), 2749-2757.

[58] Chen, Y; Kung, S-C; Taggart, DK; Halpern, AR; Penner, RM; Corn, RM. Fabricating nanoscale DNA patterns with gold nanowires. *Anal. Chem.* 2010, 82(8), 3365-3370.

[59] Sarangi, SN; Rath, S; Goswami, K; Nozaki, S; Sahu, SN. DNA template driven CdSe nanowires and nanoparticles: Structure and optical properties. *Physica E.* 2010, 42(5), 1670-1674.

[60] Wang, Z-G; Wilner, OI; Willner, I. Self-Assembly of Aptamer-Circular DNA Nanostructures for Controlled Biocatalysis. *Nano Lett.* 2009, 9(12), 4098-4102.

[61] [a] Westwater, J; Gosain, DP; Usui, S, Si Nanowires Grown via the Vapour-Liquid-Solid Reaction, *Phys. Stat. Sol.,* 1998, 165, 37-42; [b] Wagner, RS; Ellis, WC, Vapor-liquid-solid mechanism of single crystal groth, *Appl. Phys. Lett.*, 1964, 4(5), 89-90; [c] Wagner, RS; Doherty, CJ, Mechanism of branching and kinking during vapor-liquid-solid crystal growth, *J. Electrochem. Soc.,* 1968, 115(1), 93-99.

[62] [a] Qin, L; Park, S; Huang, L; Mirkin, CA, On-Wire Lithography, *Science* 2005, 309, 113-115; [b] Gates, BD; Xu, Q; Stewart, M; Ryan, D; Willson, CG; Whitesides, GM, New Approaches to Nanofabrication: Molding, Printing, and Other Techniques, *Chem. Rev.* 2005, 105(4), 1171-1196.

[63] Liusman, C; Li, S; Chen, X; Wei, W; Zhang, H; Schatx, GC; Boey, F; Mirkin, CA. Free-Standing Bimetallic Nanorings and Nanoring Arrays Made by On-Wire Lithography, *ACS Nano* 2010, 4(12), 7676-7682.

[64] [a] Liu, D; Wang, M; Deng, Z; Walulu, R; Mao, C. Tensegrity: Construction of rigid DNA triangles with flexible four-arm DNA DNA junction, *J. Am. Chem. Soc., 2004,* 126(8), 2324-2325; [b] Ding, B; Sha, R; Seeman, NC, Pseudohexagonal 2D DNA Crystals from Double Crossover Cohesion, *J. Am. Chem. Soc.*, 2004, 126(33), 10230-10231; [c] Rothemund, PWK; Papadakis, N; Winfree, E, Algorithmic self-assembly of DNA Sierpinski triangles, *PLoS Biol.,* 2004, 2, 2041-2053; [d] Malo, J; Mitchell, JC; Venien-Bryan, C; Harris, JR; Wille, H; Sherratt, DJ; Turberfield, AJ, Engineering a 2D Protein-DNA Crystal, *Angew. Chem., Int. Ed.*, 2005, 44, 3057-3061.

[65] [a] Feng, L; Park, SH; Reif, JH; Yan, H, A Two-State DNA Lattice Switched by DNA Nanoactuator, *Angew. Chem., Int. Ed.*, 2003, 42, 4342-4346; [b] Mathieu, F; Liao, S; Kopatsch, J; Wang, T; Mao, C; Seeman, NC, Six-Helix Bundles Designed from DNA, *Nano Lett.,* 2005, 5, 661-665; [c] Park, SH; Barish, R; Li, H; Reif, JH; Finkelstein, G; Yan, H; LaBean, TH, Three-helix bundle DNA tiles self-assembled into 2D lattice of 1D templates for silver nanowires, *Nano Lett, 2005,* 5, 693-696; [d] Sun, X; Ko, SH; Zhang, C; Ribber, AE; Mao, C. Surface-Mediated DNA Self-Assembly, *J. Am. Chem. Soc.*, 2009, 131, 13248-13249; [e] Singh, A; Snyder, S; Lee, L; Johnston, APR; Caruso, F; Yingling, YG, Effect of Oligonucleotide Length on the Assembly of DNA Materials: Molecular Dynamics Simulations of Layer-by-Layer DNA Films, *Langmuir, 2010,* 26(22), 17339-17347.

[66] Winfree, E; Liu, F; Wenzler, LA; Seeman, NC, Design and self-assembly of two-dimensional DNA crystals, *Nature*, 1998, 394, 539-544.

[67] Zheng, J; Constantinou, PE; Micheel, C., Alivisatos, AP; Kiehl, RA; Seeman, NC. Two-Dimensional Nanoparticle Arrays Show the Organizational Power of Robust DNA Motifs, *Nano Lett.*, 2006, 6(7), 1502-1504.

[68] [a] Braich, RS; Chelyapov, N; Johnson, C; Rothemund, PWK; Adleman, L. Solution of a 20-Variable 3-SAT Problem on a DNA Computer, *Science*, 2002, 296(5567), 499-502 ; [b] Benenson, Y; Paz-Elizur, T; Adar, R; Keinan, E; Livneh, Z; Shapiro, E. Programmable and autonomous computing machine made of biomolecules, *Nature*, 2001, 414, 430-434.

[69] Han, D; Pal, S; Nangreave, J; Deng, Z; Liu, Y; Yan, H. DNA Origami with Complex Curvatures in Three-Dimensional Space, *Science*, 2011, 334, 342-346.

[70] [a] Alivisatos, AP; Johnsson, KP; Peng, X; Wilson, TE; Lowth, CJ; Brunchez, MP; Schultz, PG. Organization of Nanocrystal Molecules using DNA, *Nature*, 1996, 382, 609-611; [b] Mirkin, CA. Programming the Asssembly of Two- and Three-Dimensional Architectures with DNA and Nanoscale Inorganic Building Blocks, *Inorg. Chem.*, 2000, 39, 2258-2272.

[71] Wang, Z; Liu, J; Zhang, K; Cai, H; Zhang, G; Wu, Y; Kong, T; Wang, X; Chen, J. Hou, J. Fabrication of Well-Aligned and Highly Dense Cadmium Sulfide Nanowires on DNA Scaffolds Using the Poly(dimethylsiloxane) Transfer Method, *J. Phys. Chem. C.*, 2009, 113, 5428-5433.

[72] [a] Fortuna, SA; Li, X. Metal-catalyzed semiconductor nanowires: a review on the control of growth directions. *Semicond. Sci. Technol.* 2010, 25, 024005/1-024005/16; [b] Pauzaukie, PJ; Yang, P. Nanowire Photonics. *Mater. Today*, 2006, 9, 36-45.

[73] Appenzeller, J; Knoch, J; Bjork, MT; Riel, H; Schmid, H; Reiss, W. Toward nanowire electronics, *IEEE Trans. Electron Devices*, 2008, 55, 2827-2845.

[74] [a] Husain, A; Hone, J; Postma, HWC; Huang, XMH; Drake, T; Barbic, M; Scherer A; Roukes, ML. Nanowire-based very-high-frequency electromechanical resonator, *Appl. Phys. Lett.*, 2003, 83, 1240-1242; [b] Li, M; Mhiladvala, RB; Morrow, TJ; Sioss, JA; Lew, K-K; Redwing, JM, Keating, CD, Mayer, TS. Bottom-up assembly of large-area nanowire resonator arrays, *Nat. Nano* 2008, 3, 88-92.

[75] Patolsky, F; Timko, BP; Zheng, G; Lieber, CM. Nanowire-based nanoelectronic devices in the life sciences, *MRS Bull.*, 2007, 32, 142-149.

[76] [a] Wright, A; Gabaldon, J; Burckel, DB; Jiang, Y.-B; Tian, ZR; Liu, J; Brinker, CJ; Fan, H, Hierarchically Organized Nanoparticle Mesostructure Arrays Formed through Hydrothermal Self-Assembly, *Chem. Mater.*, 2006, 18, 3034-3038; [b] Cao, YWC; Jin, RC; Mirken, CA, Nanoparticles with Raman spectroscopic fingerprints for DNA and RNA detection, *Science,* 2002, 297(5586), 1536-1540.

[77] [a] Saunders, AE; Shah, PS; Sigman, MB; Hanrath, T; Hwang, HS; Lim, KT; Johnston, KP; Korgel, BA, *Nano Lett.*, Inverse Opal Nanocrystal Superlattice Films, 2004, 4(10), 1943-1948; [b] Petruska, MA; Malko, AV; Voyles, PM; Klimov, VI, High-performance, quantum dot nanocomposites for nonlinear and optical gain applications, *Adv. Mater.*, 2003, 15(7-8), 610-613.

[78] [a] Andres, RP; Bielefeld, JD; Henderson, JI; Janes, DB; Kolagunta, VR; Kubiak, CP; Mahoney, WJ; Osifchin, RG, *Science*, Self-assembly of a two-dimensional superlattice of molecularly linked metal clusters, 1996, 273, 1690-1693; [b] Sato, T; Ahmed, H; Brown, D; Johnson, BFG, *J. Appl. Phys.*, Single electron transistor using a molecularly linked gold colloidal particle chain, 1997, 82, 696-701.

[79] [a] Nam, J; Won, N; Jin, H; Chung, H; Kim, S, pH-Induced Aggregation of Gold Nanoparticles for Photothermal Cancer Therapy, *J. Am. Chem. Soc.*, 2009, 131, 13639-13645; [b] Xu, X; Han, MS; Mirkin, CA, A gold-nanoparticle-based real-time colorimetric screening method for endonuclease activity and inhibition, *Angew. Chem., Int. Ed.*, 2007, 46, 3468-3470; [c] Lee, J-S; Ulmann, PA; Han, MS; Mirkin, CA, A DNA-Gold Nanoparticle-Based Colorimetric Competition Assay for the Detection of Cystein, *Nano Lett.,* 2008, 8, 529-533.

[80] [a] Podsiadlo, P; Sinani, VA; Bahng, JH; Kam, NWS; Lee, J; Kotov, NA, Gold Nanoparticles Enhance the Anti-Leukemia Action of a 6-Mercaptopurine Chemotherapeutic Agent, *Langmuir*, 2008, 24(2), 568-574; [b] Chen, Y-H; Tsai, C-Y;

Huang, P-Y; Chang, M-Y; Cheng, P-C; Chou, C-H; Chen, D-H; Wang, C-R; Shiau, A-L; Wu, C-L, Methotrexate Conjugated to Gold Nanoparticles Inhibits Tumor Growth in a Syngenic Lung Tumor Model, *Mol. Pharmaceutics*, 2007, 4(5), 713-722; [c] Paciotti, GF; Myer, L; Weinreich, D; Goia, D; Pavel, N; McLaughlin, RE; Tamarkin, L, Colloidal Gold: A Novel Nanoparticle Vector for Tumor Directed Drug Delivery, *Drug Delivery*, 2004, 11(3), 169-183.

[81] [a] Sokolov, K; Follen, M; Aaron, J; Pavlova, I; Malpica, A; Lotan, R; Richards-Kortum, R, Real-time vital optical imaging of precancer using anti-epidermal growth factor receptor antibodies conjugated to gold nanoparticles, *Cancer Res.* 2003, 63(9), 1999-2004; [b] Javier, DJ; Nitin, N; Levy, M; Ellington, A; Richards-Kortum, R, Aptamer-Targeted Gold Nanoparticles as Molecular-Specific Contrast Agents for Reflective Imaging, *Bioconjugate Chem.,* 2008, 19(6), 1309-1312; [c] Aaron, J; Nitin, N; Travis, K; Kumar, S; Collier, T; Park, SY; Jose-Yacaman, M; Coghlan, L; Follen, M; Richards-Kortum, R; Sokolov, K, Plasmon resonance coupling of metal nanoparticles for molecular imaging of carcinogenesis *in vivo*, *J. Biomed. Opt.,* 2007, 12(3), 034007/1-034007/11; [d] Lee, S; Cha, E-J; Park, K; Lee, S-Y; Hong, J-K, Sun, L-C; Kim, SY; Choi, K; Kwon, IC; Kim, K; Ahn, C-H, A near-infrared-fluorescence-quenched gold-nanoparticle imaging probe for in vivo drug screening and protease activity determination, *Angew. Chem., Int. Ed.,* 2008, 47(15), 2804-2807; [e] El-Sayad, IH; Huang, X; El-Sayad, MA, Surface Plasmon resonance scattering and absorption of anti-EGFR antibody conjugated gold nanoparticles in cancer diagnostics: Applications in oral cancer, *Nano Lett.*, 2005, 5(5), 829-834.

[82] [a] Huang, X; Jain, PK; El-Sayed, IE; El-Sayed, MA, Plasmonic photothermal therapy (PPTT) using gold nanoparticles, *Lasers Med. Sci.*, 2008, 23(3), 217-228; [b] Von Maltzahn, G; Park, J-H; Agrawal, A; Bandaru, NK; Das, SK; Sailor, MJ; Bhatia, SN, Computationally Guided Photothermal Tumor Therapy Using Long-Circulating Gold Nanorods Antennas, *Cancer Res.*, 2009, 69(9), 3892-3900; [c] Cheng, Y; Samia, AC; Meyers, JD; Panagopoulos, I; Fei, B; Burda, C, Highly Efficient Drug Delivery with Gold Nanoparticle Vectors for in Vivo Photodynamic Therapy of Cancer, *J. Am. Chem. Soc.*, 2008, 130(32), 10643-10647.

[83] [a] Liu, B; Xie, J; Lee, JY; Ting, YP; Chen, JP, Optimization of High-Yield Biological Synthesis of Single-Crystalline Gold Nanoparticles, *J. Phys. Chem. B.*, 2005, 109, 15256-15263; [b] Kim, F; Song, JH; Yang, PD, Photochemical Synthesis of Gold Nanorods, *J. Am. Chem. Soc.*, 2002, 124(48), 14316-14317; [c] Sun, Y; Xia, Y, Shape-Controlled Synthesis of Gold and Silver Nanoparticles, *Science*, 2002, 298, 2176-2179; [d] Hao, E; Bailey, RC; Schatz, GC; Hupp, JT; Li, S, Synthesis and Optical Properties of "Branched" Gold Nanocrystals, *Nano Lett*, 2004, 4(2), 327-330; [e] Suzuki, M; Niidome, Y; Kuwahara, Y; Terasaki, N; Inoue, K; Yamada, S, Surface-Enhanced Nonresonance Raman Scattering from Size-and Morphology-Controlled Gold Nanoparticle Films, *J. Phys. Chem. B.*, 2004, 108(31), 11660-11665.

[84] [a] Walker, CH; St. John, JV; Wisian-Nielson, P, Synthesis and Size Control of Gold Nanoparticles Stabilized by Poly(methylphenylphosphazene, *J. Am. Chem. Soc.*, 2001, 123, 3846-3847; [b] Slot, JW; Geuze, HJ, Sizing of protein A-colloidal gold probes for immunoelectron microscopy, *J. Cell. Biol.*, 1981, 90(2), 533-536.

[85] [a] Elghanian, R; Storhoff, JJ; Mucic, RC; Letsinger, RL; Mirkin, CA, Selective colorimetric detection of polynucleotide based on the distance-dependent optical

properties of gold nanoparticles, *Science*, 1997, 277(5329), 1078-1080; [b] Storhoff, JJ; Elghanian, R; Mucic, RC; Mirkin, CA; Letsinger, RL, One-Pot Colorimetric Differentiation of Polynucleotides with Single Base Imperfections Using Gold Nanoparticle Probes, *J. Am. Chem. Soc.*, 1998, 120(9), 1959-1964.

[86] [a] Liu, J; Xu, R; Kaifer, AE, In Situ Modification of the Surface of Gold Colloidal Particles. Preparation of Cyclodextrin-Based Rotaxanes Supported on Gold Nanospheres, *Langmuir*, 1998, 14(26), 7337-7339; [b] Labande, A; Astruc, D, Colloids as redox sensors: recognition of $H_2PO_4^-$ and HSO_4^- by amidoferrocenylalkylthiol-gold nanoparticles, *Chem. Commun.*, 2000, 12, 1007-1008.

[87] [a] Nakao, H; Hayashi, H; Yoshino, T; Sugiyama, S; Otobe, K; Ohtani, T, Development of novel polymer-coated substrates for straightening and fixing DNA, *Nano Lett.*, 2002, 2(5), 475-479; [b] Nakao, H; Gad, M; Sugiyama, S; Otobe, K; Ohtani, T, Transfer-printing of highly aligned DNA nanowires, *J. Am. Chem. Soc.*, 2003, 125(24), 7162-7163.

[88] F. Patolsky, Y. Weizmann, O. Lioubashevski, I. Willner, Au-Nanoparticle Nanowires Based on DNA and Polylysine Templates, *Angew. Chem. Int. Ed.*, 2002, 41(13), 2323-2327.

[89] Salem, AK; Searson, PC; Leong, KW, Multifunctional nanorods for gene delivery, *Nat. Mater.*, 2003, 2(10), 668-671.

[90] Berti, L; Alessandrini, A; Facci, P, DNA-Templated Photoinduced Silver Deposition, *J. Am. Chem. Soc.*, 2005, 127, 11216-11217.

[91] Hossain, Z; Huq, FJ, Studies on the interaction between Ag^+ and DNA, *Inorg. Biochem.*, 2002, 91, 398-404.

[92] Wei, G; Zhou, H; Liu, Z; Song, Y; Wang, L; Sun, L; Li, Z. One-step synthesis of Silver Nanoparticles, Nanorods, and Nanowires on the Surface of a DNA Network. *J. Phys. Chem. B.*, 2005, 109, 8738-8743.

[93] [a] Spiro, TG. *Nucleic Acid-Metal Ion Interactions*. 1st Edition. New York: Wiley Interscience; 1980; [b] Marzilli, LG; Kistenmacher, TJ; Rossi, M. An Extension of the role of O(2) of cytosine residues in the binding of metal ions. Synthesis and structure of an usual polymeric silver(I) complex of 1-methylcytosine. *J. Am. Chem. Soc.*, 1977, 99(8), 2797-2798; [c] Eichorn GL, (ed.) *Inorganic Biochemistry*, 1st Edition. (Vol. 2, Ch. 33-34), Amsterdam: Elsevier; 1973.

[94] [a] Holgate, CS; Jackson, P; Cowen, PN; Bird, CC. Immunogold-silver staining: new method of immunostaining with enhanced sensitivity. *Histochem. Cytochem.*, 1983, 31(7), 938-944; [b] Birrell, GB; Habliston, DL; Hedberg, KK; Griffith, OH. Silver-enhanced colloidal gold as a cell surface marker for photoelectron microscopy. *J. Histochem. Cytochem.*, 1986, 34(3), 339-345.

[95] Keren, K; Berman, RS; Braun, E. Patterned DNA Metallization by a Sequence-Specific Localization of a Reducing Agent, *Nano Lett.*, 2004, 4(2), 323-326.

[96] Eichen, Y; Braun, E; Sivan, U; Ben-Yoseph, G. Self-assembly of nanoelectronic components and circuits using biological templates, *Acta Polym.*, 1998, 49, 663-670.

[97] Barton, JK. *Bioinorganic Chemistry* (eds Bertini, I. *et. al.*), Ch. 8. Mill Valley, CA: University Science Book Valley; 1994.

[98] Lu, J; Yang, L; Xie, A; Shen, Y. DNA-templated photo-induced silver nanowires: Fabrication and use in detection of relative humidity. *Biophys. Chem.*, 2009, 145, 91-97.

[99] [a] Fu, XQ; Wang, C; Yu, HC; Wang, YG; Wang, TH. Fast humidity sensors based on CeO_2 nanowires. *Nanotechnology*, 2007, 18, 145503-145506; [b] Chou, KS; Lee, TK; Liu, FJ. Sensing mechanism of a porous ceramic as humidity sensors. *Sens. Actuators B.*, 1996, 56, 106-11; [c] Zhao, J; Bulduml, A; Han, J; Lu, JP. Gas molecule adsorption in carbon nanotubes and nanotube bundles, *Nanotechnology*, 2002, 13, 195-200; [d] Chen, Z; Wlodarski, W. A thin-film sensing element for ozone, humidity and temperature, *Sens. Actuators B.*, 2002, 64, 42-48.

[100] Seidel, R; Colombi Ciacchi, L; Weigel, M; Pompe, W; Mertig, M. Synthesis of Platinum Cluster Chains on DNA Templates: Conditions for a Template-Controlled Cluster Growth. *J. Phys. Chem. B.*, 2004, 108, 10801-10811.

[101] Mertig, M; Ciacchi, LC; Seidel, R; Pompe, W. DNA as a Selective Metallization Template, *Nano Lett.*, 2002, 2(8) 841-844.

[102] Floriani, C. Transition metal complexes as bifunctional carriers of polar organometallics: Their application to large molecule modifications and to hydrocarbon activation. *Pure & Appl. Chem.,* 1996, 68(1), 1-8.

[103] Metcalfe, C; Thomas, JA. Kinetically inert Transition Metal Complexes that reversibly bind to DNA. *Chem. Soc. Rev.*, 2003, 32, 215-224.

[104] Becerril, HA; Ludtke, P; Willardson, BM; Wooley, AT. DNA-Templated Nickel Nanostructures and Protein Assemblies, *Langmuir*, 2006, 22, 10140-10144.

[105] Li, G; Liu, H; Chen, X; Zhang, L; Bu, Y. Multi-Copper-Mediated DNA Base Pairs Acting as Suitable Building Blocks for the DNA-Based Nanowires, *J. Phys. Chem. C.*, 2011, 115, 2855–2864.

[106] [a] Takahara, PM; Rosenzweig, AC; Lippard, SJ. Crystal structure of double-stranded DNA containing the major adduct of the anticancer drug cisplatin. *Nature*, 1995, 377(6650), 649-652; [b] Huang, H; Zhu, L; Reld, BR; Drobny, DP; Hopkins, PB. *Science*, 1995, 270(5243), 1842-1845; [c] Herman, F; Kozelka, J; Stoven, V; Guittet, E; Girault, JP; Tam, HD; Igolen, J; Lallemand, JY; Chottard, JC. A d(GpG)-platinated decanucleotide duplex is kinked: an extended NRM and molecular mechanics study. *Eur. J. Biochem.*, 1990, 194(1), 119-133.

[107] Grabert H; Devoret, MH. *Single Charge Tunneling*, Plenum, New York, **(1992)**.

[108] Richter, J; Mertig, M; Pompe, W; Vinzelberg, H. Low-temperature resistance of DNA-templated nanowires. *Appl. Phys. A.,* 2002, 74, 725-728.

[109] [a] Thouless, DJ. Maximum metallic resistance in thin wires. *Phys. Rev. Lett.*, 1977, 39(18), 1167-1169; [b] Abrahams, E; Anderson, PW; Licciardello, DC; Ramakrishnan, TV. Scaling Theory of Localization: Absence of Quantum Diffusion in Two Dimensions. *Phys. Rev. Lett.*, 1979, 42(10), 673-676; [c] Anderson, PW; Abrahams, E; Ramakrishnan, TV. Possible Explanation of Nonlinear Conductivity in Thin-Film Metal Wires. *Phys. Rev. Lett.*, 1979, 43(10), 718-720; [d] Altshuler, BL; Aronov, AG; Lee, PA. Interaction Effects in Disordered Fermi Systems in Two Dimensions. *Phys. Rev. Lett.*, 1980, 44(19), 1288-1291.

[110] [a] Dolan, GJ; Osheroff, DD. Nonmetallic Conduction in Thin Films at Low Temperatures. *Phys. Rev. Lett.*, 1979, 43(10), 721-724; [b] Burns, MJ; McGinnis, WC; Simon, RW; Deutscher, G; Chaikin, PM. Minimum Metallic Conductivity and Thermopower in Thin Palladium Films. *Phys. Rev. Lett.*, 1981, 47(22), 1620-1624; [c] Bishop, DJ; Tsui, DC; Dynes, RC. Nonmetallic conduction in electron inversion layers at low temperature. *Phys. Rev. Lett.*, 1980, 44(17), 1153-1156.

[111] McGinnis, WC; Chaikin, PM. Electron localization and interaction effects in palladium and palladium-gold films. *Phys. Rev. B.*, 1985, 32(10), 6319-6330.

[112] Kundu, S; Wang, K; Huitink, D; Liang, H. Photoinduced Formation of Electrically Conductive Thin Palladium Nanowires on DNA Scaffolds, *Langmuir*, 2009, 25(17) 10146-10152.

[113] Dabbousi, BO; Bawendi, MG; Onitsuka, O; Rubner, MF. Electroluminescence from CdSe quantum-dot/polymer composites. *Appl. Phys. Lett.*, 1995, 66(11), 1316-1318.

[114] Schmid, G. Cluster and Colloids, From theory to Applications, VCH, Weinheim, Germany (1994).

[115] Freeman, R; Grabar, KC; Allison, KJ; Bright, RM; Davis, JA; Guthrie, AP; Hommer, MB; Jackson, MA; Smith, PC; Walter, DG; Natan, MJ. Self-assembled metal colloid monolayers: an approach to SERS substrates. *Science*, 1995, 267(5204), 1629-1631.

[116] [a] Mitchell, GP; Mirkin, CA; Lestinger, RL. Programmed Assembly of DNA Functionalized Quantum Dots, *J. Am. Chem. Soc.*, 1999, 121, 8122-8123; [b] Hu, F; Ran, Y; Zhou, Z; Gao, M. Preparation of bioconjugates of CdTe nanocrystals for cancer marker detection. *Nanotechnology*, 2006, 17, 2972-2977.

[117] Sarangi, SN, Sahu, SN. CdSe nanocrystalline thin films: composition, structure and optical properties. *Physica E*, 2004, 23, 159–167.

[118] Hannestad, JK; Sandin, P; Albinsson, B. Self-Assembled DNA Photonic Wire for Long-Range Energy Transfer. *J. Am. Chem. Soc.*, 2008, 130, 15889-15895.

[119] Dong, L; Hollis, T; Connolly, BA; Wright, NG; Horrocks, BR; Houlton, A. DNA-Templated Semiconductor Nanoparticle Chains and Wires, *Adv. Mater.*, 2007, 19, 1748-1751.

[120] Li, J; Bai, C; Wang, C; Zhu, C; Lin, Z; Li, Q; Cao, E. A convenient method of aligning large DNA molecules on bare mica surfaces for atomic microscopy. *Nucleic Acids Res.*, 1998, 26(20), 4785-4786.

[121] [a] Coffer, JL; Bigham, SR; Li, X; Pinizzotto, RF; Rho, YG; Pirtle, RM; Pirtle, IL. Dictation of the Shape of Mesoscale Semiconductor Nanoparticle Assemblies by Plasmid DNA. *Appl. Phys. Lett.*, 1996, 69, 3851; [b] Kuczynski, JP; Milosavljevic, BH; Thomas, JK. Effect of the synthetic preparation on the photochemical behavior of colloidal cadmium sulfide. *J. Phys. Chem.*, 1983, 87, 3368-3370; [c] Kuczynski, JP; Milosavljevic, BH; Thomas, JK. Photophysical properties of cadmium sulfide in Nafion film. *J. Phys. Chem.*, 1984, 88, 980-984; [d] Bagnall, DM; Ullrich, B; Sakai, H; Segawa, Y. Micro-cavity lasing of optically excited CdS thin films at room temperature. *J. Cryst. Growth*, 2000, 214, 1015-1018.

[122] Long, Y; Chen, Z; Wang, W; Bai, F; Jin, A; Gu, C. Electrical conductivity of single CdS nanowire synthesized by aqueous chemical growth. *Appl. Phys. Lett.*, 2005, 86, 153102/1-153102/3; b) Lee, J-H; Yi, S-Y; Yang, K-J; Park, J-H; Oh, R-D. *Thin Solid Films*, 2003, 431-432, 344.

[123] Levina, L; Sukhovatkin, V; Musikhin, S; Cauchi, S; Nisman, R; Bazett-Jones, DP; Sargent, EH. Efficient Infrared-Emitting PbS Quantum Dots grown on DNA and Stable in Aqueous solution and Blood Plasma. *Adv. Mater.*, 2005, 17, 1854-1857.

[124] Nakao, H; Hayashi, H; Iwata, F; Karasawa, H; Hirano, K; Sugiyama, S; Ohtani, T. Fabricating and Aligning π-Conjugated Polymer-Functionalized DNA Nanowires: Atomic Force Microscopic and Scanning Near-Field Optical Microscopic Studies, *Langmuir*, 2005, 21, 7945-7950.

[125] [a] Shirakawa, H. The discovery of polyacetylene film: The dawning of an area of conducting polymers. *Synth. Met.*, 2002, 125, 3-10; [b] MacDiarmid, AG. Synthetic metals: a novel role for organic polymers. *Synth. Met.*, 2002, 125, 11-22; [c] Heeger, AJ. Semiconducting and metallic polymers: the fourth generation of polymeric materials. *Synth. Met.*, 2002, 125, 23-42.

[126] Jager, EWH; Smela, E; Inganas, O. Microfabricating conjugated polymer actuators. *Science*, 2000, 290(5496), 1540-1546.

[127] Yu, J; Holdcroft, S. Chemically amplified soft lithography of a low band gap polymer. *Chem. Commun.*, 2001, 14, 1274-1275.

[128] [a] Martin, CR. Membrane-Based Synthesis of Nanomaterials. *Chem. Mater.,* 1996, 8(8), 1739-1746; [b] Park, S; Chung, S-W; Mirkin, CA. Hybrid Organic-Inorganic, Rod-Shaped Nanoresistors and Diodes. *J. Am. Chem. Soc.*, 2004, 126(38), 11772-11773.

[129] Maynor, BW; Filocamo, SF; Grinstaff, MW; Liu, J. Direct-Writing of Polymer Nanostructures: Poly(thiophene) Nanowires on Semiconducting and Insulating Surfaces. *J. Am. Chem. Soc.*, 2002, 124(4), 522-523.

[130] Nagarajan, R; Liu, W; Kumar, J; Tripathy, SK; Bruno, FF; Samuelson, LA. *Macromolecules*, 2001, 34(12), 3921-3927.

[131] [a] Barker, KD; Benoit, BR; Bordelon, JA; Davis, RJ; Delmas, AA; Mytykh, OV; Petty, JT; Petty, JF; Wheeler, JF; Kane-Maguire, NAP. Intercalative binding and photoredox behavior of [Cr(phen)2(dppz)3+ with D-DNA. *Inorg. Chim. Acta*, 2001, 322(1,2), 74-78; [b] Song, Y-F; Pang, P. Mononuclear tetrapyrido[3,2-a:2'3'-C:2",3"-h:3"',2"'-j]phenazine (tpphz) cobalt complex. *Polyhedron*, 2001, 20(6), 501-506; [c] Yamamoto, T. Shimizu, T; Kurokawa, E. Doping behavior of water-soluble n-conjugated polythiophenes depending on pH and interaction of the polymer with DNA. *React. Funct. Polym.*, 2000, 43(1,2), 79-84; [d] Yang, D; Strode, JT; Spielmann, HP; Wang, AH-J; Burke, TG. DNA Interactions of Two Clinical Camptothecin Drugs Stabilize Their Active Lactone Forms. *J. Am. Chem. Soc.*, 1998, 120(12), 2979-2980.

[132] Dong, L; Hollis, T; Fishwich, S; Connolly, BA; Wright, NG; Horrocks, BR; Houlton, A. Synthesis, Manipulation and Conductivity of Supramolecular Polymer Nanowires, *Chem. Eur. J.*, 2007, 13, 822-828.

[133] [a] Jennings, P; Jones, AC; Mount, AR. Fluorescence Properties of Electropolymerzed 5-Substituted Indoles in Solution, *J. Chem. Soc., Faraday Trans*, 1998, 94, 3619-3624; [b] Weiqiang, Z; Jingkun, X; Mengping, G. The Fluorescence Spectra of Polyindole and Its Derivatives, *Chem. Bull. Huaxue Tongbao*, 2008, 71, 75-79; [c] Billaud, D; Maarouf, EB; Hannecart, E. Electrochemical Polymerization of Indole, *Polymer*, 1994, 35, 2010-2011; [d] Billaud, D; Maarouf, EB; Hannecart, E. Chemical Oxidation and Polymerization of Indole, *Synth. Met.*, 1995, 69, 571-572.

[134] [a] Zotti, GA; Zecchin, S; Schiavon, G; Seraglia, RB; Berlin, A; Canavesi, A. Structure of Polyindoles from Anodic Coupling of Indoles: An Electrochemical Approach, *Chem. Mater.*, 1994, 6, 1742-1748; [b] Saraç, AS; Ozkara, S; Sezer, E. Electrocopolymerization of Indole and Thiophene: Conductivity-Peak Current Relationship and *In Situ* Spectrochemical Investigation of Soluble Co-Oligomers, *Int. J. Polym. Anal. Charact.*, 2003, 8, 395-409; [c] Tüken, T; Yazici, B; Erbil, A. Electrochemical Synthesis of Polyindole on Nickel-Coated Mild Steel and Its Corrosion Performance, *Surf. Coat. Technol.*, 2005, 200, 2301-2309; [d] Eraldemir, O; Sari, B;

Gok, A; Unal, HI. Synthesis and Characterization of Polyindole/Poly(vinyl acetate) Conducting Composites, *J. Macromol. Sci., Part A*, 2008, 45, 205-211.

[135] Liu, W.; Kumar, J.; Tripathy, S.; Senecal, K. J.; Samuelson, L. Enzymatically Synthesized Conducting Polyaniline. *J. Am.Chem. Soc.*, 1999, 121(1), 71-78

[136] [a] Huang, JX; Virji, S; Weiller, BH; Kaner, RB. Polyaniline Nanofibers: Facile Synthesis and Chemical Sensors. *J. Am. Chem. Soc.*, 2003, 125(2), 314-315; [b] Virji, S; Huang, JX; Kaner, RB; Weiller, BH. Polyaniline nanofibers gas sensors: examination of response mechanisms. *Nano Lett.*, 2004, 4, 491-496; [c] Zhang, HQ; Boussaad, S; Ly, N; Tao, NK. Magnetic-field-assisted assembly of metal/polymer/metal junction sensors. *Appl. Phys. Lett.*, 2004, 84(1), 133-135.

[137] [a] Lazzara, TD; Bourret, GR; Lennox, RB; van de Ven, TGM. Polymer Templated Synthesis of AgCN and Ag Nanowires, *Chem. Mater.*, 2009, 21, 2020-2026; [b] Metwalli, E; Moulin, J-F; Perlich, J; Wang, W; Diethert, A; Roth, SV; Müller-Buschbaum, P. Polymer-Template-Assisted Growth of Gold Nanowires Using a Novel Flow-Stream Technique, *Langmuir*, 2009, 25(19) 11815-11821.

[138] [a] Zhang, F; Nyberg, T; Inganäs, O. Conducting Polymer Nanowires and Nanodots Made with Soft Lithography, *Nano Lett.*, 2002, 2(12), 1373-1377; [b] Liu, HQ; Kameoka, J; Czaplewski, DA; Craighead, HG. Polymeric Nanowire Chemical Sensor. *Nano Lett.*, 2004, 4(4), 671-675; [c] Hernandez, SC; Chaudhuri, D; Chen, W; Myung, NV; Mulchandani, A. Single polypyrrole nanowire ammonia gas sensor. *Electroanalysis*, 2007, 19(19-20), 2125-2130; [d] Ramanathan, K; Banger, MA; Yun, M; Chen, W; Myung, NV; Mulchandani, A. Bioaffinity Sensing Using Biologically Functionalized Conducting-Polymer Nanowire. *J. Am. Chem. Soc.*, 2005, 127(2), 496-497.

[139] [a] Wanekaya, AK; Chen, W; Myung, NV; Mulchandani, A. Nanowire-based electrochemical biosensors. *Electroanalysis*, 2006, 18(6), 533-550; [b] Wanekaya, AK; Bangar, MA; Yun, M; Chen, W; Myung, NV; Mulchandani, A. Field-Effect Transistors Based on Single Nanowires of Conducting Polymers. *J. Phys. Chem. C.*, 2007, 111, 5218-5221; [c] Liu, HQ; Reccius, CH; Craighead, HG. Single electrospun regioregular poly(3-hexylthiophene) nanofiber field-effect transistor. *Appl. Phys. Lett.*, 2005, 87(25), 2531016/1-253106/3.

[140] Samitsu, S; Shimomura, T; Ito, K; Fujimori, M; Heike, S; Hashizume, T. Conductivity measurements of individual poly(3,4-ethylenedioxythiophene)/poly(styrenesulfonate) nanowires on nanoelectrodes using manipulation with an atomic force microscope. *Appl. Phys. Lett.*, 2005, 86(23), 233103/1-233103/3.

[141] He, HX; Li, CZ; Tao, NJ. Conductance of polymer nanowires fabricated by a combined electrodeposition and mechanical break junction method. *Appl. Phys. Lett.*, 2001, 78(6), 811-813.

[142] [a] Friend, RH; Gymer, RW; Holmes, AB; Burroughes, JH; Marks, RN; Taliani, C; Bradley, DDC; Dos Santos, DA; Bredas, JL; Logdlund, M; Salaneck, WR. Electroluminescence in conjugated polymers. *Nature*, 1999, 397(6715), 121-128; [b] Burroughes, JH; Bradley, DDC; Brown, AR; Marks, RN; Mackay, K; Friend, RH; Burns, PL; Homes, AB. Light-emitting diodes based on conjugated polymers. *Nature*, 1990, 347(6293), 539-541; [c] Noy, A; Miller, AE; Klare, JE; Weeks, BL; Woods, BW; DeYoreo, JJ. Fabrication of Luminescent Nanostructures and Polymer Nanowires Using Dip-Pen Nanolithography, *Nano Lett.*, 2002, 2(2) 109-112.

[143] Lipomi, DJ; Chiechi, RC; Dickey, MD; Whitesides, GM. Fabrication of Conjugated Polymer Nanowires by Edge Lithography, *Nano Lett.*, 2008, 8(7), 2100-2105.

[144] Yun, M; Myung, NV; Vasquez, RP; Lee, C; Menke, E; Penner, RM. Electrochemically Grown Wires for Individually Addressable Sensor Arrays, *Nano Lett.*, 2004, 4(3) 419-422.

[145] Walter, MJ; Borys, NJ; van Schooten, KJ. Light-Harvesting Action Spectroscopy of Single Conjugated Polymer Nanowires, *Nano Lett.*, 2008, 8(10), 3330-3335.

[146] [a] Liu, J; Lin, Y; Liang, L; Voigt, JA; Huber, DL; Tian, ZR; Coker, E; Mckenzie, B; Mcdermott, MJ. Templateless assembly of molecularly aligned conductive polymer nanowires: A new approach to oriented nanostructures. *Chem.—Eur. J.*, 2003, 9(3), 604-611; [b] Chiou, N-R; Lu, C; Guan, J; Lee, J; Epstein, AJ. Growth and alignment of polyaniline nanofibers with superhydrophobic, superhydrophillic and other properties. *Nat. Nanotechnol.*, 2007, 2(6), 354-357; [c] Tang, Q; Wu, J; Sun, X; Li, Q; Lin, J. Shape and Size Control of Oriented Polyaniline Microstructure by a Self-Assembly Method. *Langmuir*, 2009, 25(9), 5253-5257.

[147] Wang, Y; Tran, HD; Kaner, RB. Template-Free Growth of Aligned Bundles of Conducting Polymer Nanowires. *J. Phys. Chem. C.*, 2009, 113, 10346-10349.

[148] Tseng, RJ; Huang, JX; Ouyang, J; Kaner, RB; Yang, Y. Polyaniline Nanofiber/Gold Nanoparticle Nonvolatile Memory. *Nano Lett.,* 2005, 5(6), 1077-1080.

[149] Nadagouda, MN; Varma, RS. Green approach to bulk and template-free synthesis of thermally stable reduced polyaniline nanofibers for capacitor applications. *Green Chem.*, 2007, 9(6), 632-637.

[150] Ramanathan, K; Bangar, MA; Yun, M; Chen, W; Mulchandani, A; Myung, NV. Individually Addressable Conducting Polymer Nanowires Array. *Nano Lett.*, 2004, 4(7), 1237-1239.

[151] Thompson, RB; Ginzburg, VV; Matsen, MW; Balazs, AC. Predicting the mesophases of copolymer-nanoparticle composites. *Science*, 2001, 292(5526), 2469-2472.

[152] [a] Mao, C; Flynn, CE; Hayhurst, A; Sweeney, R; Qi, J; Georgiou, G; Iverson, B; Belcher, AM. Virus assembly of oriented quantum dot nanowires, *PNAS USA*, 2003, 100, 6946-6951; [b] Young, M; Willits, D; Uchida, M; Douglas, T. Plant Viruses as Biotemplates for Materials and Their use in Nanotechnology. *Annual Rev. Phytopathology*, 2008, 46, 361-384; [c] Falkner, JC; Turner, ME; Bosworth, JK; Trentler, TJ; Johnson, JE; Lin, T; Colvin, VL. Virus Crystals as Nanocomposite Scaffolds. *J. Am. Chem. Soc.*, 2005, 127, 5274-5275.

[153] [a] Smith, GP. Filamentous fusion phage: novel expression vectors that display cloned antigens on the virion surface. *Science*, 1985, 228, 1315-1317; [b] Levin, AM; Weiss, GA. Optimizing the affinity and specificity of proteins with molecular display. *Mol. Biosyst.*, 2006, 2(1), 49-57; [c] Petrenko, VA; Smith, GP. Phages from landscape libraries as substitute antibodies. *Protein Eng.*, 2000, 13, 589-592.

[154] [a] Goldman, ER; Pazirandeh, MP; Charles, PT; Balighian, ED; Anderson, GP. Selection of phage displayed peptide for the detection of 2,4,6-trinitrotoluene in seawater. *Anal. Chem. Acta*, 2002, 457(1), 13-19; [b] Rozinov, MN; Nolan, GP. Evolution of peptides that modulate the spectral qualities of bound, small-molecule fluorophores. *Chem. Biol.*, 1998, 5(12), 713-728.

[155] [a] Kehoe, JW; Kay, BK. Filamentous Phage Display in the New Millennium. *Chem. Rev.*, 2005, 105(11), 4056-4072; [b] Kay, BK; Hamilton, PT. Identification of enzyme

inhibitors from phage-displayed combinatorial peptide libraries. *Comb. Chem. High Throughput Screening*, 2001, 4(7), 535-543; [c] Scott, JK; Smith, GP. Searching for peptide ligands with an epitope library. *Science*, 1990, 249, 386-390; [d] Yang, LM; Tam, PY; Murray, BJ; McIntire, TM; Overstreet, CM; Weiss, GA; Penner, RM. Virus Electrodes for Universal Biodetection. *Anal. Chem.*, 2006, 78(10), 3265-3270; [e] Sidhu, SS; Kroide, S. Phage display for engineering and analyzing protein interaction interfaces. *Curr. Opin. Struct. Biol.*, 2007, 17(4), 481-487.

[156] Choo, Y; Klug, A. Designing DNA-binding proteins on the surface of filamentous phage. *Curr. Opin. Biotechnol.*, 1995, 6, 431-436.

[157] Arter, JA; Taggart, DK; McIntire, TM; Penner, RM; Weiss, GA. Virus-PEDOT Nanowires for Biosensing, *Nano Lett.*, 2010, 10, 4858-4862.

[158] Mao, C; Solis, DJ; Reiss, BD; Kottmann, ST; Sweeney, RY; Hayhurst, A; Georgiou, G; Inverson, B; Belcher, AM. Virus-Based Toolkit for the Direct Synthesis of Magnetic and Semiconducting Nanowires, *Science*, 2004, 303, 213-217.

[159] [a] Banfield, JF; Welch, SA; Zhang, H; Ebert, TT; Penn, RL. Aggregation-based crystal growth and microstructure development in natural iron oxyhydroxide biomineralization products. *Science*, 2000, 289(5480), 751-754; [b] Alivisatos, AP. Perspectives: Biomineralization: naturally aligned nanocrystals. *Science* 289, 736-737.

[160] Tsukamoto, R; Muraoka, M; Seki, M; Tabata, H; Yamashita, I. Synthesis of CoPt and $FePt_3$ nanowires using the central channel of tobacco mosaic virus as a biotemplate. *Chem. Mater.,* 2007, 19, 2389-2391.

[161] Nam, KT; Kim, D-W; Yoo, PJ; Chiang, C-Y; Meethong, N; Hammond, PT; Chiang, Y-M; Belcher, AM. Virus-Enabled Synthesis and Assembly of Nanowires for Lithium Ion Battery Electrodes, *Science*, 2006, 312, 885-888.

[162] Joly, S; et al, *Langmuir*, 2000, 16, 1354.

[163] Massalski, TB; Okamoto, H; Subramanian, PR; Kacprzak, L. Eds, *Binary Alloy Phase Diagrams* (ASM International, Materials Park, OH) **(1990)**.

[164] Nam, KT; Lee, YJ; Krauland, EM; Kottmann, ST; Belcher, AM. Peptide-mediated Reduction of Silver Ions on Engineered Biological Scaffolds. *ACS Nano*. 2008, 2, 1480-1486.

[165] [a] Nanoclusters within Microphase-Separated Diblock Copolymers: Sodium Carboxylate vs Carboxylic Acid Functionalization. *Supramol. Sci.,* 1998, 5, 41–48; [b] Ghosh, SK; Kundu, S; Mandal, M; Nath, S; Pal, T. Studies on the Evolution of Silver Nanoparticles in Micelle by UV- Photoactivation. *J. Nanopart. Res*. 2003, 5, 577-587; [c] Joly, S; Kane, R; Radzilowski, L; Wang, T; Wu, A; Cohen, RE; Thomas, EL; Rubner, MF. Multilayer Nanoreactors for Metallic and Semiconducting Particles. *Langmuir,* 2000, 16, 1354–1359.

[166] Cao, G. Nanostructures and Nanomaterials: Synthesis, properties and applications. London: Imperial College Press; 2004.

[167] Reches, M; Gazit, E. Designed aromatic homo-dipeptides: formation of ordered nanostructures and potential nanotechnological applications. Phys. Biol. 2006, 3, S10-9

[168] Scanlon, S; Aggeli, A. Self-assembling Peptide Nanotubes. *Nano Today*, 2008, 3, 22-30.

[169] Matsui, H; Douberly, GE. Organization of Peptide Nanotubes into Macroscopic Bundles. *Langmuir*, 2001, 17, 7918-7922.

[170] Yemini, M; Reches, M; Gazit, E; Rishpon, J. Peptide nanotube-modified electrodes for enzyme-biosensor applications. *Anal. Chem.* 2005, 77, 5155–5159
[171] Rajagopal, K; Schneider, JP. Self-assembling peptides and proteins for nanotechnological applications. *Curr. Opin. Struct. Biol.* 2004, 14, 480–486.
[172] [a] Main, ERG; Lowe, AR; Mochrie, SGJ; Jackson, SE; Regan, L. A recurring theme in protein engineering: the design, stability and folding of repeat proteins. *Curr. Opin. Struct. Biol.* 2005, 15, 464-471; [b] Kajander, T; Cortajarena, AL; Main, ERG; Mochrie, SGJ; Regan, L. A new folding paradigm for repeat proteins. *J. Am. Chem. Soc.* 2005, 127, 10188-10190.
[173] Tsai, CJ; Maizel, JV; Nussinov, R. Anatomy of protein structures: Visualizing how a one-dimensional protein chain folds into a three-dimensional shape. *Proc. Natl. Acad. Sci.* 2000, 97, 12038–12043.
[174] Tsai, CJ; Ma, B; Sham, YY; Kumar, S; Wolfson, HJ; Nussinov, R. A hierarchical building-block-based computational scheme for protein structure prediction. *IBM J. Res. & Dev.* 2001, 45, 513-523.
[175] Baldwin, RL; Rose GD. Is protein folding hierarchic? Local structure and peptide folding. *Trends Biochem. Sci.* 1999, 24, 26–33.
[176] Tsai, CJ; Zheng, J; Zanuy, D; Haspel, N; Wolfson, H; Aleman, C; Nussinov, R. Principles of Nanostructure Design With Protein Building Blocks. *PROTEINS: Structure, Function, and Bioinformatics*, 2007, 68, 1–12.
[177] Zheng, J; Zanuy, D; Haspel, N; Tsai, CJ; Aleman, C; Nussinov, R. Nanostructure Design Using Protein Building Blocks Enhanced by Conformationally Constrained Synthetic Residues. *Biochemistry*, 2007, 46, 1205-1218.
[178] Kojima, S; Kuriki,Y; Yoshida, T; Yazaki, K; Miura, K. Fibril formation by an amphipathic alpha-helix-forming polypeptide produced by gene engineering. *Proc. Jpn. Acad. Ser. B-Phys. Biol. Sci.* 1997, 73, 7-11.
[179] Pandya, MJ; Spooner, GM; Sunde, M; Thorpe, JR; Rodger, A; Woolfson DN. Sticky-end assembly of a designed peptide fiber provides insight into protein fibrillogenesis. *Biochemistry,* 2000, 39, 8728-8734.
[180] 180 Fairman, R; Akerfeldt, KS. Peptides as novel smart materials. *Curr. Opin. Struct. Biol.* 2005, 15, 453-463.
[181] Woolfson, DN; Ryadnov, MG. Peptide-based fibrous biomaterials: some things old, new and borrowed. *Curr. Opin. Chem. Biol.* 2006, 10, 559-567.
[182] [a] Reches, M; Gazit, E. Casting metal nanowires within discrete self-assembled peptide nanotubes. *Science*, 2003, 300, 625-627; [b] Scheibel, T; Parthasarathy, R; Sawicki, G; Lin, XM; Jaeger, H; Lindquist, SL. Conducting nanowires built by controlled self-assembly of amyloid fibers and selective metal deposition. *Proc. Natl. Acad. Sci.* 2003, 100, 4527-4532.
[183] Bekele, H; Fendler, JH; Kelly, JW. Self-assembling peptidomimetic monolayer nucleates oriented CdS nanocrystals. *J. Am. Chem. Soc.* 1999, 121, 7266-7267.
[184] Borgia, JA; Fields, GB. Chemical Synthesis of Proteins. *Tibtech*, 2000,18, 243-251.
[185] Archibald, DD; Mann, S. Template Mineralization of Self-Assembled Lipid Microstructures. *Nature,* 1993, 364, 430-433.
[186] Whaley, SR; English, DS; Hu, EL; Barbara, PF; Belcher, AM. Selection of peptides with semiconductor binding specificity for directed nanocrystal assembly. *Nature,* 2000, 405, 665-668.

[187] Hartgerink, JD; Beniash, E; Stupp, SI. Self-assembly and mineralization of peptide-amphiphile nanofibers. *Science,* 2001, 294, 1684-1688.

[188] Lutolf, MP; Hubbell, JA. Synthetic biomaterials as instructive extracellular microenvironments for morphogenesis in tissue engineering. *Nature Biotechnology*, 2005, 23, 47-55.

[189] Shakesheff, K; Cannizzaro, S; Langer, RS. Creating biomimetic micro-environments with synthetic polymer-peptide hybrid molecules. *J. Biomater. Sci. Polym. Ed.* 1998, 9, 507–518.

[190] Lanza, RP; Langer, R; Chick, WL. Principles of Tissue Engineering. Texas: *RG Landes and Academic Press*; 1997.

[191] Zhang, S; Holmes, T; Lockshin, C; Rich, A. Spontaneous assembly of a self-complementary oligopeptide to form a stable macroscopic membrane. *Proc. Natl. Acad. Sci.* 1993, 90, 3334-3338.

[192] Kleinman, HK; Philp, D; Hoffman, MP. Role of the extracellular matrix in morphogenesis. *Curr. Opin. Biotechnol.* 2003, 14, 526–532.

[193] Kenawy, ER; Layman, JM; Watkins, JR; Bowlin, GL; Matthews, JA; Simpson, DG. Wnek, GE. Electrospinning of poly(ethylene-co-vinyl alcohol) fibers. *Biomaterials,* 2003, 24, 907–913.

[194] Menger, FM. Supramolecular chemistry and self-assembly. *Proc. Natl. Acad. Sci.* 2002, 99, 4818–4822.

[195] Kopecek, J; Yang, J. Peptide-directed self-assembly of hydrogels. *Acta Biomater.* 2009, 5, 805-816.

[196] Holmes, TC; de Lacalle, S; Su, X; Liu, G; Rich, A; Zhang, S. Extensive neurite outgrowth and active synapse formation on self-assembling peptide scaffolds. *Proc. Natl. Acad. Sci.* 2000, 97, 6728-6733.

[197] Zhang, S; Lockshin, C; Herbert, A; Winter, E; Rich, A. Zoutin, a Z-DNA binding protein in Saccharomyces cerevisiae. *The EMBO Journal*, 1992, 11, 3787-3796.

[198] Zhang, S; Holmes, T; DiPersio, M; Hynes, RO; Su, X; Rich, A. Self-complementary oligopeptide matrices support mammalian cell attachment. *Biomaterials*, 1995, 16, 1385-1393.

[199] Kisiday, J; Jin, M; Kurtz, B; Hung, H; Semino, C; Zhang, S; Grodzinsky, AJ. Self-assembling peptide hydrogel fosters chondrocyte extracellular matrix production and cell division: Implications for cartilage tissue repair. *Proc. Natl. Acad. Sci.* 2002, 99, 9996–10001.

[200] Semino, CE, Merok, JR; Crane, GG; Panagiotakos, G; Zhang, S. Functional differentiation of hepatocyte-like spheroid structures from putative liver progenitor cells in three-dimensional peptide scaffolds. *Differentiation,* 2003, 71, 262–270.

[201] Nowak, AP; Breedveld, V; Pakstis, L; Ozbas, B; Pine, DJ; Pochan, D; Deming, TJ. Rapidly recovering hydrogel scaffolds from self-assembling diblock copolypeptide amphiphiles. *Nature*, 2002, 417, 424–428.

[202] Pakstis, LM; Ozbas, B; Hales, KD; Nowak, AP; Deming, TJ; Pochan, D. Effect of chemistry and morphology on the biofunctionality of self-assembling diblock copolypeptide hydrogels. *Biomacromolecules,* 2004, 5, 312–318.

[203] Niece, KL; Hartgerink, JD; Donners, JJ; Stupp, SI. Self-assembly combining two bioactive peptide-amphiphile molecules into nanofibers by electrostatic attraction. *J. Am. Chem. Soc.* 2003, 125, 7146–7147.

[204] Baldwin, SP; Saltzman, WM. Materials for protein delivery in tissue engineering. *Adv. Drug Del. Rev.* 1998, 33, 71-86.

[205] Vauthey, S; Santoso, S; Gong, H; Watson, N; Zhang, S. Molecular self-assembly of surfactant-like peptides to form nanotubes and nanovesicles. *Proc. Natl. Acad. Sci.* 2002, 99, 5355-5360.

[206] Von Maltzahn, G; Vauthey, S; Santoso, S; Zhang, S. Positively Charged Surfactant-like Peptides Self-assemble into Nanostructures. *Langmuir*, 2003, 19, 4332-4337.

[207] Scanlon, S; Aggeli, A. Self-assembling peptide nanotubes. *Nano Today*, 2008, 3, 22-30.

[208] Ghadiri, MR; Granja, JR; Milligan, RA; Mcree, DE; Khazanovich, N. Self-assembling organic nanotubes based on a cyclic peptide architecture. *Nature*, 1993, 366, 324-327.

[209] [a] Hartgerink, JD; Clark, TD; Ghadiri, MR. Peptide Nanotubes and Beyond. *Chem. Eur. J.* 1998, 4, 1367-1372; [b] Horne, WS; Stout, CD; Ghadiri, MR. A Heterocyclic Peptide Nanotube. *J. Am. Chem. Soc.* 2003, 125, 9372-9376.

[210] Block, MAB; Hecht, S. Wrapping Peptide Tubes: Merging Biological Self-Assembly and Polymer Synthesis. *Angew. Chem., Int. Ed.* 2005, 44, 6986-6989.

[211] Zhang, S; Holmes, T; Lockshin, C; Rich, A. Spontaneous assembly of a self-complementary oligopeptide to a stable macroscopic membrane. *Proc. Natl. Acad. Sci.* 1993, 90, 3334-3338.

[212] Halverson, K; Fraser, PE; Kirschner, DA; Lansbury, PT. Molecular Determinants of Amyloid Deposition in Alzheimer's Disease: Conformational Studies of Synthetic P-Protein Fragments. *Biochemistry*, 1990, 29, 2639-2644.

[213] Kirschner, DA; Inouye, H; Duffy, LK; Sinclair, A; Lind, M; Selkoe, DJ. Synthetic peptide homologous to beta protein from Alzheimer disease forms amyloid-like fibrils in vitro. *Proc. Natl. Acad. Sci.* 1987, 84, 6953-6957.

[214] Whitesides, GM; Mathias, JP; Seto, CT. Molecular self-assembly and nanochemistry: a chemical strategy for the synthesis of nanostructures. *Science*, 1991, 254, 1312-1319.

[215] Campbell, GR; Campbell, JH. Development of Tissue Engineered Vascular Grafts. *Curr. Pharm. Biotechnol.* 2007, 8, 43-50.

[216] Weinberg, CB; Bell, E. Regulation of proliferation of bovine aortic endothelial cells, smooth muscle cells, and adventitial fibroblasts in collagen lattices. *J. Cell Physiol.* 1985, 122, 410-414.

[217] Ziegler, T; Alexander, RW; Nerem, RM. An endothelial cell-smooth muscle cell co-culture model for use in the investigation of flow effects on vascular biology. *Ann. Biomed. Eng.* 1995, 23, 216-225.

[218] Hirai, J; Matsuda, T. Venous reconstruction using hybrid vascular tissue composed of vascular cells and collagen: tissue regeneration process. *Cell Transplant,* 1996, 5, 93-105.

[219] Narmoneva, DA; Oni, O; Sieminski, AL; Zhang, S; Gertler, JP; Kamm, RD; Lee, RT. Self-assembling short oligopeptides and the promotion of angiogenesis. *Biomaterials*, 2005, 26, 4837-4846.

[220] L'Heureux, N; Stoclet, JC; Auger, FA; Lagaud, GJ; Germain, L; Andriantsitohaina, R. A human tissue-engineered vascular media: a new model for pharmacological studies of contractile responses. *FASEB J.* 2001, 15, 515-524.

[221] L'Heureux, N; Paquet, S; Labbe, R; Germain, L; Auger, FA. A completely biological tissue-engineered human blood vessel. *FASEB J.* 1998, 12, 47-56.

[222] Zhou, WY; Guo, B; Liu, M; Liao, R; Rabie, ABM; Jia, D. Poly(vinyl alcohol)/Halloysite nanotubes bionanocomposite films: Properties and in vitro osteoblasts and fibroblasts response. *J. Biomed. Mater. Res.* 2009, 93A, 1574-1587.

[223] Nerem, RM; Seliktar, D. Vascular Tissue Engineering. *Annu. Rev. Biomed. Eng.* 2001, 3, 225-243.

[224] Schmidt, CE; Baier, JM. Acellular vascular tissues: natural biomaterials for tissue repair and tissue engineering. *Biomaterials*, 2000, 21, 2215-2231.

[225] Arieta, AH; Dermitzakis, K; Damian, D; Lungarella, M; Pfeifer, R. (2008). Sensory-motor coupling in rehabilitation robotics. In: Takahashi, Y (Ed.), Handbook of Service Robotics (pp. 21-36). Vienna, Austria: I-Tech Education and Publishing.

[226] Someya, T; Sekitani, T; Iba, S; Kato, Y; Kawaguchi, H; Sakurai, T. A large-area flexible pressure sensor matrix with organic field-effect transistors for artificial skin applications. *Proc. Natl Acad. Sci., USA* 2004, 101, 9966-9970.

[227] Someya, T; Sekitani, T; Iba, S; Kato, Y; Sakurai, T; Kawaguchi, H. Organic Transistor Integrated Circuits for Large-Area Sensors. *Mol. Cryst. Liq. Cryst.* 2006, 444, 13-22.

[228] Fan, Z; Ruebusch, DJ; Rathore, AA; Kapadia, R; Ergen, O; Leu, PW; Javey, A. Challenges and Prospects of Nanopillar-Based Solar Cells. *Nano Res.* 2009, 2, 829-843.

[229] Takei, K; Takahashi, T; Ho, JC; Ko, H; Gillies, AG; Leu, PW; Fearing, RS; Javey, A. Nanowire active-matrix circuitry for low-voltage macroscale artificial skin. *Nature Mater*, 2010, 9, 821-826.

[230] Fan, Z; Ho, JC; Jacobson, ZA; Razavi, H; Javey, A. Large scale, heterogeneous integration of nanowire arrays for image sensor circuitry. *Proc. Natl. Acad. Sci.* 2008, 105, 11066-11070.

[231] Someya, T; Sekitani, T; Iba, S; Kato, Y; Kawaguchi, H; Sakurai, T. Comfortable, flexible, large-area networks of pressure and thermal sensors with organic transistor active matrixes. *Proc. Natl. Acad. Sci.* 2005, 102, 12321-12325.

[232] [a] Birbaumer, N; Murguialday, AR; Cohen, L. Brain-computer interface in paralysis. *Curr. Opin. Neurol.* 2008, 21, 634–638; [b] Rutten, WLC. Selective electrical interfaces with the nervous system. *Annu. ReV. Biomed. Eng.* 2002, 4, 407–452.

[233] Linsmeier, CE; Prinz, CN; Pettersson, LME; Caroff, P; Samuelson, L; Shouenborg, J; Montelius, L; Danielsen, N. Nanowire Biocompatibility in the Brain – Looking for a Needle in a 3D Stack. *Nano Lett.* 2009, 9, 4184-4190.

[234] Keefer, EW; Botterman, BR; Romero, MI; Rossi, AF; Gross, GW. Carbon nanotube coating improves neuronal recordings. *Nat. Nanotechnol.* 2008, 3, 434–439.

[235] Oberpenning, F; Meng, J; Yoo, JJ; Atala, A. De novo reconstitution of a functional mammalian urinary bladder by tissue engineering. *Nat. Biotechnol.* 1999, 17, 149-155.

[236] Ourednik, V. Segregation of human neural stem cells in the developing primate forebrain. *Science*, 2001, 293, 1820-1824.

[237] Park, KI; Teng, YD; Snyder, EY. The injured brain interacts reciprocally with neural stem cells supported by scaffolds to reconstitute lost tissue. *Nat. Biotechnol.* 2002, 20, 1111–1117.

[238] Mahoney, MJ; Saltzman, WM. Transplantation of brain cells assembled around a programmable synthetic microenvironment. *Nat. Biotechnol.* 2001, 19, 934–939.

[239] Mahoney, MJ; Saltzman, WM. Cultures of cells from fetal rat brain: Methods to control composition, morphology, and biochemical activity. *Biotechnol. Bioeng.* 1999, 62, 461–467.

[240] Cui, X; Wiler, J; Dzaman, M; Altschuler, RA; Martin, DC. In vivo studies of polypyrrole/peptide coated neural probes. *Biomaterials*, 2003, 24, 777-787.

[241] Kim, D.H., Richardson-Burns, S., Povlich, L., Abidian, M.R., Spanninga, S., Hendricks, J.L., Martin, D.C. (2008). Soft, Fuzzy, and Bioactive Conducting Polymes for Improving the Chronic Performance of Neural Prosthetic Devices. In W. M. Reichert (Ed.) Indwelling Neural Implants: Strategies for Contending with the *In Vivo* Environment (Chapter 7). Florida, USA: CRC Press.

[242] Hallstrom, W; Martensson, T; Prinz, C; Gustavsson, P; Montelius, L; Samuelson, L; Kanje, M. Gallium Phosphide Nanowires as a Substrate for Cultured Neurons. *Nano Lett.* 2007, 7, 2960–2965.

[243] Patolsky, F; Timko, BP; Yu, GH; Fang, Y; Greytak, AB; Zheng, GF; Lieber CM. Detection, Stimulation, and Inhibition of Neuronal Signals with High-Density Nanowire Transistor Arrays. *Science,* 2006, 313, 1100–1104.

[244] Linsmeier, CE; Prinz, CN; Pettersson, LME; Caroff, P; Samuelson, L; Shouenborg, J; Montelius, L; Danielsen, N. Nanowire Biocompatibility in the Brain – Looking for a Needle in a 3D Stack. *Nano Lett.* 2009, 9, 4184-4190.

[245] Rosengren, A; Wallman, L; Danielsen, N; Laurell, T; Bjursten, LM. Tissue reactions evoked by porous and plane surfaces made out of silicon and titanium. *IEEE Trans. Biomed. Eng.* 2002, 49, 392–399.

[246] Linsmeier, CE; Wallman, L; Faxius, L; Schouenborg, J; Bjursten, LM; Danielsen, N. Soft tissue reactions evoked by implanted gallium phosphide. *Biomaterials*, 2008, 29, 4598–4604.

[247] Park, I; Li, Z; Li, X; Pisano, AP; Williams; RS. Towards the silicon nanowire-based sensor for intracellular biochemical detection. *Biosens. Bioelectron.* 2007, 22, 2065-2070.

[248] [a] Chen, RJ; Bangsaruntip, S; Drouvalakis, KA; Kam, NWS.; Shim, M; Li, YM; Kim, W; Utz, PJ; Dai, HJ. Noncovalent Functionalization of Carbon Nanotubes for Highly Specific Electronic Biosensors. *Proc. Natl. Acad. Sci.* 2003, 100, 4984–4989; [b] Star, A; Gabriel, JCP; Bradley, K; Gruner, G. Electronic Detection of Specific Protein Binding Using Nanotube FET Devices. *Nano Lett.* 2003, 3, 459–463.

[249] Besteman, K; Lee, JO; Wiertz, FGM; Heering, HA; Dekker, C. Enzyme-Coated Carbon Nanotubes as Single-Molecule Biosensors. *Nano Lett.* 2003, 3, 727–730.

[250] [a] Patolsky, F; Timko, BP; Yu, GH; Fang, Y; Greytak, AB; Zheng, GF; Lieber, CM. Detection, Stimulation, and Inhibition of Neuronal Signals with High-Density Nanowire Transistor Arrays. *Science,* 2006, 313, 1100–1104; [b] Stern, E; Steenblock, ER; Reed, MA; Fahmy, TM. Label-Free Electronic Detection of the Antigen-Specific T-Cell Immune Response. *Nano Lett.* 2008, 8, 3310–3314; [c] Wang, CW; Pan, CY; Wu, HC; Shih, PY; Tsai, CC; Liao, KT; Lu, LL; Hsieh, WH; Chen, CD; Chen, YT. In Situ Detection of Chromogranin A Released from Living Neurons with a Single-Walled Carbon-Nanotube Field-Effect Transistor. *Small,* 2007, 3, 1350–1355.

[251] Ishikawa, FN; Curreli, M; Chang, HK; Chen, PC; Zhang, R; Cote, RJ; Thompson, ME; Zhou, C. A Calibration Method for Nanowire Biosensors to Suppress Device-to-Device Variation. *ACS Nano,* 2009, 3, 3969-3976.

[252] Sonnichsen, C; Alivisatos, AP. Gold nanorods as novel nonbleaching plasmon-based orientation sensors for polarized single-particle microscopy, *Nano Lett.* 5 (2005) 301–304.

[253] Huang, XJ; Choi, YK. Chemical sensors based on nanostructured materials. *Sensors and Actuators B: Chemical,* 2007, 122, 659-671.

[254] Hahm, JI; Lieber, CM. Direct Ultrasensitive Electrical Detection of DNA and DNA Sequence Variations Using Nanowire Nanosensors. *Nano Lett.* 2004, 4, 51-54.

[255] Wang, WU; Chen, C; Lin, KH; Fang, Y; Lieber, CM. Label-free detection of small-molecule – protein interactions by using nanowire nanosensors. *Proc. Natl. Acad. Sci.* 2005, 102, 3208-3212.

[256] Cui, Y; Wei, QQ; Park, HK; Lieber, CM. Nanowire Nanosensors for Highly Sensitive and Selective Detection of Biological and Chemical Species. *Science,* 2001, 293, 1289–1292

[257] Fraden, J. Handbook of Modern Sensors: Physics, Designs, and Applications. New York: Springer; 2004.

[258] McAlpine, MC; Agnew, HD; Rohde, RD; Blanco, M; Ahmad, H; Stuparu, AD; Goddard, WA; Heath, JR. Peptide-Nanowire Hybrid Materials for Selective Sensing of Small Molecules. *J. Am. Chem. Soc.* 2008, 130, 9583-9589.

[259] Zhang, D.; Liu, Z.; Li, C.; Tang, T.; Liu, X.; Han, S.; Lei, B.; Zhou, C. Detection of NO_2 down to ppb levels using individual and multiple In_2O_3 nanowire devices. *Nano Lett.* 2004, 4, 1919–1924.

[260] Stella, R; Barisci, JN; Serra, G; Wallace, GG; De Rossi, D. Characterisation of olive oil by an electronic nose based on conduting polymer sensors. *Sens. Actuators B* 2000, 63, 1–9.

[261] Freund, MS; Lewis, NS. A chemically diverse conducting polymer-based "electronic nose". *Proc. Natl. Acad. Sci.* 1995, 92, 2652–2656.

[262] Qi, P; Vermesh, O; Grecu, M; Javey, A; Wang, Q; Dai, H; Peng, S; Cho, K. J. Toward large arrays of multiplex functionalized carbon nanotube sensors for highly sensitive and selective molecular detection. *Nano Lett.* 2003, 3, 347–351.

[263] Snow, E. S.; Perkins, F. K.; Houser, E. J.; Badescu, S. C.; Reinecke, T. L. Chemical detection with a single-walled carbon nanotube capacitor. *Science,* 2005, 307, 1942–1945.

[264] Zhang, GJ; Chua, JH; Chee, RE; Agarwal, A; Wong, SM; Buddharaju, KD; Balasubramanian, N. Highly sensitive measurements of PNA-DNA hybridization using oxide-etched silicon nanowire biosensors. *Biosens. Bioelectron.* 2008, 23, 1701-1707.

[265] Sysoev, VV; Button, BK; Wepsiec, K; Dmitriev, S; Kolmakov, A. Toward the Nanoscopic "Electronic Nose": Hydrogen vs Carbon Monoxide Discrimination with an Array of Individual Metal Oxide Nano- and Mesowire Sensors. *Nano Lett.* 2006, 6, 1584–1588.

[266] McAlpine, MC; Ahmad, H; Wang, D; Heath, JR. Highly Ordered Nanowire Arrays on Plastic Substrates for Ultrasensitive Flexible Chemical Sensors. *Nature Mater.* 2007, 6, 379- 384.

[267] Heath, JR. Superlattice nanowire pattern transfer (SNAP). *Acc. Chem. Res.* 2008, 41, 1609-1617.

[268] Lin, Z; Strother, T; Cai, W; Cao, X; Smith, LM; Hamers, RJ. DNA Attachment and Hybridization at the Silicon (100) Surface. *Langmuir,* 2002, 18, 788-796.

[269] McAlpine, MC; Friedman, RS; Jin, S; Lin, KH; Wang, WU; Lieber, CM. High-Performance Nanowire Electronics and Photonics on Glass and Plastic Substrates. *Nano Lett.* 2003, 3, 1531-1535.

[270] Di Natale, C; Macagnano, A; Martinelli, E; Paolesse, R; D'Arcangelo, G; Roscioni, C; Finazzi-Agro, A; D'Amico, A. Lung cancer identification by the analysis of breath by means of an array of non-selective gas sensors. *Biosens. Bioelectron.* 2003, 18, 1209-1218.

[271] Li, C; Curreli, M; Lin, H; Lei, B; Ishikawa, FN; Datar, R; Cote, RJ; Thompson, ME; Zhou, CW. Complementary Detection of Prostate-Specific Antigen Using In2O3 Nanowires and Carbon Nanotubes. *J. Am. Chem. Soc.* 2005, 127, 12484–12485.

[272] Alon, U. (2007). An introduction to Systems Biology: Design Principles of Biological Circuits. London, UK: CRC Press.

[273] Porrata, P; Goun, E; Matsui, H. Size-controlled peptide nanotube fabrication using polycarbonate membranes as templates. *Chem. Mater.* 2002, 14, 4378-4381.

[274] Schnur, JM. Lipid Tubules: A Paradigm for Molecularly Engineered Structures. *Science,* 1993, 262, 1669-1676.

[275] Zhou, Y; Ji, Q; Shimizu, Y; Koshizaki, N; Shimizu, T. One-Dimensional Confinement of CdS Nanodots and Subsequent Formation of CdS Nanowires by Using a Glycolipid Nanotube as a Ship-in-Bottle Scaffold. *J. Phys. Chem. C.* 2008, 112, 19412-18416.

[276] Xia, Y; Yang, P; Sun, Y; Wu, Y; Mayers, B; Gates, E; Yin, Y; Kim, F; Yan, H. One-Dimensional Nanostructures: Synthesis, Characterization, and Applications. *AdV. Mater.* 2003, 15, 353-389.

[277] [a] Shimizu, T. Bottom-Up Synthesis and Structural Properties of Self-Assembled High-Axial-Ratio Nanostructures. *Macromol. Rapid Commun* 2002, 23, 311-331; [b] Iwaura, R; Hoeben, FJM; Masuda, M; Schenning, APHJ; Meijer, EW; Shimizu, T. Molecular-level helical stack of a nucleotide-appended oligo (p-phenylenevinylene) directed by supramolecular self-assembly with a complementary oligonucleotide as a template. *J. Am. Chem. Soc.* 2006, 128, 13298-13304.

[278] Kamiya, S; Minamikawa, H; Jung, JH; Yang, Bl; Masuda, M; Shimizu, T. Molecular Structure of Glucopyranosylamide Lipid and Nanotube Morphology. *Langmuir*, 2005, 21, 1401-1443.

[279] Spector, MS; Selinger, JV; Singh, A; Rodriguez, JM; Price, RR; Schnur, JM. Controlling the Morphology of Chiral Lipid Tubules. *Langmuir*, 1998, 14, 3493-3500.

[280] Kameta, N; Masuda, M; Minamikawa, H; Goutev, NV; Rim, JA; Jung, JH; Shimizu, T. Selective Construction of Supramolecular Nanotube Hosts with Cationic Inner Surfaces. *AdV. Mater.* 2005, 17, 2732-2736.

[281] Matsui, H; Porrata, P; Douberly, GEJ. Protein tubule immobilization on self- assembled mono-layers on Au. *Nano Lett.* 2001, 1, 461-464.

[282] Mann, S. Self-assembly and transformation of hybrid nano-objects and nanostructures under equilibrium and non-equilibrium conditions. *Nat. Mater.* 2009, 8, 781-792.

[283] Zhou, Y; Shimizu, T. Lipid Nanotubes: A Unique Template To Create Diverse One-Dimensional Nanostructures. *Chem. Mater.* 2008, 20, 625–633.

[284] Masuda, M; Shimizu, T. Lipid Nanotubes and Microtubes: Experimental Evidence for Unsymmetrical Monolayer Membrane Formation from Unsymmetrical Bolaamphiphiles. *Langmuir,* 2004, 20, 5969-5977.

[285] Karlsson, A; Karlsson, R; Karlsson, M; Cans, AS; Stromberg, A; Ryttsen, F; Orwar, O. Networks of nanotubes and containers. *Nature,* 2001, 409, 150-152.

[286] Claussen, RC; Rabatic, BM; Stupp, SI. Aqueous Self-Assembly of Unsymmetric Peptide Bolaamphiphiles into Nanofibers with Hydrophilic Cores and Surfaces. *J. Am. Chem. Soc.* 2003, 125, 12680-12681.

[287] MacGilivray, L.R., & Atwood, J. L. (2000). Spherical Molecular Containters: From Discovery to Design. In Gokel, G. W (Ed*.), Advances in Supramolecular Chemistry* (157-183). Stamford: JAI Press.

[288] Yang, B; Kamiya, S; Yoshida, K; Shimizu, T. Glycolipid Nanotube Hollow Cylinders as Substrates: Fabrication of One-Dimensional Metallic-Organic Nanocomposites and Metal Nanowires. *Chem. Commun.* 2004, 16, 2826-2831.

[289] Sott, K; Lobovkina, T; Lizana, L; Tokarz, M; Bauer, B; Konkoli, Z; Orwar, O. Controlling Enzymatic Reactions by Geometry in a Biomimetic Nanoscale Network. *Nano Lett.* 2006, 6, 209-214.

[290] Schnur, JM; Price, R; Rudolph, AS. Biologically engineered microstructures – controlled-release applications. *J. Controlled Release,* 1994, 28, 3-13.

[291] Reches, M; Gazit, E. Casting Metal Nanowires Within Discrete Self-Assembled Peptide Nanotubes. *Science*, 2003, 300, 625-627.

[292] Djalali, R; Chen, Y; Matsui, H. Au Nanocrystal Growth on Nanotubes Controlled by Conformations and Charges of Sequenced Peptide Templates. *J. Am. Chem. Soc.* 2003, 125, 5873-5879.

[293] Schnur, JM. Lipid Tubules: A Paradigm for Molecularly Engineered Structures. *Science,* 1993, 262, 1669-1676.

[294] Small, D.M. (1986). *The Physical Chemistry of Lipids*. New York: Plenum Press.

[295] Kameta, N; Masuda, M; Minamikawa, H; Mishima, Y; Yamashita, I; Shimizu, T. Functionalizable Organic Nanochannels Based on Lipid Nanotubes: Encapsulation and Nanofluidic Behavior of Biomacromolecules. *Chem. Mater.* 2007, 19, 3553-3560.

[296] Kameta, N; Masuda, M; Minamikawa, H; Shimizu, T. Self-assembly and thermal phase transition behavior of unsymmetrical bolaamphiphiles having glucose- and amino-hydrophilic headgroups. *Langmuir,* 2007, *23*, 4634-4641.

[297] Yui, H; Shimizu, Y; Kamiya, S; Yamashita, I; Masuda, M; Ito, K; Shimizu, T. Encapsulation of ferritin within a hollow cylinder of glycolipid nanotubes. *Chem. Lett.* 2005, 34, 232-233.

[298] Tien, H., Ottova-Leitmannova, A. (2003). Planar lipid bilayers (BLMs) and Their Applications. Amsterdam: Elsevier Science.

[299] Roux, A; Cappello, G; Cartaud, J; Prost, J; Goud, B; Bassereau, P. A minimal system allowing tubulation with molecular motors pulling on giant liposomes. *Proc. Natl. Acad. Sci.* 2002, 99, 5394-5399.

[300] Huang, SCJ; Artyukhin, AB; Martinez, JA; Sirbuly, DJ; Wang, Y; Ju, JW; Stroeve, P; Noy, A. Formation, Stability, and Mobility of One-Dimensional Lipid Bilayer on High Curvature Substrates. *Nano Lett.* 2007, 7, 3355-3359.

[301] Sackmann, E. Supported Membranes: Scientific and Practical Applications. *Science*, 1996, 271, 43-48.

[302] Artyukhin, AB; Shestakov, A; Harper, J; Bakajin, O; Stroeve, P; Noy, A. Functional One-Dimensional Lipid Bilayers on Carbon Nanotube Templates. *J. Am. Chem. Soc.* 2005, 127, 7538-7542.

[303] Smith, HL; Jablin, MS; Vidyasagar, A; Saiz, J. Watkins, E; Toomey, R; Hurd, AJ; Majewski, J. Model Lipid Membranes on a Tunable Polymer Cushion. *Phys. Rev. Lett.* 2009, 102, 1-4.

[304] Sirbuly, DJ; Fischer, N; Huang, SCJ; Tok, J; Bakajin, O; Noy, A. Biofunctional subwavelength optical waveguides for biodetection. *ACS Nano*, 2008, 2, 255-262.

[305] Zhou, X; Moran-Mirabal, JM; Craighead, HG; McEuen, PL. Supported lipid bilayer/carbon nanotube hybrids. *Nat. Nanotechnnol.* 2007, 2, 185-190.

[306] Noy, A; Misra, N; Martinez, J. Lipid-coated nanowires enable small-scale bioelectronics. *SPIE Newsroom*, 2009, 1-2.

[307] Noy, A. Bionanoelectronics. *Adv. Mater.* 2011, 23, 807–820.

[308] Misra, N; Martinez, JA; Huang, SCJ; Wang, Y; Stroeve, P; Grigoropoulos, CP; Noy, A. Bioelectronic silicon nanowire devices using functional membrane proteins. *Proc. Natl. Acad. Sci.* 2009, 106, 13780-13784.

[309] Langridge-Smith, J; Dubinsky, W. Donnan equilibrium and pH gradient in isolated tracheal apical membrane vesicles. *Am. J. Physiol.* 1985, 249, 417– 420.

[310] Noy, A; Artyukhin, AB; Misra, N. Bionanoelectronics with 1D materials. *Mater. Today,* 2009, 12, 22-31.

[311] Merkel, TJ; Herlihy, KP; Nunes J; Orgel, RM; Rolland, JP; DeSimone, JM. Scalable, shape-specific, top-down fabrication methods for the synthesis of engineered colloidal particles. *Langmuir*, 2010, 26(16), 13086–130896

[312] Li, Y; Yin, X-F; Chen, F-R; Yang, H-H; Zhuang, Z-X; Wang, X-R. Synthesis of Magnetic Molecularly Imprinted Polymer Nanowires Using a Nanoporous Alumina Template, *Macromolecules*, 2006, 39, 4497-4499.

[313] [a] Haupt, K; Mosbach, K. Molecularly Imprinted Polymers and Their Use in Biomimetic Sensors. *Chem. Rev.*, 2000, 100(7), 2495-2504; [b] Haupt, K. Molecularly imprinted polymers: the next generation. *Anal. Chem.*, 2003, 75(17), 376A-383A; c) Wulff, G. Enzyme-like Catalysis by Molecularly Imprinted Polymers. *Chem. Rev.*, 2002, 102(1), 1-27.

[314] Yue, F; Ngin, T; Hailin, G. A novel paper pH sensor based on polypyrrole. *Sens. Actuators, B*, 1996, 32, 33-39.

[315] Beh, WS; Kim, IT; Qin, D; Xia, YN; Whitesides, GM. Formation of patterned microstructures of conducting polymers by soft lithography and applications in microelectronic device fabrication. *Adv. Mater.*, 1999, 11(12), 1038-1041.

[316] Granlund, T; Nyberg, T; Roman, LS; Svensson, M; Inganäs, O. Patterning of polymer light-emitting diodes with soft lithography. *Adv. Mater.*, 2000, 12(4), 269-273.

[317] [a] Roman, LS; Mammo, W; Pettersson, L; Andersson, MR; Inganäs, O. High quantum efficiency polythiophene/C60 photodiodes. *Adv. Mater.*, 1998, 10(10), 774-777; [b] Roman, LS; Berggren, M; Inganäs, O. Polymer diodes with high rectification. *Appl. Phys. Lett.*, 1999, 75(22), 3557-3559.

[318] Andersson, M; Ekebad, PO; Hjertberg, T; Wennerström, P; Inganäs, O. Polythiophene with a free amino acid side chain. *Polym. Commun.*, 1991, 32(18), 546-548.

[319] Xu, QB; Rioux, RM; Whitesides, GM. Fabrication of Complex Metallic Nanostructures by Nanoskiving. *ACS Nano*, 2007, 1(3), 215-227.

[320] Lim, JH; Mirkin, CA. Electrostatically driven dip-pen nanolithography of conducting polymers. *Adv. Mater.*, 2002, 14(20), 1474-1477.

[321] McQuade, DT; Pullen, AE; Swagger, TM. Conjugated polymer-based chemical sensors. *Chem. Rec.*, 2000, 100(7), 2537-2574.

[322] Hong, SH; Zhu, J; Mirkin, CA. Multiple ink nanolithography: Toward a multiple-pen nano-plotter. *Science*, 1999, 286(5439), 523-525.

[323] Li, J; Qin, DJ; Baker, JR; Tomalia, DA. The characterization of high generation poly(amidoamine) G9 dendrimers by atomic force microscopy (AFM). *Macromol. Symp.*, 2001, 167, 257-269.

[324] Maynor, BW; Li, Y; Liu, J. Au "Ink" for AFM "Dip-Pen" Nanolithography. *Langmuir*, 2001, 17, 2575-2578.

[325] Demers, LM; Ginger, DS; Park, S-J; Li, Z; Chung, S-W; Mirkin, CA. Direct Patterning of Modified Oligonucleotide on Metals and Insulators by Dip-Pen Nanolithography. *Science*, 2002, 296, 1836-1838.

[326] Lee, K-B; Park, S-J; Mirkin, CA; Smith, JC; Mrksich, M. Protein Nanoarrays Generated By Dip-Pen Nanolithography. *Science* 2002, 295, 1702-1705.

[327] Wu, CG; Bein, T. Conducting polyaniline filaments in a mesoporous channel host. *Science*, 1994, 264(5166), 1757-1759.

In: Nanowires: Properties, Synthesis and Applications
Editor: Vincent Lefevre
ISBN: 978-1-61470-129-3

Chapter 2

SEMICONDUCTOR NANOWIRES AND HETEROSTRUCTURES BASED GAS SENSORS

N. S. Ramgir, N. Datta, M. Kaur, A. K. Debnath, D. K. Aswal and S. K. Gupta

Technical Physics Division, Bhabha Atomic Research Centre
Mumbai, India

ABSTRACT

In recent years, significant interest has emerged in the synthesis of nanoscale materials owing to their superior and enhanced functional properties. The most attractive class of materials for functional nanodevices are based on semiconductors, in particular metal oxides based nanostructures. For application as gas sensor, nanostructures offer several advantages including high surface area-to-volume ratio, dimensions comparable to the extension of the surface charge region, relatively simple preparation methods allowing large-scale production and sensors that are convenient to use. Of the various nanostructures, nanowires (NWs) are particularly useful for gas sensing application as they offer various advantages. These include pathway for electron transfer (length of NWs), enhanced and tunable surface reactivity implying possible room temperature operation, faster response and recovery time and ease of fabrication and manipulation. The smaller dimension further enhances the possibility of high integration density thereby leading to smaller size of the actual sensor device and low power consumption. All the above mentioned features definitely make NWs a potential candidate for the development and realization of next generation sensing devices. This chapter deals with the progress made towards the effective use of semiconducting NWs for achieving superior sensing performance has been critically addressed. In particular, different sensor configurations like single-NW based, multiple-NW based, NW films and as grown films have been investigated in detail. Besides, the result obtained using the investigations of doping element, incorporation into the polymer matrix and use of heterostructures on improvement in the sensing characteristics has been elaborated with examples. Steps taken towards commercialization of ultimate sensor device and the major obstacles involved are also discussed.

Keywords: NWs, nanosensors, ZnO, CuO, SnO_2, polymer composites, heterostructures

1. INTRODUCTION

In recent years, significant interest has emerged in the synthesis of nanoscale materials [1]. One of the most attractive classes of materials for functional nanodevices is semiconductors or in particular metal oxides [2]. Metal oxide nanostructures offers several advantages including high surface-to-volume ratio, dimensions comparable to the extension of surface charge region, superior stability owing to the high crystallinity and relatively simple preparation methods allowing large-scale production [3]. All the above mentioned features make nanostructures very promising for the development of a new generation gas sensors. Accordingly, various semiconducting oxide nanostructures like nanoparticles, nanorods, nanowires (NWs), nanobelts, nanopencils and 3-D heterostructures like nanorings, nanohelixes, with promising sensing characteristics have been reported [4]. Among these, NWs in particular is looked upon as a potential candidate for the realization of the next generation of sensors. NWs are crystalline structures with precise chemical composition, surface terminations, and are dislocation-defect free. NWs exhibit physical properties, which are significantly different from their coarse-grained polycrystalline counterpart because of their nanosized dimensions. Their use as gas-sensing materials should reduce instabilities, suffered from their polycrystalline counterpart, associated with grain coalescence and drift in electrical properties. High degree of crystallinity and atomic sharp terminations make them very promising for better understanding of sensing principles and for development of a new generation gas sensors. They offer various advantages including high surface to volume ratio, Debye length comparable to target molecule, fast response and recovery time, enhanced and tunable surface reactivity implying possible room temperature operation, ease of fabrication and manipulation and above all the ease of incorporation into microelectronic devices [5, 6]. Also, surface effects dominate in these, which leads to the enhancement of the surface related properties, such as catalytic activity or surface adsorption: key properties for superior chemical sensors production [7].

Providing electrical contacts to NWs is often considered as a tedious and complex process. Usually "pick and place" approach is used to realize a sensor using NWs. And depending upon the configuration in which NWs are being used the sensors are classified as either single, multiple or Mat-type and thin film sensors. To realize an electrical contact NWs are first dispersed into solvents like methanol, ethanol or isopropanol in very small concentrations. Isolated single NWs (Figure 1 (a)) are then located under optical microscope or SEM and contacts are provided by depositing an electrode layer on top of NW using Focused Ion Beam (FIB) deposition. Alternatively, NWs are drop casted onto substrates containing predefined electrodes and are aligned using dielectrophoresis technique. For multiple NWs or mat-type film configuration, (Figure 1 (b)) highly concentrate NWs suspension is either drop casted or spin coated onto the substrate and then contact electrodes are deposited over them. NW samples are then subjected to annealing for better adherence of the sensor film. For field effect transistor configuration heavily doped Si, acting as a back gate, is used. The schematic of a FET based sensor is shown in Figure 1(c).

In 1-D disordered network of NW, only a small fraction of the adsorbed species adsorbed near the grain boundaries is active in modifying the device electrical transport properties. In the sensors based on single-crystalline nanostructures, almost all of the adsorbed species are active in producing a surface depletion layer. Free carriers should cross the single-crystalline nanostructures bulk along the axis in a FET channel-like way. A change in electrical resistance of the NW is measured as a function of time for known amount of exposed gas. The resistance can be measured by a simple ohmmeter. Alternatively, a constant potential is applied to the sensor and change in current is measured. Most often the resistance of the sensor is measured as a function of chemical interaction occurring on its surface and is commonly referred to as "*Chemiresistive sensor*". The electrical resistance of the sensor in particular, is very sensitive to the chemical environment. Afterwards contacts are made on the film by thermally evaporating metals (e.g. Au, Ag etc) and Si substrate operates as back contact.

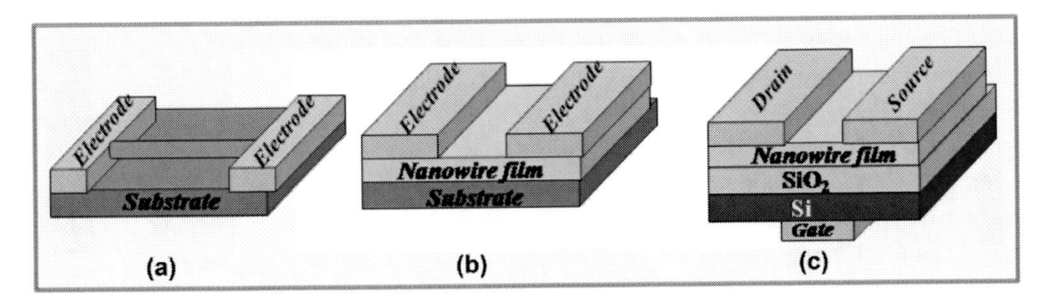

Figure 1. Schematic showing various configurations of a chemiresistive sensor (a) Single NWs (b) Mat-type NWs thick film and (c) FET based chemi-sensor.

The present article represents a review of recent research on NWs for gas sensing application that has been carried out in our laboratory. This review is focused on the description of the metal oxide based NWs namely ZnO, SnO_2, and CuO, for possible gas sensing application. Additionally, approaches to improve the sensor performance like incorporation of sensitizers, and possible polymer composite and heterostructures are also elaborated. The present review has been divided into five sections for clarity and simplicity. Section 1 gives a brief introduction to the NW based sensors such as single NW, multiple NW or Mat-type and chemiresistive field effect transistor (FET) sensors. Section 2-4 discusses in detail the growth and gas sensing properties exhibited by ZnO, SnO_2, and CuO, respectively. In particular, they give an overview of the different techniques used for the growth of NWs and heterostructures and relative discussion of different growth mechanisms involved. Techniques like thermal evaporation, vapor phase deposition and hydrothermal growth were used effectively to realize NWs of these materials. For example, ZnO and SnO_2 NWs have been prepared by thermal evaporation at atmospheric pressure. CuO NWs have been grown by heating of copper sheets in oxidizing atmosphere. Besides, these sections also discusses in detail all the possible ways (single, multiple or mat, thick films, embedded in polymer matrix, surface functionalized) in which NWs have been used for gas sensing application and results of respective gas sensing characteristics are conferred. Finally, the review ends briefing the important steps taken towards the commercialization of sensors based on these nanostructures and highlighting the loopholes that are still to be addressed.

2. ZnO Nanowires

Among various semiconducting oxide materials, ZnO has generated a great deal of interest due to its direct wide band gap of 3.37 eV, large exciton binding energy of 60 meV, and processing advantages for its nanostructures [8]. This is besides the intrinsic defects existing in ZnO with characteristic emission of a wide band(s) covering a large part of the visible range. In addition, ZnO is biosafe (non-toxic) and possesses piezoelectric properties. Accordingly, different nanoforms of ZnO have been investigated for their possible device applications. Preparation and gas sensing performances of ZnO 1D nanostructures has been reviewed by a number of authors [9].

ZnO has a hexagonal structure (space group *C6mc*) with lattice parameters a = 0.3296 and c = 0.5206 nm. The structure of ZnO can be simply described as a number of alternating planes composed of tetrahedrally coordinated O^{2-} and Zn^{2+} ions, stacked alternately along the c-axis [10] (Figure 2). The tetrahedral coordination in ZnO results in noncentral symmetric structure and consequently is responsible for the exhibited piezoelectricity and pyroelectricity. Another important characteristic of ZnO is its polar surfaces. The most common polar surface is the basal plane. The oppositely charged ions produce positively charged Zn-(0001) and negatively charged O-$(000\bar{1})$ surfaces, resulting in a normal dipole moment and spontaneous polarization along the c-axis as well as a divergence in surface energy. The other two most commonly observed facets for ZnO are $(2\bar{1}\bar{1}0)$ and $(0\bar{1}10)$, which are non-polar surfaces and have lower energy than the (0001) facets.

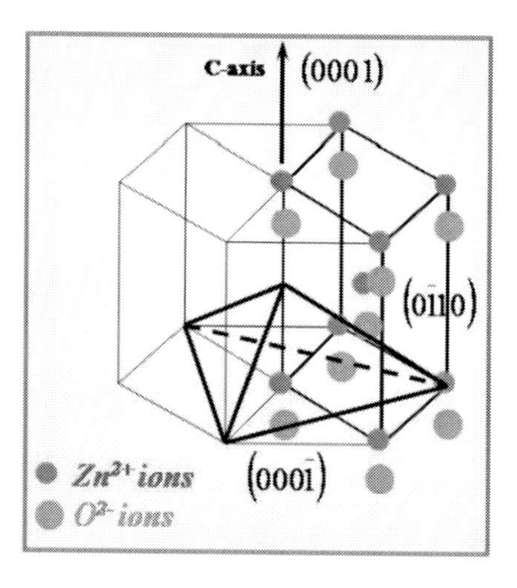

Figure 2. ZnO crystal structure.

Structurally, ZnO has three types of fast growth directions: $\langle 2\bar{1}\bar{1}0\rangle$ ($\pm[2\bar{1}\bar{1}0]$, $\pm[\bar{1}2\bar{1}0]$, $\pm[\bar{1}\bar{1}20]$); $\langle 0\bar{1}10\rangle$ ($\pm[0\bar{1}10]$, $\pm[10\bar{1}0]$, $\pm[1\bar{1}00]$); and $\pm[0001]$. Together with the polar surfaces (due to atomic terminations) ZnO exhibits a wide range of novel structures that can be grown by tuning the growth rates along these directions. One of the most profound factors determining the morphology involves the relative surface activities of various growth facets under given conditions. Macroscopically, a crystal has different kinetic parameters for different crystal planes, which are emphasized under controlled growth conditions [11]. Thus, after an initial period of nucleation and incubation, a crystallite will commonly develop into a three dimensional object with well-defined, low index crystallographic faces. Different nanoforms of ZnO have been realized commonly using vapor phase or chemical approaches. We have mostly investigated the gas sensing properties of ZnO NWs which were grown using vapor phase deposition and hydrothermal approach.

2.1. NW deposition and growth mechanisms

There are two different approaches to the production of 1D structures: top-down and bottom up technologies [12]. The first one is based on standard micro fabrication methods with deposition, etching and ion beam milling on planar substrates in order to reduce the lateral dimensions of the films to the nanometer size. Electron beam, focused ion beam, X-ray lithography, nano-imprinting and scanning probe microscopy techniques can be used for the selective removal processes [13]. The advantages are the use of the well developed technology of semiconductor industry and the ability to work on planar surfaces, while disadvantages are their extremely elevated costs and preparation times. In the top-down approach highly ordered NWs can be obtained, but at the moment this technology does not fulfill the industrial requirements for the production of low cost and large numbers of devices [14]. Furthermore the 1D nanostructures produced with these techniques are in general not single-crystalline.

The second approach, bottom-up, consists of the assembly of molecular building blocks or chemical synthesis by vapor phase transport, electrochemical deposition, solution-based techniques or template growth [15]. Its advantages are the high purity of the nano-crystalline materials produced, their small diameters, the low cost of the experimental set ups together with the possibility to easily vary the intentional doping and the possible formation of junctions. The main disadvantage is their integration on planar substrates for the exploitation of their useful properties, for example transfer and making contacts can be troublesome. The bottom-up approach allows low cost fabrication although it could be very difficult to get them well arranged and patterned.

In the past years the number of synthesis techniques has grown exponentially. Numerous one-dimensional nanostructures with useful properties, compositions, and morphologies have recently been fabricated using bottom-up synthetic routes. Some of these structures could not have been created easily or economically using top-down technologies. We can divide these growth widely in two different categories, vapor and solution phase growth. Before going into the depth of gas sensing applications, it will be appropriate and crucial to know first the basic properties of different semiconducting oxides namely, zinc oxide (ZnO), tin oxide (SnO_2), and copper oxide (CuO) and different deposition methods employed with corresponding NWs growth mechanism. Additionally, the techniques used to improve both the sensitivity and selectivity is also elaborated.

2.1.1. Thermal evaporation method

In this method pure Zn metal powder was heated in a quartz boat placed inside a quartz tube. The quartz boat was heated to the desired temperature between 600 and 900°C under Ar atmosphere at a heating rate of 6°C/min. On stabilizing the temperature at the desired value, the gas atmosphere was switched to 95% Ar and 5% O_2 at a flow rate of 500 sccm. The material was maintained under these conditions for 1 to 5 h and then cooled to room temperature. In this process, growth of nanostructures on Zn metal powder was observed at temperatures between 600 and 800°C. Evaporation of Zn in this temperature range was found to be negligible. At a temperature of 900°C, Zn was found to evaporate and deposit in the quartz tube in the direction of the gas flow. The local temperature at different places in the quartz tube was measured and deposits from these places were collected and analyzed.

NWs with diameter of <100 nm were observed growing radially from the grains at 600°C. Both the density and diameter of the NWs was found to increase with the deposition temperature. Figure 3 shows the corresponding SEM image of NWs grown at 700°C. The NW growth in this case takes place by a self catalytic growth mechanism [16, 17]. In brief, when the zinc powder is heated to temperatures higher than its melting point (419.5°C), Zn droplets are formed and, when oxygen is introduced in the chamber, it reacts with the outer surface of the previously formed Zn droplets and forms nanosized ZnO nuclei on the surface of these droplets. These ZnO nuclei grow outward in the form of ZnO NWs. At a temperature of 800°C, thick nanorods of nearly 500 nm diameter were obtained and the entire Zn source was found converted to ZnO. On heating Zn metal to 900°C, Zn gets evaporated and different nanoforms were found deposited in the form of powder on the quartz tube downstream. More specifically, at a temperature of 200°C, NWs with diameters in the range of 100 to 200 nm were collected.

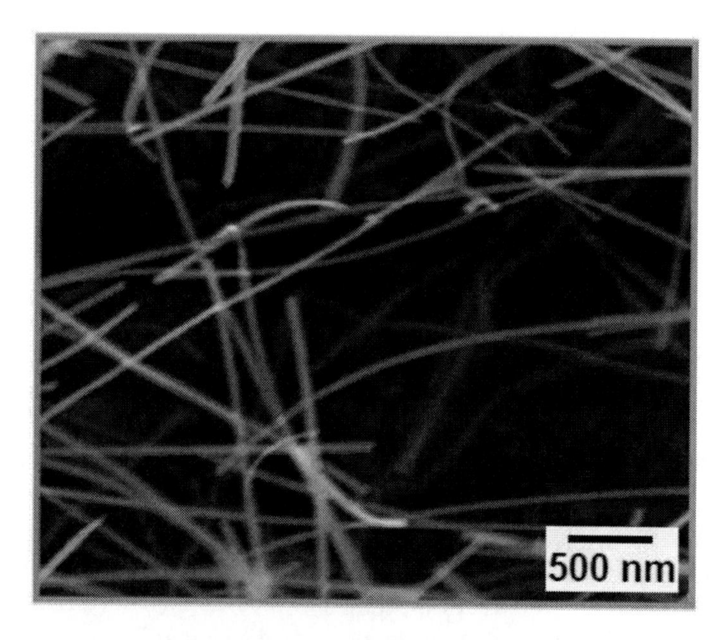

Figure 3. ZnO NWs grown at 700°C after heating Zn powder for 5 h.

2.1.2. Hydrothermal Growth

ZnO NWs were grown on oxidized silicon substrates of 10×10 mm^2 size by hydrothermal technique [18]. Hydrothermal growth was carried out by suspending silicon wafers upside-down in an open beaker filled with an aqueous equimolar (0.025 M) solution of zinc nitrate hydrate and hexamine at 90°C. NWs growth was carried out on plain Si/SiO_2 substrates as well as those coated with ZnO NPs of 10 nm size (to act as seed). For this ZnO NPs were synthesized by chemical process using zinc acetate dihydrate as precursor [19]. The NaOH solution in ethanol (5 to 15 mM concentration) was slowly added (over a period of 15 min) to the solution of zinc acetate dihydrate in ethanol (30 to 45 nM concentration) kept at a temperature of $60-75°C$ (the volume ratio of the two solutions was maintained at 2:1). Growth was carried out for few hours under constant stirring. Growth at temperatures above 60°C ensures that precipitation of hydroxides does not take place. NPs with different sizes were obtained by controlling the precursor concentration and the growth temperatures. For NWs growth NPs with 10 nm diameters were spin casted several times (4-6 times) on oxidized silicon wafers (100 nm SiO_2 layer) to result in a continuous film. After growth, the wafers were rinsed thoroughly with de-ionized water, dried under Ar flow and used for gas sensing studies and further characterization.

The results of NWs grown on substrates without seed particles and with seed particles for different growth time are shown in Figure 4. It is seen that the NWs grown on substrates with seed particles are more uniform and have lower diameter of 50-100 nm compared to ~500 nm for those grown on substrates without seed particles. Length of NWs was found to increase with increase in growth time. NWs grown on substrates with seed particles were used for further studies.

Figure 4. SEM images of NW-films grown on silicon substrates: (a) without seed particles and with seed particles for (b) 6 h and (c) 21 h growth times.

In aqueous solution, zinc(II) is solvated by water and gives rise to aquo ions including $ZnOH^{+}_{(aq)}$, $Zn\text{-}(OH)_{2(aq)}$, $Zn(OH)_{2(s)}$, $Zn(OH)_{3\text{-}(aq)}$, and $Zn(OH)_4^{2-}{}_{(aq)}$ [20]. For given zinc(II) concentration, the stability of these complexes is defined by the pH and the temperature. Dehydration of these hydroxyl species results in the formation of solid ZnO nuclei. And ZnO crystal can grow continuously by the condensation of the surface hydroxyl groups with the zinc-hydroxyl complexes. These hydrolysis and condensation reactions of zinc salts result in one-dimensional ZnO crystals. Hexamine (HMTA) which is a nonionic cyclic tertiary amine that can act as a Lewis base to metal ions and also as a bidentate ligand capable of bridging two zinc(II) ions in solution is used very often to promote one-dimensional ZnO precipitation [21, 22]. In particular, it functions in part, by decomposing during the reaction and increasing the pH to above ~9 at the crystal surface. It is also known to hydrolyze, producing formaldehyde and ammonia in the pH and temperature range of the ZnO NW reaction. It acts as a pH buffer by slowly decomposing to provide a gradual and controlled supply of ammonia, which can form ammonium hydroxide as well as complex zinc(II) to form $Zn(NH_3)_4^{2+}$ [23]. Because dehydration of the zinc hydroxide intermediates control the growth of ZnO, the slow release of hydroxide may have a profound effect on the kinetics of the reaction. Additionally, ligands such as HMTA and ammonia can kinetically control species in solution by coordinating to zinc(II) and keeping the free zinc ion concentration low. Additionally, they can also coordinate to the ZnO crystal, hindering the growth of certain surfaces. However, the exact role of HMTA in NW growth is still under a debate and demands detailed investigations [21].

2.2. Gas sensing measurements

2.2.1. Drop casted mat-type NW film

Thick films were made out of the NWs by making a paste of NWs in methanol and painting them on alumina substrate followed by annealing at 500°C. Two Au contacts (120 nm) were thermally evaporated on these films. Silver wires were attached to these contact pads to measure the resistance change on exposure to the desired gases. All the sensor films were subjected to vacuum annealing at 400°C for 1 h prior to testing. The sensors were investigated for their response to several toxic gases (H_2S, Cl_2, NH_3, NO, CO and CH_4) at various temperatures. One of the important drawbacks of "Pick and Place approach" is the difficulty in dictating the exact placement of the laterally dispersed nanostructure on the device substrate. However, some control over the alignment of the NWs between electrodes can be achieved using alignment techniques.

In principle dielectrophoresis (DEP), which is the electrokinetic motion of dielectrically polarized materials in non-uniform electric fields, could be a powerful tool for self alignment of NWs into a well defined space region [24, 25]. In this technique an external electric field is induced, without mechanical nanomanipulation techniques [26]. Suspended NWs can then be trapped in the microelectrode gap. First of all, microelectrode gap is made by depositing gold electrodes with a required spacing of 25 μm on alumina substrates, figure 5 (a). Above technique is then used to align and isolate NWs. NWs dispersed in methanol were aligned between the microelectrode gap under the application of electric field of 20 Vpp at 100 KHz, across two electrodes. Consequently, NWs were trapped and aligned along the electric field lines bridging the electrode gap where the electric field becomes higher. These samples were then viewed one by one, grid by grid, under SEM to reveal presence of few isolated and aligned 1-D nanostructures in each sample. Figure 5 shows the corresponding SEM images of the NWs trapped and aligned between 25 μm electrodes. These samples were tested for their behavior towards various gases. The geometry of the electrodes as well as the magnitude and frequency of the applied electric field were proven to be significantly affecting the density of the NWs assembled between the electrodes [27].

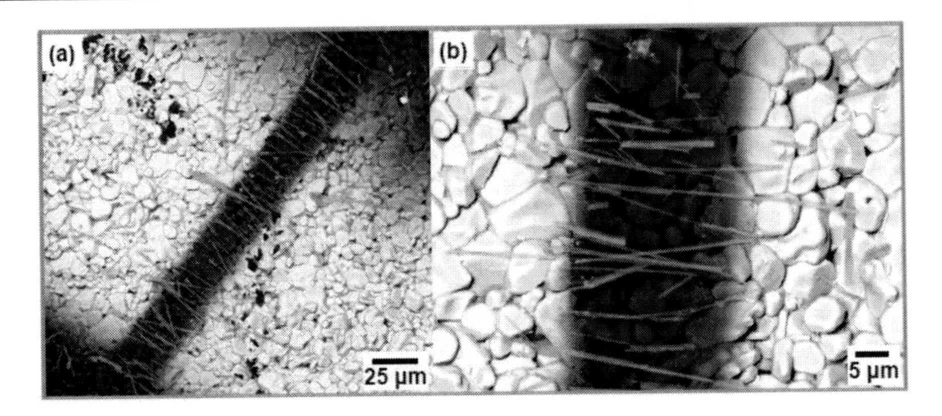

Figure 5. (a) Multiple NWs aligned between two gold electrodes with 25 μm gap, and (b) magnified view of aligned NWs.

Figure 6 shows the set-up used for the gas sensing measurements. Briefly, the sensor was mounted upside down in a leak tight stainless steel chamber (volume: 250 cm^3) equipped with temperature control unit. A desired concentration of the test gas in the chamber is achieved by injecting a known quantity of gas using a micro-syringe. The response data was acquired by using a personal computer equipped with Labview software. Once a steady state was achieved, recovery of sensors was recorded by exposing the sensors to air, which is achieved by opening the lid of the chamber. A change in resistance of the sensor film as a function of time (response curve) was recorded at different operating temperatures and concentration of different gases (i.e. Cl_2, H_2S, NH_3, CH_4, CO and NO), which were commercially procured.

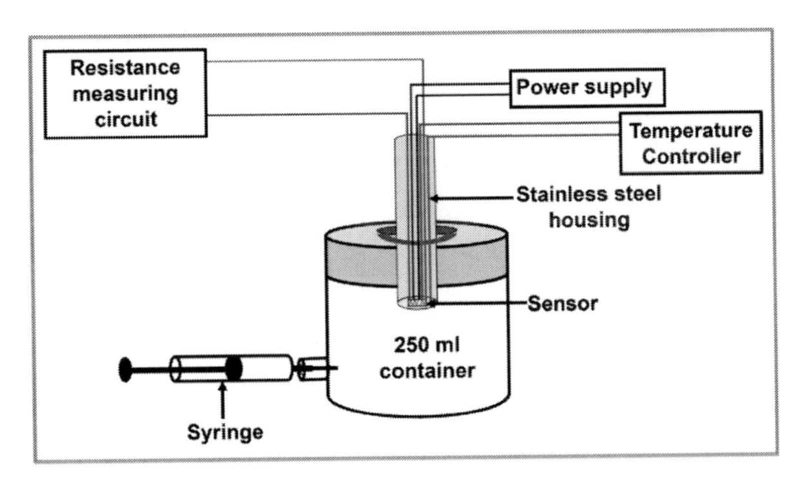

Figure 6. Static gas sensing measurement unit.

A typical sensor is characterized by following five parameters: (i) sensitivity/sensors response, (ii) response time, (iii) recovery time, (iv) selectivity, and (v) long term stability. The sensor response (*S*) can be defined by many ways, including

(a) A ratio of resistance in air to that in gas i.e.

$$S = \frac{R_a}{R_g}, \tag{1}$$

(b) $$S\,(\%) = 100 \times \frac{(R_a - R_g)}{R_g}, \quad (2)$$

A positive value of S implies film resistance decreases on gas exposure and vice versa

(c) $$S = 100 \times \frac{(C_g - C_a)}{C_a}, \quad (3)$$

(d) $$\mathrm{S} = \frac{|I_g - I_a|}{I_a} \quad (4)$$

where, C_g and C_a are the conductance in test gas and air respectively, I_a and I_g are current in air and test gas, respectively. The response time is the time interval over which resistance attains a fixed percentage (usually 90%) of final value when the sensor is exposed to full-scale concentration of the gas. Recovery time is the time interval over which sensor resistance reduces to 10% of the saturation value when the sensor is exposed to full-scale concentration of the gas and then placed in the clean air. A good sensor should have a small response and recovery times so that sensor can be used over and over again. Moreover, the sensor should be selective to a particular gas only and it should not degrade on continuous operations for long durations.

Sensor films exhibited good sensitivity towards H_2S and NO gases and very little or no response towards other gases. Response of a typical film to different concentration of H_2S and NO is shown in Figure 7. It is seen that the films have good response to both gases at room temperature. Typical response and recovery times for 10 ppm concentration are 250 and 700 s for H_2S and 250 and 150 s for NO, respectively. The response and recovery times were found to increase with concentration.

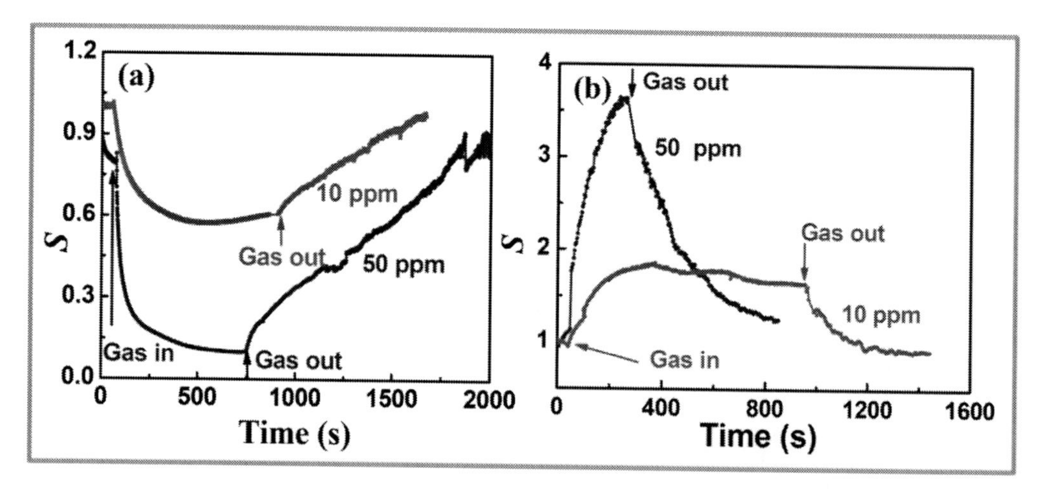

Figure 7. Response (S) of a typical ZnO NW film to different concentrations of (a) H_2S and (b) NO. The measurements have been carried out at room temperature.

ZnO is a well known sensor material and gas sensing properties of nanostructures of ZnO are well reported [28]. Both NO [29] as well as H_2S [30] gas sensing properties of ZnO are well known. The fundamental sensing mechanism of ZnO based gas sensors relies on a change in electrical conductivity due to the process of interaction between the surface complexes such as adsorbed oxygen species, H^+ and OH^- reactive chemical species and the gas molecules to be detected. Generally sensing mechanism of all gases is explained by adsorption-desorption mechanism. Oxygen vacancies present in ZnO, acts as electron donors to make it an n-type semiconductor. Oxygen molecules from the ambient are adsorbed at the grain boundaries, which in turn capture electrons from the conduction band forming adsorbed oxygen ion. This causes a decrease in carrier concentration and increase in resistance of the sample. When the sensors are exposed to a reducing gas, the gas reacts with the adsorbed oxygen resulting in release of the trapped electrons back into the conduction band. This leads to an increase in carrier concentration and decrease in resistance of the sensor. However on exposure to oxidizing gas, reaction take place directly with the oxide surface rather than with the oxygen chemisorbed at the surface.

2.2.2. Hydrothermally grown NW film

We have demonstrated a direct deposition of NW-films on substrate [17, 31] that yields adherent films with high response and recovery times towards H_2S than that of reported earlier [32, 33]. Response of NW-films was investigated on exposure to 30 ppm of H_2S as a function of temperature and the results for a typical sensor are shown in Figure 8, along with the corresponding response and recovery times. Sensor response is found to increase with temperature with maximum at 350°C. As expected both the response and recovery times were found to decrease with increase in the temperature. For optimum temperature of 350°C, the response and recovery times were 11 and 65 s, respectively. Annealing the films upto 900°C for 1 h did not have any significant effect on the sensing characteristics. Response curves of NW-films towards different concentrations of H_2S and at optimum operating temperature of 350°C are shown in Figure 8 (c) and concentration dependence of response is shown in Figure 8 (d). These characteristics were obtained at fixed applied voltage of 1 V. It is seen that the NW-films can reliably detect H_2S at 1 ppm concentration. Sensor response is also observed to increase with concentration saturating at around 15 ppm. It exhibits a power law dependence on concentration given as: [34, 35]

$$S = A C^{\alpha} \quad (5)$$

where, A is constant. From the fit of Figure 8 (d) the value of α was found to be 0.7. The power law dependence arises from receptor and transducer functions i.e., the adsorption or interaction of H_2S with the sensor surface and the change of surface potential, respectively.

The sensor response was also compared with that on other test gases and the results on exposure to 30 ppm of different gases are depicted in Figure 9 (a). It is seen that the NW-films have negligible response to other oxidizing and reducing gases except Cl_2 indicating partial selectivity. For comparison a typical response curve on exposure to Cl_2 at 5 ppm concentration is shown in Figure 9 (b). Exposure to high concentrations of Cl_2 (>20 ppm) was observed to deteriorate the sensor film resulting in incomplete recovery.

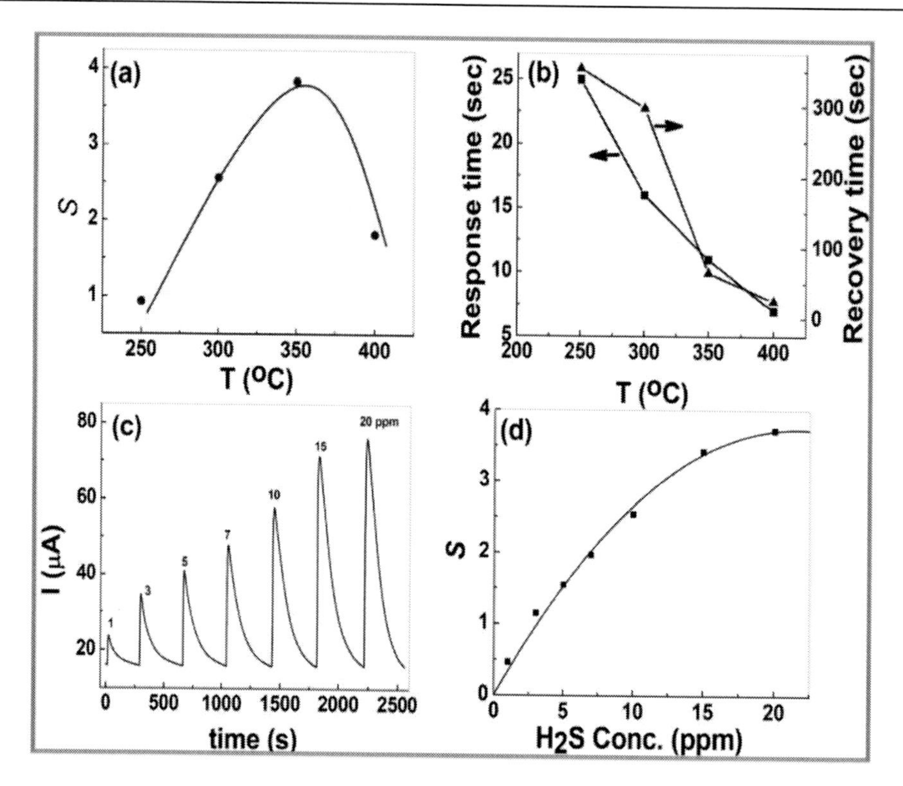

Figure 8. (a) Response of typical NW-film on exposure to 30 ppm of H_2S gas as a function of temperature, (b) the corresponding response and recovery times, (c) Response curves of ZnO NW-film towards different concentrations of H_2S at an optimum operating temperature of 350°C and (d) corresponding concentration dependence of sensor response.

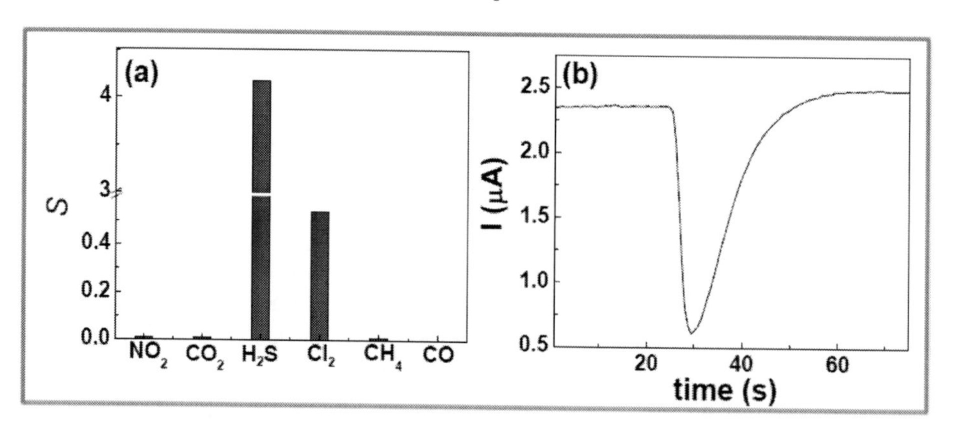

Figure 9. (a) Selectivity histogram of ZnO NW-film towards 30 ppm of different gases (except Cl_2 at 5 ppm) and (b) sensors response curve towards 5 ppm of Cl_2.

2.2.2.1. Cu modified ZnO NWs as H_2S sensor

Pure ZnO NWs were first grown onto Si/SiO_2 substrate using hydrothermal technique as described earlier. Secondly, Cu with different thicknesses (10 – 1000 nm) was vacuum deposited onto these NWs at a base pressure of 2×10^{-5} mbar. The sensor film was subjected to annealing at 600°C for 1 h prior to investigations of their gas sensing characteristics.

Cu is known to be an excellent promoter for H_2S. It readily interacts with oxygen forming CuO. CuO is a p-type semiconductor and forms random p-n junctions with the host n-type semiconductor matrix. The depletion region formed at these junctions act as a potential barrier for charge transfer and hence the host material exhibit very high resistance. H_2S exposure of films at >160°C, leads to a conversion of CuO into CuS. CuS being metallic makes potential barrier to disappear, resulting in a sharp drop in sensor resistance. Re-exposure to fresh air leads to the formation of CuO, and the original resistance value is regained. This implies that use of Cu causes enhancement of both sensitivity and selectivity of the sensor towards H_2S. We have investigated the effect of Cu on H_2S sensing properties of ZnO NWs [36]. The sensor modified with 10 nm of Cu responded with maximum sensitivity at an operating temperature of 250°C. Figure 10 shows the response and recovery curve recorded towards increasing concentration of H_2S. The sensor exhibited a fast response and recovery times for example, towards 5 ppm of H_2S response and recovery times were 62 and 150 s, respectively. This fast response kinetics could be attributed to the effective electron conduction pathway provided by NWs.

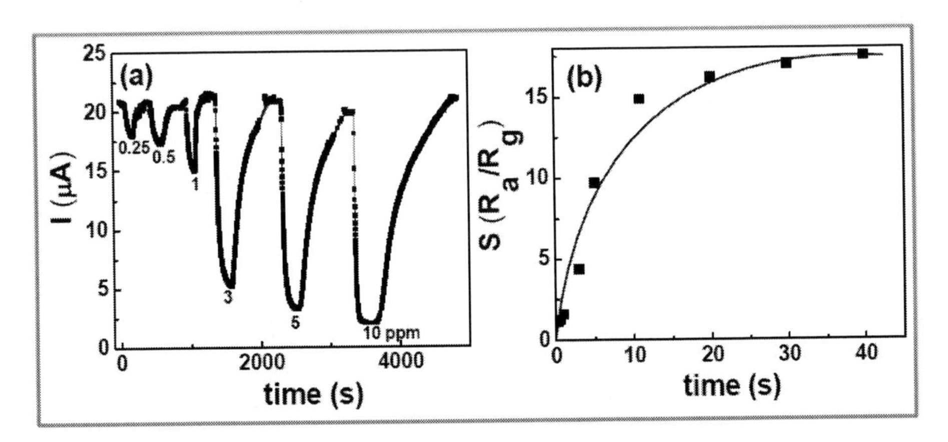

Figure 10. (a) Sensors response towards increasing concentration of H_2S at an operating temperature of 250°C (b) corresponding variation of sensors response with gas concentration.

2.2.2.2. Surface functionalized ZnO NWs as H_2S sensor

$CuSO_4$ (aq.) solution was prepared in different mM concentrations. Freshly grown ZnO NWs was then dipped into these solutions for 1 h and then subjected to annealing at 600°C for 1 h. Annealing assures the formation of CuO on the surface which leads to the formation of random p-n junctions between p-type CuO and n-type ZnO.

We have also investigated the effect of surface modification using chemical self organization method. Figure 11 shows the response and recovery curves recorded for sample modified with Cu species using dip coating method. For this 1 mM concentration of (copper sulfate) $CuSO_4$ (aq.) was used. The sensor responded with maximum sensitivity at 350°C. Besides, it exhibited a reasonable response and recovery time for example, towards 5 ppm of H_2S response and recovery times were 90 and 240 s, respectively.

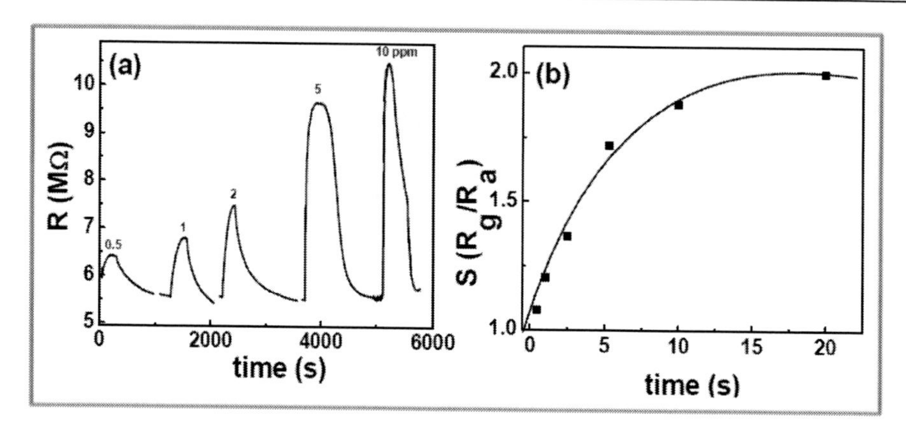

Figure 11. (a) Sensors response towards increasing concentration of H_2S at an operating temperature of 350°C (b) corresponding variation of sensors response with gas concentration.

2.2.3. NWs and polymer composites

2.2.3.1. ZnO: Poly(3-hexylthiophene) as NO_2 sensors

Thin films of conducting polymers have advantages related to easy processing, low cost and room temperature operation, however they often suffer from the lack of specificity, sluggish response and recovery. Therefore, a concept of organic-inorganic hybrid emerges, which results in altogether new sensors. Here we present chemiresistor sensors based on composite films of poly(3-hexylthiophene) [P3HT] and ZnO NWs. The hybrid P3HT: ZnO NW films were deposited on glass substrate by drop casting method [37]. In order to make the hybrid films of different compositions, first the following two solutions were prepared: P3HT in chloroform and ZnO NWs dispersed in ethanol. Then solutions are mixed in such a way that weight ratio of P3HT: ZnO in different solutions corresponds to 4:1, 2:1, 1:1, and 0.5: 1. On mixing the solution colloidal particles are seen to form in the solution. The formation of stable colloidal particle on adding alcohol to the P3HT solutions is reported due to the π stacking of the P3HT molecules. These colloidal solutions of different compositions were drop cast on the glass substrate (1 cm^2) followed by drying at room temperature. The nominal thickness of the films was found to be 1.0 ± 0.1 μm. The room temperature response of films towards 4 ppm concentration of different gases (NO_2, H_2S, NH_3, CH_4, H_2 and CO) was measured by recording the change in film resistance. The sensitivity of the film was defined by: $S(\%) = 100 \times (R_{air} - R_{gas})/R_{air}$, where R_{air} and R_{gas} are resistances in air and gas, respectively. A positive value of S implies film resistance decreases on gas exposure and vice versa. The resistance of composite P3HT: ZnO-NWs films, as shown in Figure 12 (a), increase monotonically with increasing ZnO-NWs content. As shown in Figure 12 (b), pure P3HT and ZnO-NWs films exhibit high sensitivity for NO_2 and H_2S, and a moderate sensitivity for CO. However, all the films were nearly insensitive ($|S| <1\%$) for NH_3 and CH_4. An interesting feature presented in Figure 12 (b) is that with increasing ZnO-NWs content in composite films the sensitivity for NO_2 increases gradually, while it decreases for H_2S. The films of 1:1 composition are highly selective for NO_2. However, for higher ZnO-NWs contents, the selectivity for NO_2 vanishes.

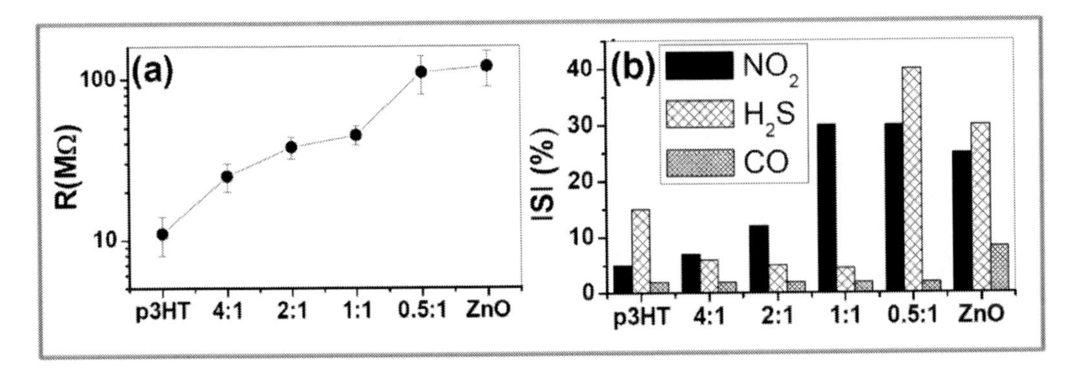

Figure 12. Room temperature resistance value for various films (Error bar indicates variation in resistance of 5 samples of same composition. A variation within ±10% indicates very good reproducibility), and (b) Bar chart showing absolute value of sensitivity, |S|, of composite films of different compositions for different gases. The gas concentration in all the cases was 4 ppm.

In Figure 13, the response curves of 1:1 composite film for 4 ppm NO_2 and 4 ppm H_2S are compared with those of pure P3HT and ZnO-NWs films. It may be noted that for pure P3HT and 1:1 composite films, the resistance increases for NO_2 exposure, while it decreases for H_2S. Exactly opposite behavior is observed for ZnO-NWs films. For pure P3HT film, the response time (time required to attain 90% of the saturation value of resistance on exposure to a gas) is ~8-10 min for both NO_2 and H_2S and, the recovery time is 30-50 min. ZnO-NWs films show a faster response (4-6 min) and faster recovery (2-4 min) for both NO_2 and H_2S. These data clear show that both pure P3HT and ZnO-NWs films lack specificity. On the other hand, 1:1 composite film shows a very fast response (~1.2 min) and high sensitivity (~32%) for NO_2, while sluggish response (~7 min) and poor sensitivity (~3.5%) for H_2S. Further studies have shown that the sensitivity of 1:1 composite films increases linearly from 0 to 70% for the NO_2 concentration range 0-10 ppm and saturates at ~75% for concentrations ≥20 ppm. The |S| of these sensors for 10-ppm concentration of other oxidizing gases, by namely, CO_2 and Cl_2, was found to be < 5%. Thus, 1:1 composite films can be utilized for fabrication of NO_2 sensors with high specificity. The stability tests of these films under ambient conditions (temperature 25-32°C and relative humidity ~65%) for a period of one month revealed no significant drift in base resistance, response and selectivity.

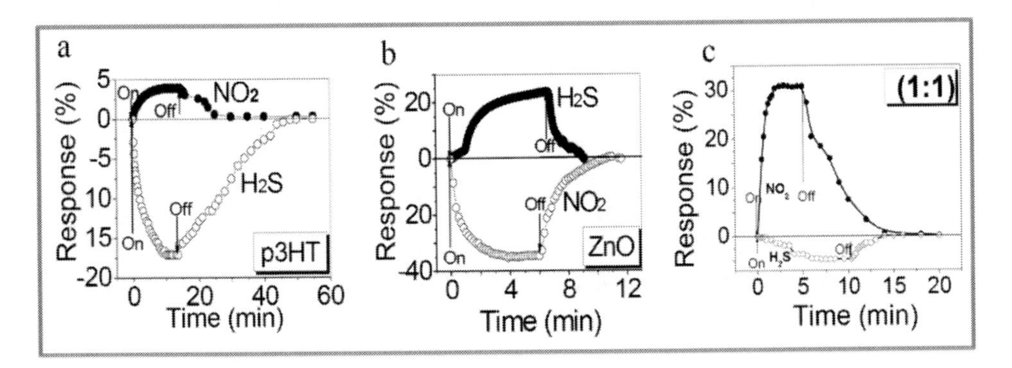

Figure 13. Response curves of (a) pure P3HT, (b) ZnO-NWs and (c) P3HT:ZnO-NWs (1:1 composition) composite films for NO_2 and H_2S.The gas concentration in all the cases was 4 ppm.

Figure 14. SEM micrographs of P3HT, composite of different compositions, and ZnO-NWs thin films.

In order to gain insight how ZnO-NWs influence the morphology and structure of P3HT films, SEM, grazing angle incidence x-ray diffraction (GIXRD), X-ray photoelectron spectroscopy (XPS) and Fourier transform infrared (FTIR) spectroscopy measurements were carried out, and some important results are shown in Figs. 14 and 15. The SEM micrographs reveal that up to a composition of 1:1, films are fairly uniform at the micro-scale and ZnO-NWs are uniformly dispersed in a P3HT, indicating that the charge transport is predominantly governed by P3HT. For higher ZnO-NWs contents, NWs get agglomerated and dominate the charge transport. The EDX analysis revealed that S/Zn ratio of P3HT:ZnO-NWs composite films are within 10% of their nominal composition. The GIXRD pattern of ZnO-NWs films was indexed to wurtzite structure having cell parameter a=3.81 Å and c =5.43 Å, in agreement with reported values [38, 39]. The GIXRD pattern of pure P3HT film show a high degree of molecular ordering at the surface, and the data was found to fit with monoclinic structure with lattice parameters a (in-plane interchain distance) = 16.6 Å, b (interplanar distance) = 4.86 Å, and c (repeating height of polymer) = 7.75 Å. For 1:1 composite film, absence of diffraction peaks corresponding to ZnO indicated that the NWs are fully covered with a P3HT layer of thickness >70 Å (the sampling depth of GIXRD). Lattice parameters a, b, and c calculated for composite films were respectively, 16.91, 4.89 and 7.88 Å. Appreciable changes in lattice parameters a and c indicate a chemical reaction between P3HT and oxygen deficient ZnO-NWs.

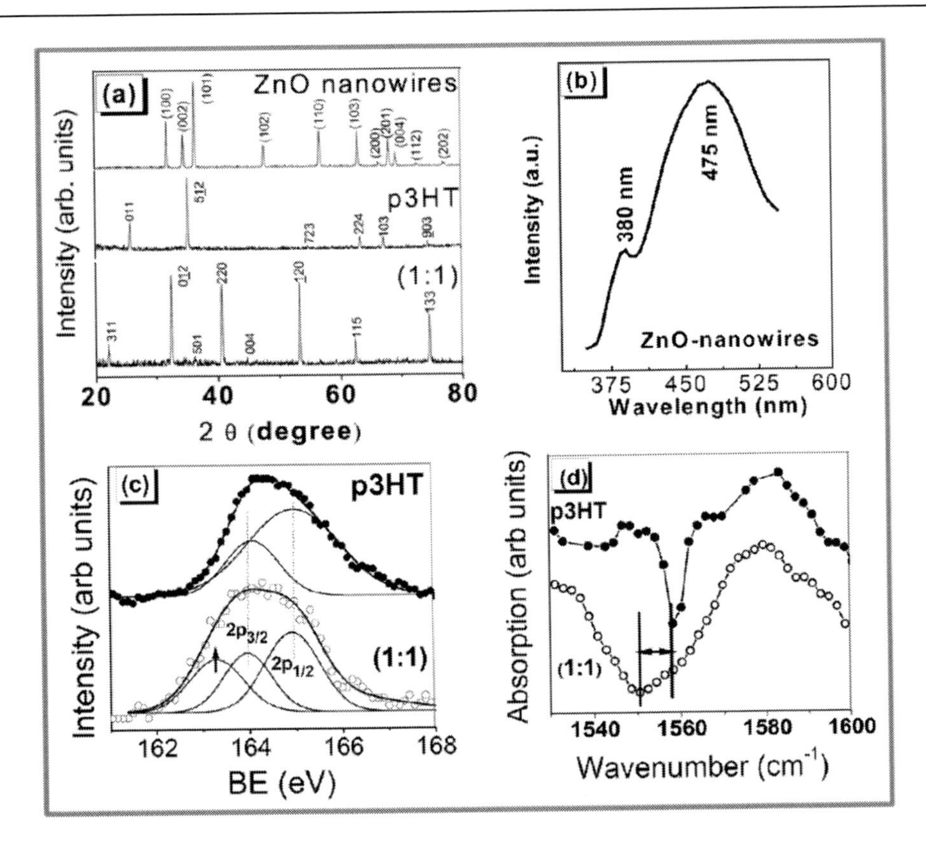

Figure 15. (a) GIXRD, (b) PL, (c) S-2p XPS and (d) FTIR spectra (in the vicinity of -CS stretching vibrations of thiophene ring) recorded for different films.

The S-2p XPS spectrum of P3HT is deconvulated into two peaks, 164.3 and 165.5 eV, corresponding to $2p_{3/2}$ and $2p_{1/2}$ doublet of S of thiophene ring [40]. However, for composite film, in addition to this doublet, a peak at ~163.4 eV, marked by arrow in Figure 18 (c), appears, which is a clear indication of S undergoing a chemical interaction. This inference is further supported by FTIR data shown in Figure 15 (d). It is observed that characteristics peaks for –CH on aliphatic chain remained unchanged in composite films; while peaks at 471, 653 and 1557 cm^{-1} associated with –CS stretching vibrations of thiophene ring are shifted to lower wavenumbers, confirming that S group is involved in the chemical reaction. It may be noted such a reaction is specific to ZnO-NWs arguably due oxygen deficiency and large surface to volume ratio, as no changes in XPS and FTIR data were observed when a composite films was prepared using ZnO microparticles. Also, PL spectrum of ZnO microparticles exhibited only one intense peak at 380 nm, indicating absence of oxygen vacancies [41].

Now we present possible sensing mechanism(s) of the two highly sensitive gases, that is, NO_2, an oxidizing (electron-acceptor) gas and H_2S, a reducing (electron-donor) gas. (i) Pure P3HT is a p-type semiconductor and thereby its resistance falls on exposure to NO_2. This is because adsorption of NO_2 at P3HT surface traps electrons, forming NO_2^- ions, and leading to increased hole carrier concentration in the polymer. Converse is applicable for H_2S, which reduces P3HT by donating electron, and thereby increasing its resistance. (ii) When an n-type ZnO-NWs thin film is exposed to air, oxygen molecules adsorb on the surface of the ZnO-

NWs and form O_2^- ions by capturing electrons from the conduction band. Thus ZnO-NWs films show a high resistance state in air ambient. Its resistance falls down on exposing to a reductive H_2S because H_2S reacts with the surface O_2^- species, which results in an enhancement of electron concentration in the ZnO-NWs. The oxidizing gas NO_2 behaves exactly opposite in manner. Adsorbed NO_2 picks up electrons from ZnO to form NO_2^- ions at the surface, and thereby increases its resistance. (iii) For P3HT:ZnO-NWs composite films an increase in resistance of P3HT on addition of ZnO-NWs, as shown in Fig. 18(a), suggests that the major role of ZnO-NWs is to reduce P3HT by donating electrons. Understandably, highly reduced P3HT will be very sensitive towards oxidizing NO_2, as it can easily pickup electrons from P3HT to get adsorbed as NO_2^- ions at its surface. On the other hand, it would be extremely difficult for reducing H_2S to further reduce P3HT, and therefore, composite films exhibit very weak sensitivity and sluggish response for H_2S. Another possible mechanism of enhanced NO_2 sensitivity of composite films could be associated to the formation of donor-acceptor like complexes between ZnO-NWs and P3HT, and this opens up a room for further research in this field.

2.2.3.2. ZnO: polypyrrole as Cl_2 sensor

Cl_2 gas is widely used in various processes useful to mankind, such as, water purification, bleaching of pulp in paper mills, treatment of sewage effluents and as insecticides [42]. However, Cl_2 is very toxic gas and an exposure to few ppm is hazardous to human health. Cl_2 causes discomfort in respiration as it interacts with mucous membrane of lungs and can be fatal after a few breaths at 1000 ppm [43]. Moreover, if Cl_2 is discharged in aquatic systems, it interacts with other industrial effluents to produce a host of chlorinated organics such as dioxin. Dioxin persists in the environment for prolonged periods and has a tendency to bioaccumulate in the food chains, which causes toxic effects to humans, such as skin infection, psychological disorders and even liver damage. Therefore, it is needed to monitor Cl_2 gas at ppm levels at room temperature. Here we demonstrate that by modifying the PPy films with ZnO-NW, one can make highly sensitive and selective Cl_2 sensors operating at room temperature. ZnO-NW modified PPy films of ~1 μm thickness were prepared by drop-casting glass substrates (10 mm×10 mm) using colloidal solutions containing polypyrrole and ZnO-NW in different weight percentage. The colloidal solutions were prepared by mixing 0.1 M aqueous solution of pure distilled pyrrole, 0.05 M aqueous solution of ammonium persulphate (to initiate the polymerization) and a suspension of ZnO-NWs in ethanol (99.9% pure). The mixed solutions were ultrasonicated for 15 minutes at room temperature for uniform dispersion of ZnO-NWs in the PPy. After drop-casting the colloidal solution onto the substrate, it was dried at room temperature to obtain a film [44, 45].

Typical response curves of pure PPy and ZnO-NW (50 wt%): PPy films for 5 ppm Cl_2 are shown in Figure 16 (a) and (b), respectively. The sensitivity (defined as $S = \rho_a/\rho_g$, where ρ_a and ρ_g are resistivity of film in air and gas, respectively) of PPy films for 5 ppm Cl_2 is 1.36 in the first exposure cycle. Though the response time is very fast (~22 s), a full recovery is not achieved even after 24 h. S keeps reducing in subsequent exposure cycles e.g. in second cycle the value of S is 1.17. Moreover, sensitivities for various tested gases were nearly same, indicating unsuitability of pure PPy films for Cl_2 gas. On the other hand, ZnO-NW(50 wt.%):PPy composite films exhibited S = 15.8 for 5 ppm Cl_2 (more than an order of magnitude higher as compared to the PPy films!) with a response time of ~55 s. Other advantageous features of the composite film are: (i) a full recovery of base resistivity in ~800 s, (ii) reproducible value of S in repeated exposure cycles, (iii) highly selective to Cl_2, Figure 16 (c), as values of S for all other gases are <1.2, and (iv) S varies linearly between 1.2 and 40 for Cl_2 in a range of 1-10 ppm, Figure 16 (d). Moreover, composite films were stable in atmospheric conditions (temperature 30-38 °C and relative humidity 70-85 %) for several months. These features qualify composite films for room temperature operating Cl_2 sensors in the range of 1-10 ppm.

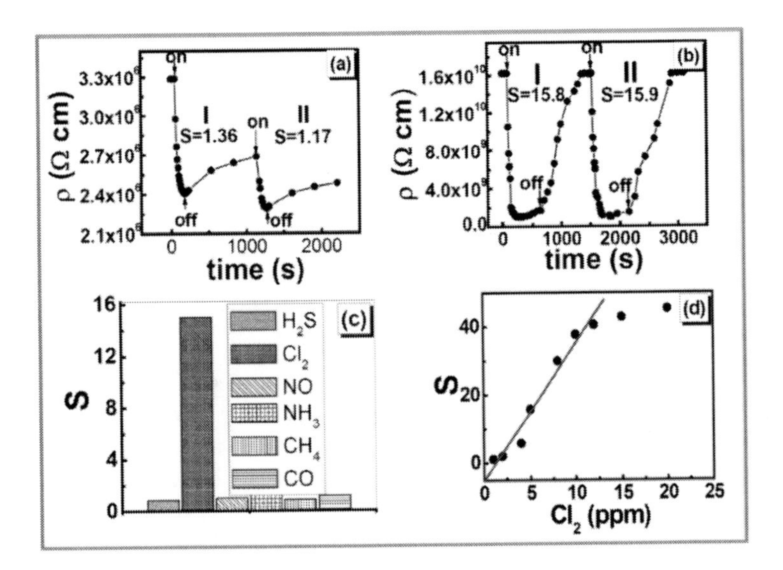

Figure 16. (a) Response curve for the pure PPy films on exposure to the 5 ppm Cl_2. The base resistivity is never recovered and the value of S decreases in the subsequent exposure cycle. (b) Response curve for the composite films on exposure to the 5 ppm Cl_2. The recovery takes place in ~800 s and S remains nearly same subsequent exposure cycles A little recovery of resistance in the saturation region (e.g. in the time period 250-1750 s) is due to leakage of Cl_2 gas from the test chamber. (c) Bar chart showing the S values of composite film for 5 ppm of various gases. (d) Variation in S of composite film as a function of Cl_2 concentration. All the measurements were carried out at room temperature.

In the following we investigate why composite films are highly sensitive as well as selective to Cl_2. One of the intriguing observation in the Figure 16 (a) and (b) is that the base resisitivity of composite film is higher by nearly four orders in magnitude as compared to the pure PPy film. In Figure 17, we plot the resistivity of composite films as a function of ZnO-NW content. It is seen that 2 wt.% ZnO-NW abruptly increases the resistivity by nearly two orders in magnitude. Subsequently, the resistivity increases exponentially with ZnO-NW content. The granular morphology of the polymer films (as shown in Figure 20) indicates that the total resistivity (ρ) can be expressed as:

$$\rho = \rho_{Intra} + \rho_{Inter} + \rho_{Ionic} \tag{6}$$

ρ_{Intra} is the contribution from the intragrain region and depends on the conjugation lengths (l) over which charge carriers are localized. In the case of unoriented polymer, the ρ_{Intra} varies as $1/l^n$ (n= 3.2-3.5), indicating that shorter is the conjugation length higher is the resistivity [46].

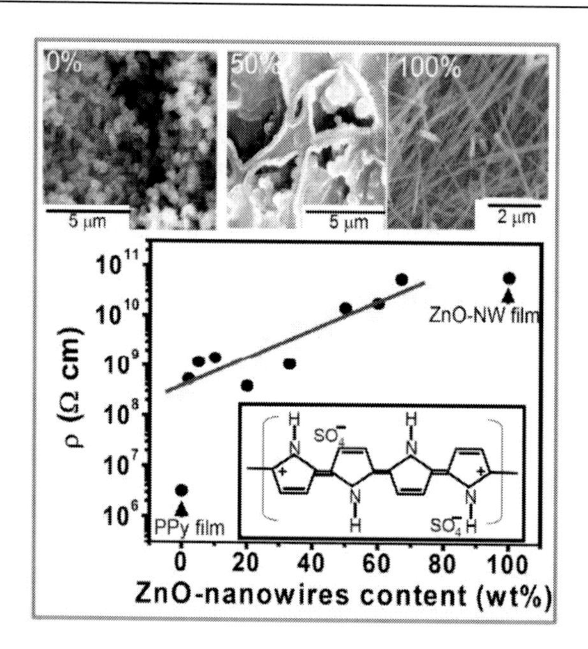

Figure 17. Variation in resistivity of composite films as a function of ZnO-NW content. The straight line is a linear fit of the semilog data. Also shown are the scanning electron micrographs of the pure PPy (%), composite (50%) and pure ZnO-NW (100%) films. The inset shows the molecular structure of the polypyrrole (SO_4^- is the counter ion), which is conductive due to long conjugation.

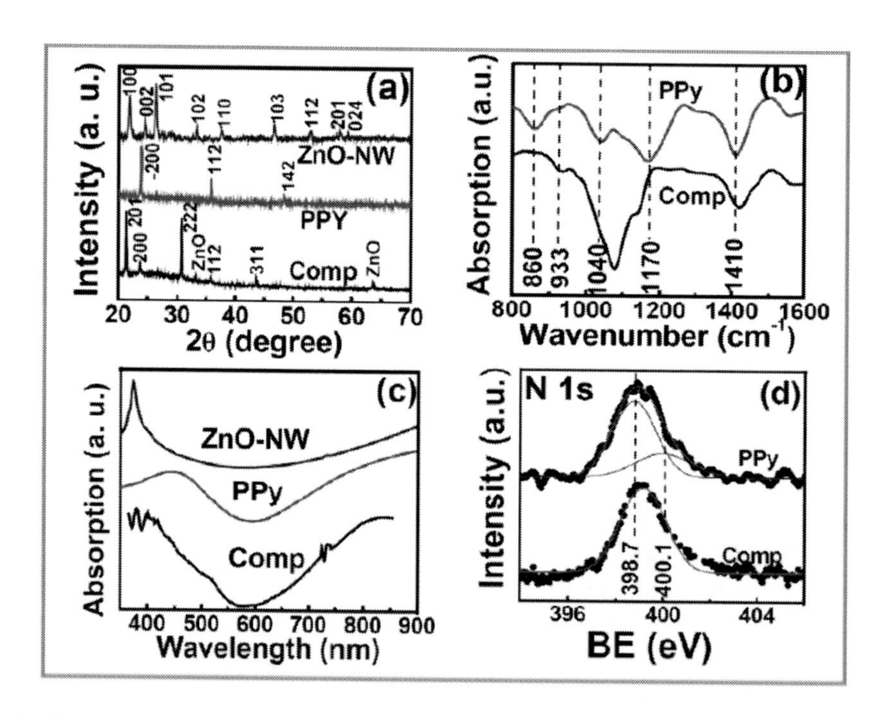

Figure 18. (a) XRD, (b) FTIR (c) UV/Vis and (d) N 1s XPS spectra recorded for ZnO-NW, PPy and composite films.

If l is too short, then the charge transport is limited by interchain hopping. In fact, l can be altered by improving crystallinity and changing doping levels of conducting polymer. ρ_{Inter} is the intergrain resistance and depends on the height of the potential barrier formed at the grain boundaries. In general, smaller is the grain size, higher is the value of ρ_{Inter}. ρ_{Ionic} is controlled by mobility of the counter ions. In the present case, as shown by SEM images, ZnO-NWs get coated with polymer and are randomly distributed in the composite film. Also, the grain size of the polymer is significantly enlarged indicating that conduction in composite films is largely due to the polymer. Despite of an enhanced grain growth, the resistivity of the composite films increases dramatically, indicating that contribution form intergrain region is negligible. Since the ρ_{Ionic} contribution is usually small, therefore an increase in resisitivity of composite films can be attributed solely due to increase in ρ_{Intra}.

In order to gain insight how ZnO-NW modify the ρ_{Intra} of PPy, we have measured XRD, FTIR, UV/Vis and XPS spectra and the results are shown in Figure 18 (a) - (d). The XRD pattern of ZnO-NW film was indexed to a wurtzite structure having cell parameters a = 3.253 Å and c = 5.21 Å. The XRD pattern of pure PPy film shows a high degree of molecular ordering, and the data were found to fit to a triclinic structure of oxidized PPy with lattice parameters a = 8.628 Å, b = 7.954 Å, and c = 7.110 Å with angles α = 59.03°, β= 116.40° and γ= 91.93° and a volume of 362.536 $Å^3$ [47]. For composite films, presence of faint ZnO diffraction peaks indicates that the ZnO-NW are fully covered with a PPy. The calculated lattice parameters for PPy in the composite films are a =8.584 Å, b=8.095 Å, and c =7.082 Å with angles α = 59.05°, β= 115.880° and γ= 91.64° and a volume of 367.30 $Å^3$. Appreciable changes in lattice parameters, particularly an increase in b (corresponding to interchain spacing), and increase in volume indicate a strong interaction between PPy and ZnO-NW. FTIR data, as shown in Figure 21(b), further confirms this inference. For pure PPy film, all the characteristic absorption peaks of PPy are observed, i.e. 890 cm^{-1} (=C—H out-of-plane vibration), 1040 cm^{-1} (=C—H in-plane vibration), 1170 cm^{-1} (N—C stretch bending) and 1410 cm^{-1} (pyrrole ring vibration) [48] However, for the composite film a prominent shift in these peak positions as well as intensities (with respect to the intensity of pyrrole ring vibration at 1410 cm^{-1}) have been observed. The most prominent changes are (i) a shift of =C—H in-plane vibration peak to high value i.e. from 1040 cm^{-1} to 1070 cm^{-1} with increase in its intensity and (ii) a shift of N—C stretch bending to a lower value i.e. from 1170 cm^{-1} to 1135 cm^{-1}. Since the frequency of a vibration is directly proportional to the strength of the bond (i.e. force constant), an increase in =C—H in-plane vibration peak indicates that the electrons are more localized in the pyrrole ring. This also indicates the shortening of N—C bonds, suggesting localization of p-orbital electrons on N. In addition, enhancement of the intensity of =C—H in-plane vibration peak indicates increase in the dipole moment. More evidence of interaction between PPy and ZnO-NW is seen from the UV/Vis spectra, Figure 21 (c). ZnO-NW and PPy films exhibit absorption peak at 372 nm (3.33 eV) and 444 nm (2.79 eV), respectively. In case of the composite film, peaks due to ZnO-NW and PPy shift, respectively to 379nm (3.27 eV) and 415 nm (2.98 eV). Shifting of absorption peak to lower wavelengths is indicative of poor conjugation. N1s XPS spectra of PPy and composite films, as shown in Figure 21 (d), also confirms the interaction between ZnO-NW and PPy. The spectrum for PPy film reveals two chemically different nitrogens. The stronger peak at 398.7 eV can be assigned to neutral (N) whereas the higher binding energy peak at 400.1 eV is assigned to the oxidized (N^+) moieties. The observed ratio of N^+/N = 0.36 correlates well with the reported value in the conducting PPy [49]. However, for composite film only one peak corresponding to (N) is observed, indicating that PPy is in highly reduced form.

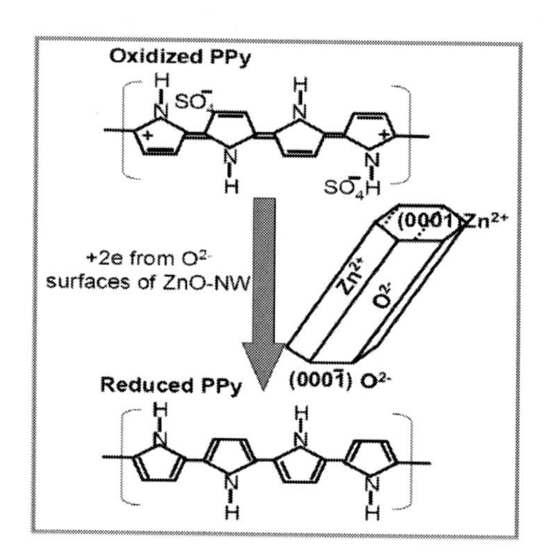

Figure 19. Reduction of oxidized-PPy due to electron transfer from O^{2-} terminated polar surfaces of ZnO NWs, which results in a loss of conjugation and, hence conductivity.

Based on XRD, FTIR, XPS and UV/Vis data, we conclude that the role of ZnO-NW is to change polypyrrole from oxidized (i.e. quinoid structure with large conjugation) to reduced (i.e. benzoid structure with almost no conjugation) state and, this process is schematically depicted in Figure 19. The structure of ZnO can be described as a number of alternating planes composed of tetrahedrally coordinated O^{2-} and Zn^{2+} ions, stacked alternately along the c-axis. An important characteristic of ZnO-NW is its polar surfaces. The oppositely charged ions produce Zn^{2+}-terminated (0001) and O^{2-} terminated $(000\bar{1})$ end polar surfaces, resulting in a normal dipole moment and spontaneous polarization along the c-axis. The six side surfaces of the ZnO-NW are: two non-polar $\pm(01\bar{1}\bar{2})$, Zn^{2+}-terminated $(0\bar{1}01)$ and $(\bar{1}011)$, and O^{2-} terminated $(\bar{1}10\bar{1})$ and $(10\bar{1}\bar{1})$. Thus in a composite film, an interaction of PPy with O^{2-} terminated ZnO-NW surfaces results in transfer of electrons to PPy, which highly reduces PPy. In a highly reduced PPy charges are strongly localized in PPy ring and the conjugation is completely lost. This causes in a dramatic enhancement of intrachain resistivity as the charge transport can only occur via hopping mechanism. A strong localization of charges in highly reduced PPy also makes it insensitive towards most of the gases. However, Cl_2 being highly oxidizing gas can pick up electrons from reduced PPy to form Cl_2^- counter-ions, which results in a resistivity decrease of composite films

3. SnO₂ Nanowires

SnO_2 an n-type wide band gap (3.6 eV) semiconductor because of its unique conductance properties has been employed in various opto-electronic applications. It has a tetragonal/rutile structure with lattice parameters a = 0.480 nm and c = 0.308 nm [50]. Two unique characteristics of tin oxide, i.e., the variation of valence state and oxygen vacancy, make them useful for applications including gas sensors, catalyst, electrode material, antireflecting coating, and several other smart and functional devices [51]. Additionally, the ability to control stoichiometry, ease of fabrication into different forms, high chemical and thermal stability, and availability of various precursors has lead to its widespread use in gas sensing applications. However, it suffers from various drawbacks including poor selectivity, ageing and humidity induced effects involving grain growth and surface poisoning causing poor reliability. Besides, widespread utility for commercial application requires a reliable commercial low cost sensor. All the above mentioned drawbacks can easily be overcome by employing NWs of SnO_2 with the main advantage of high surface to volume ratio holding the promise for the same.

3.1. NWs growth using thermal evaporation method

We have synthesized SnO_2 NWs using a thermal evaporation method [52, 53]. In brief, a high purity Sn powder (99.99% pure) was loaded in an alumina boat kept inside a small 1 inch diameter quartz tube [54]. This tube was then loaded inside a 2.5 cm diameter and 1 m long quartz tube placed inside a horizontal tubular furnace. The temperature of the furnace was ramped to 900°C in the presence of Ar (500 sccm) and O_2 (~0.1 %) and maintained for 1 h. It may be noted that the melting point of Sn is 232°C, but it needs temperature of 997°C to attain a vapor pressure of 10^{-4} mbar, while SnO_2 (melting point 1127°◦C) sublimes with a vapor pressure of 10^{-4} mbar at 350°C. Therefore, the evaporation of Sn occurs after its oxidation. After growth, the furnace was cooled to room temperature. SnO_2 NWs were observed deposited on the upper side walls of the alumina boat. NWs in the present case grow by evaporation of metals that get oxidized and deposit on sides of alumina crucible. The growth was seen to initiate on edges of alumina boat and is attributed to Vapor-Solid (VS) growth mechanism. Figure 20 shows the corresponding SEM and TEM image of as grown NWs having a diameter between 70 and 150 nm and length extending up to few tens of microns. The higher aspect ratio of these NWs ~1000 coupled with smaller diameters is very promising and crucial for realizing sensors with faster reaction kinetics.

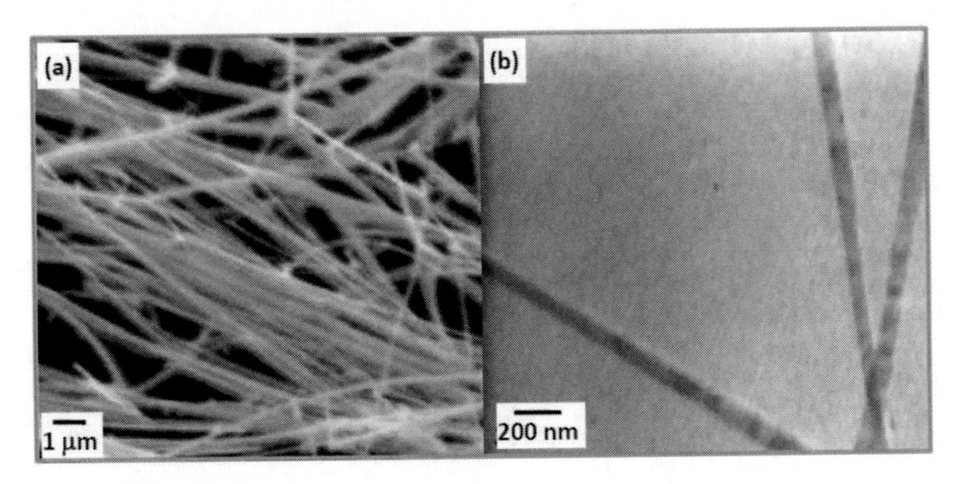

Figure 20. (a) SEM and (b) TEM image of SnO_2 NWs

3.2. Gas sensing properties

3.2.1. Single NW as Cl_2 and H_2S sensor

The "pick and place" approach was used to fabricate sensor devices using single NWs. Single wire sensors were prepared by alignment of a metal mask (12 μm spacing) on to a single NW under an optical microscope. For this, a drop of dilute suspension of wires in methanol was placed on a glass substrate. Single wires were then located under an optical microscope and carefully, W – wire (of 12 μm diameter) as a metal mask was aligned on the NWs and locked. The alignment was made possible by long length of NWs (100–500 μm) and solution sticking to wires by surface tension. For measurements, Au contacts (120 nm) were vacuum evaporated. Miniature Pt heaters were attached to backside of substrates using alumina paint.

The gas sensing characteristics of the Single SnO_2 NW was investigated in a static system at various temperatures [50]. Figure 21 shows the room temperature response of a typical sensor towards 2 ppm H_2S and 6 ppm Cl_2, respectively. It is interesting to note that upon exposure to both reducing as well as oxidizing gas the resistance is observed to decrease.

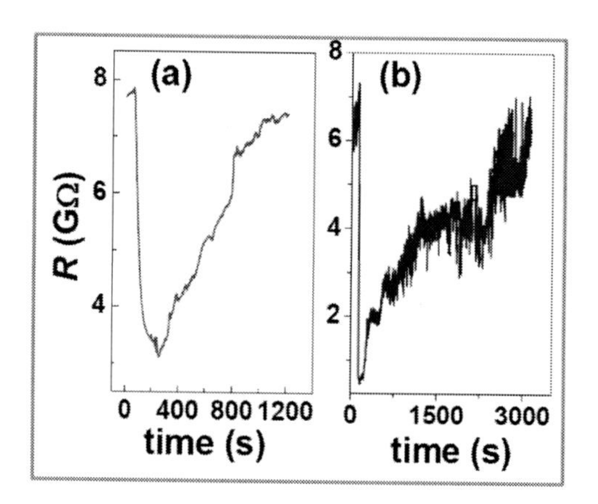

Figure 21. Room temperature response of single NW sensor towards (a) 2 ppm H_2S and (b) 6 ppm Cl_2, respectively.

The anomalous response towards Cl_2 could probably be assigned to the strong interaction with the lattice and adsorbed oxygen species. We have calculated earlier the typical depletion layer width on SnO_2 surface and is found to be ~900 nm which is much bigger than the NW diameter [11]. Hence it is expected that the single NWs are encapsulated in an electron depleted environment. Thus, a small perturbation arising from the interaction with gas molecules will be directly reflected as a change in the bulk conductance. At room temperature and on single NW which is nearly a perfect monocrystal, Cl_2 interacts predominantly with lattice oxygen and adsorbed oxygen species as per equation; [55]

$$\frac{1}{2}Cl_2 + O^{2-}{}_{(ad)} \rightarrow Cl^-{}_{(ad)} + \frac{1}{2}O_2 + e^- \quad (7)$$

$$\frac{1}{2}Cl_2 + O_o^{2-}{}_{(ad)} \rightarrow Cl_o^- + \frac{1}{2}O_2 + e^- \quad (8)$$

where, subscripts (ad) and o indicates adsorbed oxygen and lattice oxygen species, respectively. This results in release of free charge carriers in the bulk of NWs and accordingly, a reduction in resistance is observed. The response towards Cl_2 is much higher than that of H_2S. Single NWs could detect H_2S reversibly with response and recovery time of 170 and 550 s, respectively. On the other hand for Cl_2 (6 ppm) although the response time is 100 s, the sensors recovery is very slow ~ 40 min. It was observed that with increase in temperature the sensitivity decreased and became quite low at 50°C (data not shown).

3.2.2. Mat-type NW thick film as Cl_2 and H_2S sensor

Mat-type thick film based sensor samples were prepared by pressing the as grown NWs on alumina substrate. These sensors on contrary are expected to show a behavior dominant by inter-grain resistance. Accordingly, the gas sensing properties of sensor films were investigated. Figure 22 shows the effect of operating temperature on the sensitivity of a typical Mat-type films towards Cl_2 and H_2S, respectively. In these sensors the resistance was found to decrease on exposure to H_2S and increased on exposure to Cl_2. This result is as expected for an n-type SnO_2, on interaction with reducing (H_2S) and oxidizing (Cl_2) gases. Interestingly, for both the cases the operating temperature for maximum sensitivity was found to be 150°C. Importantly, at all the operating temperatures the sensor exhibited higher sensitivity towards 500 ppb of Cl_2 in comparison with that of 7 ppm H_2S.

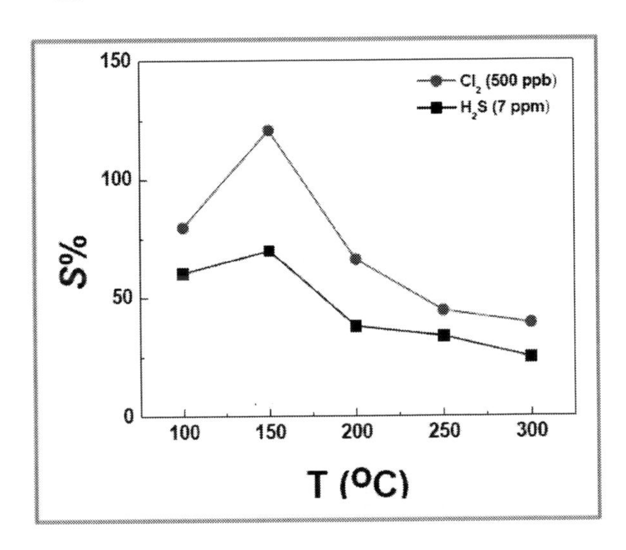

Figure 22. Sensitivity dependence of mat film based sensor on operating temperature towards Cl_2 and H_2S.

The sensors robustness was confirmed by exposing the sensor towards increasing concentrations of test gases. Accordingly, Figure 23 shows the response of sensor towards increasing concentration of Cl_2. The response was found to increase with concentration as illustrated in Figure 23 (b). The lowest detectable concentration was found to be 100 ppb. Interestingly, both the response and recovery times were very fast and increases with an increase in gas concentration For example, the response time towards 1 ppm of Cl_2 was 14 s, while recovery time was 135 s, respectively.

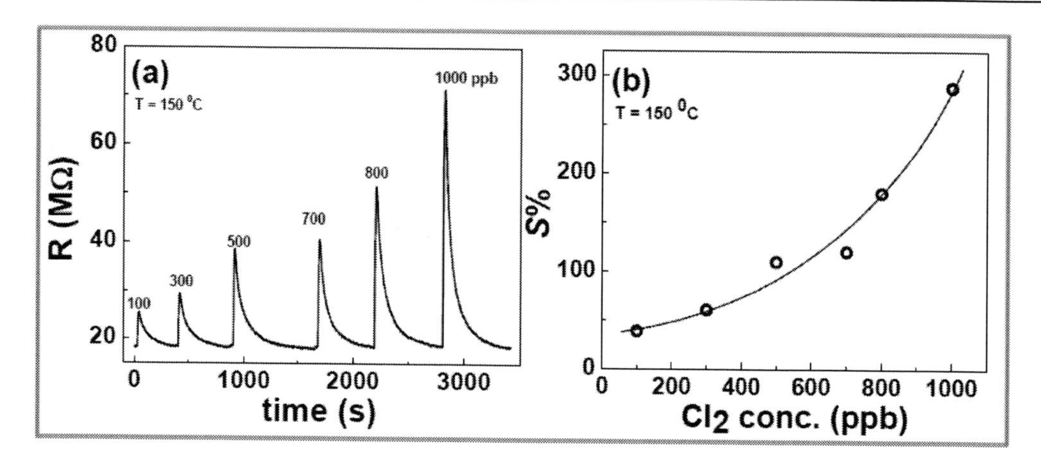

Figure 23. (a) Sensors response towards increasing concentration of Cl_2 and (b) corresponding dependence of sensitivity on gas concentration.

Contrary to Cl_2, the sensor responded only to higher concentration of H_2S as shown in Figure 24 (a). The sensitivity was found to saturate at around 10 ppm (Figure 24 (b)). Sensor exhibited faster reaction kinetics as evident from the response and recovery time behavior. For example, the response times towards 1 ppm H_2S was 6 s while recovery time was meager 70 s.

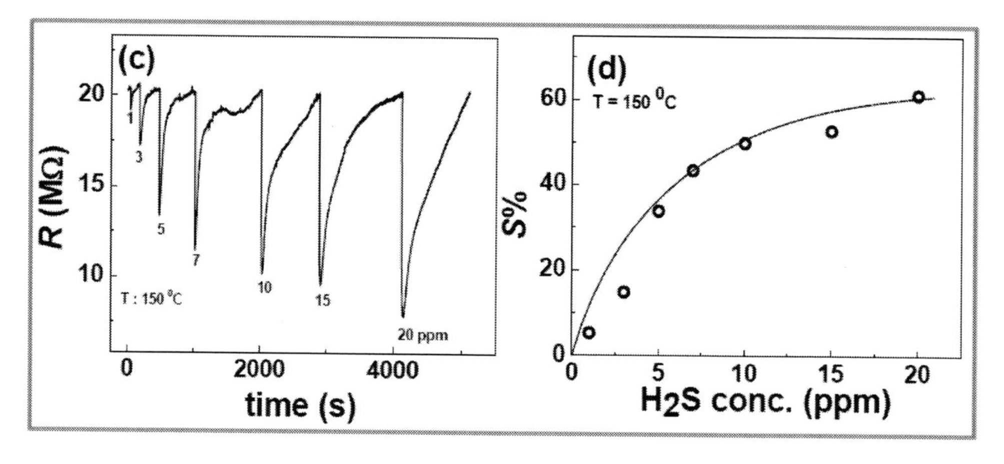

Figure 24. (a) Sensors response towards increasing concentration of H_2S and (b) corresponding dependence of sensitivity on gas concentration.

As observed from the above data, the two sensor configurations have a varied response towards the two test gases. In case of single NW sensors because of the high crystalline nature of NWs, the observed sensitivity is attributed solely to the bulk changes. In case of Mat-type sensors the effective resistance is considered to have contribution arising from two components namely bulk and intergrain resistance. Firstly the bulk resistance which is due to the change occurring on the NW surface and secondly the grain boundary resistance arising from the inter-grain (boundary between two neighboring NWs) region. Our results indicate that the single wire sensor is more appropriate for detection of Cl_2 and a Mat-type sensor could be used for both Cl_2 as well as H_2S.

4. CuO NWs

CuO is a p-type semiconductor with a narrow band gap of about 1.36 eV and has been a candidate material for photothermal and photoconductive applications [56]. It is used in fabrication of lithium–copper oxide electrochemical cells and is the basis of several high-Tc superconductors [57]. It exhibits a monoclinic structure with lattice constants: $a = 0.468$ nm, $b= 0.343$ nm, $c = 0.517$ nm and $\beta = 99.54°$ [58]. CuO NWs can be potentially applicable in gas sensing [59], magnetic storage media [60], catalysis [61] and field emitters [62]. They can be synthesized by using precursors [63], hydrothermal decomposition route [64], self-catalytic growth processes [65] and solvothermal route [66]. A few studies on growth of CuO NWs by thermal annealing copper foils in oxygen atmosphere [67] have also been reported. In comparison to complex chemical methods, thermal annealing of copper foil provides simple, convenient and fast method for synthesizing CuO NWs. NWs synthesized by this method are always found to grow perpendicular to the surface of copper foil. It has been reported that [68] diameter and number density of NWs are a critical function of oxygen flow rate and annealing temperature. The length of NWs depends mainly on the annealing time.

4.1. NWs growth using thermal oxidation of Cu

We have synthesized CuO NWs using thermal oxidation of high-purity Copper (Cu) sheets. In brief, Cu sheet of 1 mm thickness and 10×10 mm^2 sizes was first cleaned in dil. HNO_3 to remove the native oxide layer and adsorbed impurities. The foil was then thoroughly rinsed with deionized water followed by ultrasonication in acetone for 5 min. Thermal oxidation of Cu sheet was carried out in a resistively heated furnace at different temperatures (between 400 and 800°C) and times under flowing O_2 atmosphere. The sample temperature was monitored by placing a thermocouple in vicinity of the sample. In all the experiments, the rate of heating of Cu sheet was maintained at 6°C/min and after oxidation samples were quenched by removing from furnace. A constant O_2 flow rate (50 sccm) was maintained throughout the cycle of heating, oxidation and quenching of the samples. An optimum growth of NWs was achieved at 675°C. Accordingly, figure 29 (a) shows the SEM images of CuO NWs after oxidation of Cu strips for 22 h. The length of the NWs was found to be >20 μm. In the present case, the NWs growth occurs via relaxation of stresses [69, 70]. Stress-induced growth mechanism occurs mainly via oxidation of Cu to Cu_2O and subsequently to CuO that provides the energy required to form additional surface [71] of NWs. In particular, oxidation causes stress in the foil due to volume and structural changes [72, 73]. And sufficiently high rate of oxidation coupled with low mobility of atoms in solid leads to relaxation of stress by formation of small crystallites of CuO from which NWs are observed to grow. NWs growth, therefore, occurs to reduce the stress generated during oxidation of Cu. This mechanism also explains the observations that the NWs do not grow

(a) during oxidation of Cu_2O pellet, although in case of copper strip, Cu_2O is first formed and

(b) at temperatures above 800°C.

Cu_2O pellets are sufficiently porous and therefore the stresses due to change in volume are easily accommodated and in case of high temperatures, high mobility of atoms helps in reducing the stress.

4.2. Gas sensing properties

4.2.1. Single NW as H_2S and NO sensor

Figure 25 shows the gas sensing properties of sensor sample with a single CuO NW aligned between two Au electrodes with 25 μm spacing. These NWs exhibited good sensitivity towards both H_2S (~200% for 10 ppm) and NO (~200% for 200 ppm). Exposure to H_2S (Figure 25 a) showed slight increase in resistance followed by decrease in resistance with very fast response (30 s) and recovery times (60 s). On exposure to NO gas (Figure 25 b), contrary to expectations, increase in resistance was observed with fairly high sensitivity and small response and recovery times. Mechanism of increase in resistance on exposure to NO instead of decrease as expected for p-type semiconductor is not understood and needs further investigations.

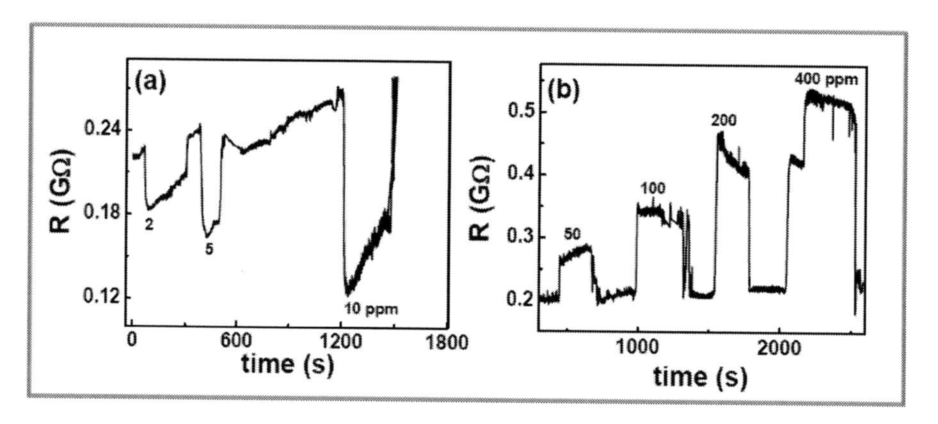

Figure 25. Room temperature gas sensing characteristics of isolated CuO NWs on exposure to (a) H_2S, and (b) NO.

4.2.2. Drop casted Mat-type NW film as H_2S sensor

CuO NW films were prepared by dispersing them in methanol and dropping this solution between the electrodes to form a mesh of NWs. These films were exposed to different concentration of H_2S at room temperature. In this case increase in resistance was observed for exposure upto 200 ppm, in accordance with the p-type behavior of films. The recovery in this case was found to be very slow. For comparison gas sensing properties were studied for CuO thin films made by oxidation of thermally evaporated copper films. In this case also only increase in resistance was observed, figure 26 (b).

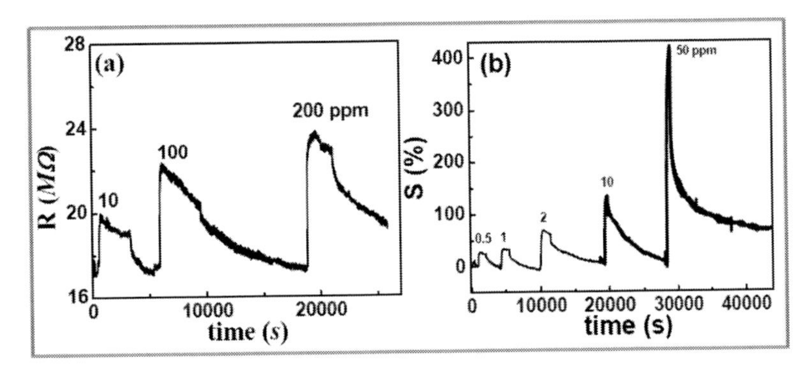

Figure 26. H_2S gas sensing characteristics at room temperature for CuO (a) NW film and (b) thin film.

Careful investigations at 200°C indicates that NW films were sensitive to H_2S concentrations, as low as 200 ppb, with maximum sensitivity of 150% (Figure 27 (a)) and fast response (~1 min) and recovery (~4 min) times [74]. Since CuO is a p-type semiconductor therefore on exposure to reducing gas (H_2S) its resistance increases. Increase in resistance is followed by decrease in resistance at higher concentrations (≥10 ppm) because the surface of CuO on reacting with H_2S forms CuS (Figure 27 (b)) [3], which is metallic in nature. This reaction is very feeble and takes place at high concentrations and temperature. The sensor films when exposed to very high dosage (e.g. 400 ppm) became unrecoverable. EDAX (Figure 27 (c)) and XPS (Figure 27 (d)) studies were carried out to confirm the formation of CuS. Presence of sulphur has been confirmed by EDAX and XPS. The binding energy of S 1s was at ~163 eV, which corresponds to CuS.

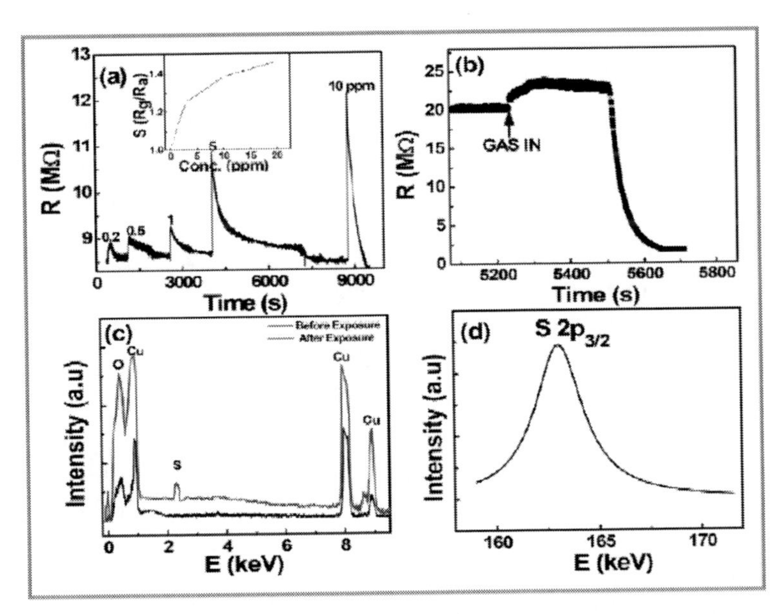

Figure 27. (a) Response curves towards different concentrations of H_2S (b) response curve at 50 ppm H_2S confirming CuS formation, (c) EDAX spectra before and after H_2S exposure and (d) S 2p spectra recorded using XPS measurements.

4.2.3. As grown CuO flakes containing NWs as H_2S and NO sensor

The flake with NWs easily gets detached from the remaining Cu foil due to stress after cooling. Au contacts were deposited on as grown CuO flakes (with NWs) and effect of exposure to different gases such as H_2S, NH_3, NO, CO and CH_4 was studied. At room temperature, the flakes showed good response to H_2S (Figure 28) and very small response to NO, inset of figure 30. The response to H_2S, in this case consisted of increase in resistance at lower concentration and an increase in resistance followed by decrease at higher concentrations. Recovery was slow at 10 ppm and incomplete on exposure to 50 ppm gas.

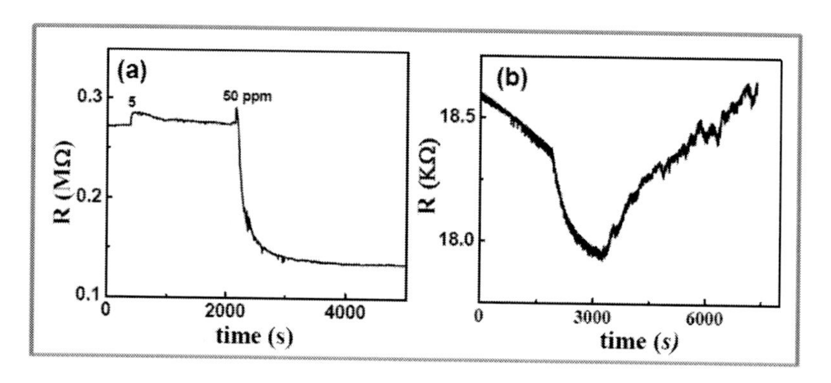

Figure 28. (a) Resistance change in a CuO flake on exposure to H_2S gas, (b) Change in resistance in another flake on exposure to 50 ppm NO gas.

4.2.4. Hierarchical CuO: $W_{18}O_{49}$ heterostructures as H_2S sensor

The growth of CuO: $W_{18}O_{49}$ hierarchical heterostructure NWs (HSNW) was carried out in two steps. In the first step, CuO NWs were grown by thermal oxidation of Cu in a horizontal tubular furnace as described earlier. In the second step, tungsten oxide NWs were grown on CuO NWs by thermal evaporation of tungsten in a vacuum deposition system in the presence of air at a pressure of 2.5×10^{-4} mbar [79, 75]. A collection of CuO NWs was placed at a distance of 2 cm from tungsten filament and the system was evacuated to 2×10^{-5} mbar. The depositions were carried out for 30 min at a source temperature of ~1965°C. The temperature of the filament was measured using a pyrometer (RAYTEK make). The pressure inside the chamber was maintained by controlled flow of air through a needle valve. The deposition occurs by slow oxidation of tungsten filament followed by evaporation, as oxides of tungsten are volatile. For example, a vapor pressure of 10^{-4} mbar occurs at 980°C for WO_3 compared to 2760°C for W metal (melting point of W is 3422°C and that of WO_3 is 1473°C). The temperature of substrates containing CuO NWs increased to 700°C during deposition. The growth of heterostructures is attributed mainly to VS mechanism. To study effect of growth parameters, the depositions of $W_{18}O_{49}$ were also carried out at different filament temperatures between 1850°C and 2050°C and at pressures between 5×10^{-5} and 7×10^{-4} mbar.

Initially effect of growth conditions on the nature of $W_{18}O_{49}$ NWs was investigated and the results are shown in Figure 29. NW density and diameter was found to increase with increase in pressure and change in oxygen partial pressure. For example, at a pressure of 2×10^{-4} mbar in air, thickness of CuO: $W_{18}O_{49}$ increased to around 1 μm. This is understandable as increase in pressure leads to higher oxidation rate and thereby increased evaporation rate. In heterostructures, a dense growth of tungsten oxide shell with protruding NWs was observed.

Figure 29. SEM micrograph of as grown CuO NWs grown at 675°C for 22 h, and CuO:$W_{18}O_{49}$ heterostructures (b) in air at 2×10^{-4} mbar and (c) in air at 2×10^{-3} mbar pressure, (d) in oxygen at 2×10^{-4} mbar and (e) in oxygen at 2×10^{-3} mbar.

Isolated heterostructure samples were highly selective towards H_2S, no response towards other gases was observed. Figure 30 shows the response and recovery curves towards H_2S as a function of operating temperature. Exposure to gas resulted in resistance decrease implying gas interaction mainly with the $W_{18}O_{49}$ NW shell [76]. Besides no major variation in the sensors response was observed with increase in the temperature.

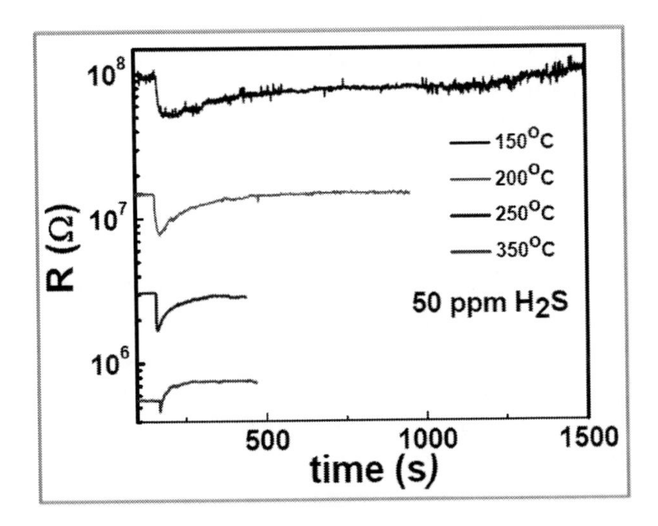

Figure 30. Sensor response as a function of temperature towards 50 ppm H_2S.

5. A Step towards Commercialization of Sensor Devices

Previous section clearly emphasizes NWs as a potential material for the realization of next generation sensor devices. However, there are certain points that need to be addressed before their widespread use for the same. One of the important requirements is the understanding of the growth mechanism. This is specifically crucial to realize NWs with desired crystal quality, aspect ratio in the predefined position for optimum sensor performance. We have targeted some of the issues which directly ease out the process towards commercialization of sensor devices based on NWs using ZnO as a model system. Some of the achievements realized toward these goals are listed below:

5.1. Control over growth region and mechanism:

5.1.1. Tuning growth position: A simulation based study

Vapor-phase synthesis inside a horizontal tube furnace is the common method for the growth of NWs. The growth is known to be very sensitively dependent on parameters like source type, substrate orientation, size of the catalyst, temperature or temperature profile of the furnace, partial pressure of reactants and gas flow and last, not least on the geometry of the tube furnace [77, 78, 79]. However, complex modeling of the NWs growth and its dependence on the various parameters and furnace geometries were rarely done up to now [80, 81]. Mostly an empirical approach is used to determine the suitable growth conditions in the tube furnaces. Very recently, we have targeted the fundamental concepts involved in vapor phase deposition [51]. A systematic study demonstrating the necessity of an optimum O_2 concentration for successful NW growth and influence of both $CO_{(g)}$ and $O_{2(g)}$ on the growth mechanism was performed. The gas flow inside the tube was designed and simulated using ANSYS CFX 11.0 software. Figure 36 shows the simulated stream line of the mixed gases inside the tube at 10 min for different flow rate of the carrier gas (5 s/step). When the mass flow is low, for example at 5 sccm, the stream line is coiling inside the cylindrical tube surface. Once the flow rates are increased, the coiling of stream lines reduces and the stream lines get straighter. This can be explained in term of the continuity equation (which is one of the governing equations in ANSYS CFX)

$$\frac{\partial \rho}{\partial t} + \frac{\partial}{\partial x_j}(\rho \vec{u}_j) = 0 \tag{9}$$

If the mass flow rate is reduced then the second term is increased, thereby also increasing the velocity of the components. This result indicates that each gas molecule need more space for moving inside the tube due to increase in velocity with a limited volume. Consequently, gas molecules circulate to a very nearby area. Hence the sticking probability increases because gas molecules repeatedly pass on the sample, and the deposition is higher.

The net transports of Zn estimated from this simulation are shown in Figure 31. In case of a mass flow rate of 5 sccm, the $Zn_{(g)}$ distributes into both directions of the tube (left and right). The Zn concentration is higher on the right hand side (downstream) because the carrier gas ($Ar_{(g)}$) is set to flow from left to right. $Zn_{(g)}$ is able to reach the left end of the tube in < 1

min. The simulation result also corresponds to the experimental result, because ZnO NWs can be grown in the whole range of the left hand side (upstream) of the tube within only 5 min or even less. With the increase of mass flow, the active area for growing ZnO NWs gets shifted to the right hand side of the tube. The higher flow rate pushes the $Zn_{(g)}$ and therefore it can reach the left side only partially. In other words, Zn transport against the gas flow becomes difficult with the increase in carrier gas flow rate. For example, when the mass flow rate is 50 sccm, only a very small part of Zn is transported to the left hand side of the powder container as observed in Figure 32 (c). Performing the experiment under the same conditions the NWs were actually observed to grow in the region as predicted. This implies that the simulation result can help to predict and tailor the growth of NWs at different positions inside the system. Further, depending on the controlled mass flow, $Zn_{(g)}$ can even be transported to an up-stream position.

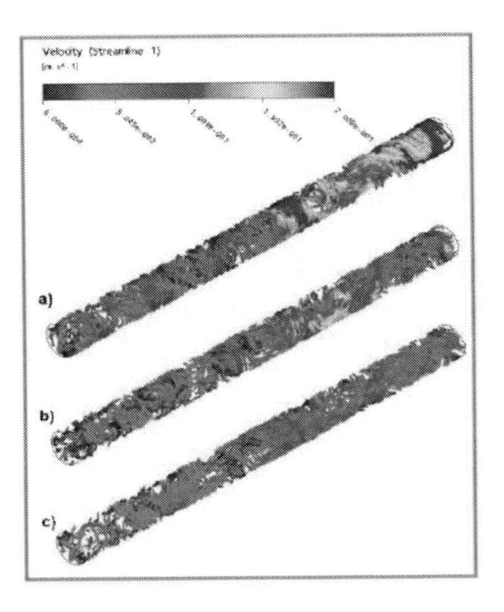

Figure 31. Stream line of the mixed gases inside the tube at 10 min for different flows of the carrier gas (a) 5 sccm, (b) 20 sccm, and (c) 50 sccm.

5.1.2. Control of ZnO NW diameter and crystal quality: Reversible VS to VLS transition

Besides the ability to control over the growth region and place inside the growth furnace the control over growth mechanism is very crucial for sensor application. A complete understanding of the growth process which enables a control of crystal quality (defects, charge carrier concentration) by tuning process parameters is a key towards direct integration of NWs into device application. Upto now NWs are reported *either grown by VS or VLS*, but a systematic transition of the growth process by means of controlling specific process parameters has not been demonstrated so far. Also the reason for the different role of the Au i.e., as a catalyst for VLS or as a surface defect for VS growth is not completely understood so far, especially in case of ZnO.

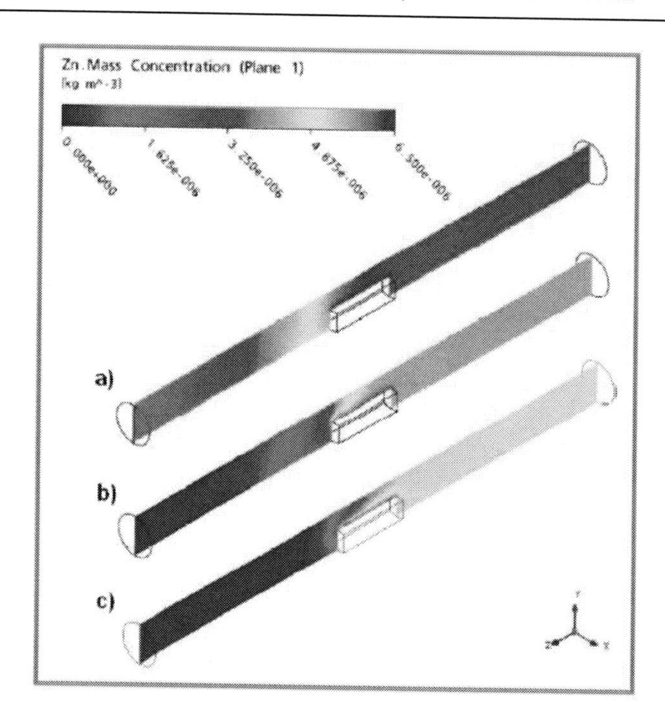

Figure 32. Simulated zinc distribution as a function of the gas flow estimated after the experiment proceeds for 10 min for Ar transport flow of (a) 5 sccm, (b) 20 sccm, and (c) 50 sccm.

The vapor phase growth has commonly been described using VLS and VS mechanism [82, 83, 84]. The use of metal nanoparticles as catalysts is specific for both VLS as well as VS processes, where the catalyst initiates and guides the growth. An epitaxial relationship between the substrate and the NWs is needed to grow NWs in a patterned and aligned arrangement. In principal, the VLS growth process is characterized by the presence of a metal particle at the NW tip and offers the possibility for diameter and arrangement control via the patterned arrays of catalyst. A carrier gas (usually Ar) transports the vaporized source material to the catalyst droplet (Au, Ag, and Ni) [85]. The vapor source material adsorbs and diffuses into the metal catalyst forming a *liquid* alloy. With time, the concentration will exceed the solubility and precipitation in the form of NWs is observed. This supersaturation and precipitation process continues until the source material is exhausted or the growth conditions are no longer preserved (lowering of pressure or temperature). Preferential 1D growth occurs in the presence of a reactant as long as the catalyst remains on top of the growing wire [86].

On the other hand, VS process involves the transport of vapor species from the source to the substrate and their subsequent (enhanced) condensation on the still solid (metal) catalyst [87, 88]. Surface defects, dislocations or metal patterns on the substrate can provide energetically favored nucleation sites for the incoming vapor. At these sites, Zn solidifies quickly to ZnO by oxidation when oxygen is either intentional or un-intentional in the reaction chamber, especially at places with inhomogeneties or defects, and creates ZnO wires or vortex like nanostructures [89]. VS NWs are observed to grow slower (10 to 100 times) than VLS NWs because of the lower diffusivity, the contributions of desorption, and the weaker surface reactivity in the solid phases. These NWs show a clear hexagonal cross section unlike VLS NWs.

We have demonstrated the controlled and reversible transition of the growth mode from VS to VLS by simply applying ionic liquids (IL) namely butyl methyl imadazolium tetrafluoroborate ($BMIm^+BF_4^-$) and butyl methyl imadazolium hexafluorophosphate ($BMIm^+PF_6^-$) as additional carbon source [90]. Figure 33 schematically illustrates the steps involved in both kinds of growth processes. The carbothermal reduction takes place via the reaction:

$$ZnO_{(s)} + C_{(s)} \rightarrow Zn_{(g)} + CO_{(g)} \quad (10)$$

providing the Zn source material for the NW growth. The backward reaction of equation (10) is also thermodynamically viable but its contribution for the NW growth can be considered as negligible. In case of VS, the growth is initiated by the formation of ZnO nuclei via an oxidation reaction:

$$2\,Zn_{(g)} + O_{2(g)} \rightarrow 2\,ZnO_{(s)} \quad (11)$$

where, Au only acts as surface inhomogeneity serving as preferential defect site for subsequent vapor adsorption and O_2 is provided externally along with the Ar carrier gas. Zn oxidation could also result from the reaction containing CO and CO_2 i.e., [91]

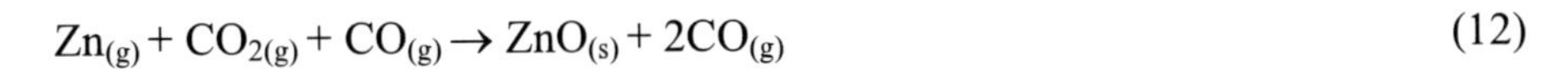

$$Zn_{(g)} + CO_{2(g)} + CO_{(g)} \rightarrow ZnO_{(s)} + 2CO_{(g)} \quad (12)$$

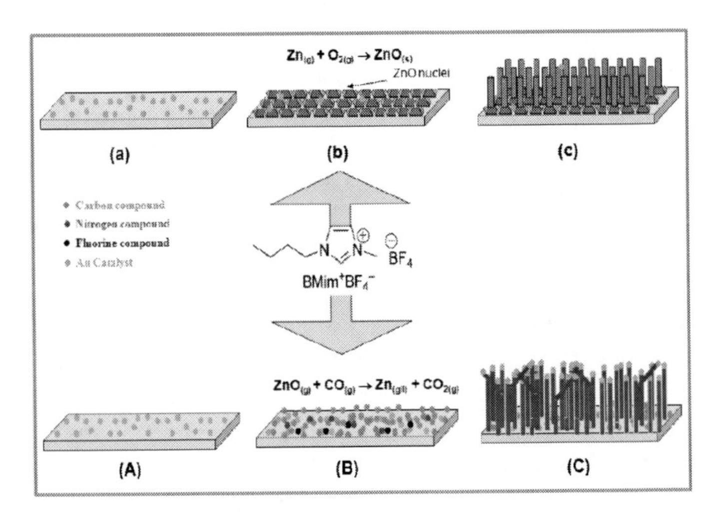

Figure 33. Steps involved in the VS (top images) and the VLS (bottom images) growth mechanism on GaN/Si substrates. (a) Substrate with Au catalyst seeds, (b) Ramping-up to higher temperatures leads to the formation of ZnO nuclei with the Au dots as favored growth sites. (c) ZnO NWs grow from these nuclei. Upon the use of IL the growth process transforms into a reactive VLS: (A) Au-Zn catalyst seed layer, (B) deposition of hydrogenated C (and N or F) on the substrate when the furnace is ramped at higher temperature, (C) NWs growing on the substrate surface with kinks and different preferred growth planes.

However, equation (11) is still expected to be the dominant reaction governing the growth. VS growth mechanism involves a crystallographic-direction-dependent anisotropic growth [92]. Au serves as surface inhomogeneity or "defect" and provides the energetically favored nucleation site for the condensing vapor. The precipitation and oxidation of Zn takes place into (0001)-oriented quasi-hexagonal ZnO nuclei. Finally, ZnO nucleate (bigger dimension) at the top of the pyramids, and grow homoepitaxially into faceted short pillars/NWs. The surface diffusion on solid surfaces is slow and hence the resulting NWs are of shorter length.

Now, in case of VLS, the NWs grow in the region where the substrate temperature is homogeneous and almost equal to the source temperature. The decomposition of IL results in release of (semi static condition) hydrogenated carbon, fluorine and nitrogen compounds (impurities) inside of the tube [93]. Every time the furnace is ramped at high temperatures the contaminations desorbs from the wall and adsorb on the Au containing substrates forming an additional source of C and O. Recently, it was demonstrated that the presence of an ultrathin carbon layer (2 nm) pre-deposited on the substrate surface strongly enhances the NW growth [94, 95, 96]. We note here that the experiments carried out with the patterned Au using photoresist techniques (without plasma treatment and annealing) also exhibited a VLS type of growth mechanism further supporting the role of C. Of the various gases, in particular CO is known to adsorb strongly on quartz and could even result in the formation of Zn following equation:

$$ZnO_{(s)} + CO_{(g)} \rightarrow Zn_{(g/l)} + CO_{2(g)} \qquad (13)$$

For conditions similar to VS, reaction 13 implies that at initial stages most of the ZnO formed via reaction 11 might be converted back to Zn on the substrate containing Au and thereby increasing the Zn concentration. The metallic Zn subsequently forms a liquid alloy with the Au present on the substrate and at the here used deposition temperatures (below the boiling point of Zn and well in the eutectic range of the Zn-Au system), these alloy droplets are highly mobile on the substrate surface and, hence, result in the formation of a film. Please note that for the temperature available in the VLS region (875-930°C) already the inclusion of 15 at.% Zn into Au is enough to result in a melted Au-Zn droplet [97]. If the droplet gets more and more Zn and no other process takes place, the size of the dot gets bigger. On a certain point, a segregation of liquid Zn might take place and a crawling of the droplet on the surface takes place leaving a track of Zn behind. If now oxygen is available, then that out-sourced liquid Zn oxidizes forming solid ZnO and hence starting the ZnO wire. The still liquid Au-Zn alloy at the top governs the efficient adsorption of the Zn gas. It is also interesting to note here that some of the wires in VS dominant condition do exhibit a sidewise crawling of NWs starting from the root. This is probably due to contribution arising from CO reduction process. The difference in the VS and the VLS process involving the liquid Au-Zn phase is the amount of Zn segregating-out and the O available to oxidize that Zn immediately. The whole process is actually controlled by the carbon species in the furnace controlling the amount of Zn gas, by the flow conditions and by the availability of oxygen. By following this route consequently the process should be called a *reactive VLS mode*. The solid ZnO phase is based on the Zn reaction with a reliable O source. ZnO is not soluble in Au; the ZnO and Au phases exist independently from each other.

Here, it is also significant to know that the choice of process parameter can effectively switch a growth mechanism from VLS to VS. Importantly, by optimizing the O_2 inside the system it is possible to switch the equilibrium either towards reaction 18 or reaction 16. This further implies that at temperatures above boiling point of Zn (907°C) and with a high O_2 flow rate it is possible to suppress the VLS mode completely, and reverse the growth mechanism back to VS mode (ZnO melting point 1975 °C). Besides the VS state of the growth tube is also easily recovered by cleaning with concentrated hydrochloric acid (HCl) thereby implying the complete reversible switching between VS and VLS NW growth modes.

With this understanding we were able to control the diameter of NWs. As shown in figure 34, VS growth results in NWs with dimension between 40 -150 nm, while employing VLS, NWs as small as 9 nm in diameter were possible. Figure 35 shows the PL spectra of the NWs grown via VS and VLS mechanisms. The VLS grown NWs show a strong and sharp UV emission at 378 nm (3.28 eV) (Figure 35) attributed to near band edge emission, arising from the recombination of free or bound excitons of ZnO. The broad green emission peak around 2.4 eV (516 nm) appears for both the VS and the VLS NWs, however, with different ratios. This green emission is discussed as deep-level or trap-state emission, caused by defects of the crystal or crystal surface [98]. The involvement of oxygen (O) vacancies, interstitial O, Zn vacancies, Zn interstitials, and other extrinsic impurities has also been discussed [99]. Surface O deficiencies are discussed as electron capture centers, which can reduce the recombination rate of electrons and holes [100].

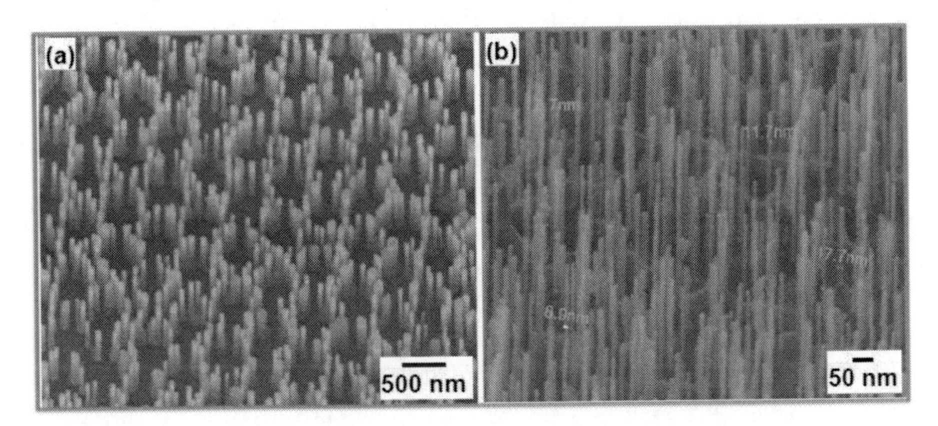

Figure 34. SEM images of VS (a) and VLS (b) NWs grown using vapor phase deposition

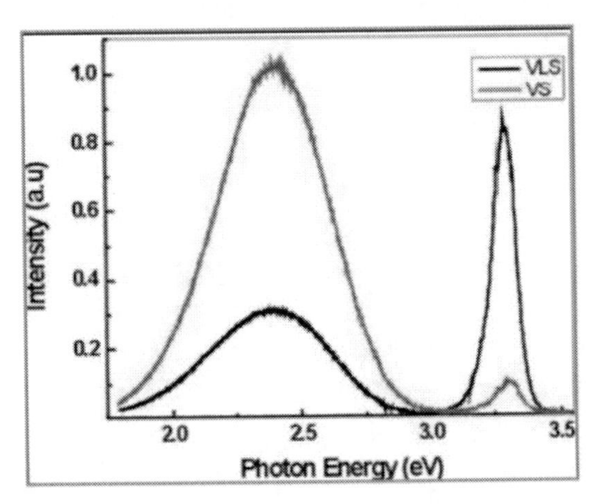

Figure 35. PL spectra recorded for as-synthesized NWs grown for 20 min on Si (100) wafers at 930°C.

In summary, a reversible switching between the VS to the reactive VLS growth mechanism has been demonstrated for the first time by adding or removing an ionic liquid and by a careful control of process parameters. Formation of liquid Zn and a subsequent liquid Au-Zn alloy due to an additional carbon source in the system are mandatory for the VLS transition. Our results clarify, for the first time, the reason for the different roles of Au, which is, acting as surface defects or as a catalyst in VS and VLS growth modes, respectively. Thus, our results demonstrate the possibility of controlling not only the growth mode and region but also the crystal quality of the NWs by using ILs as an additional carbon source. The present work is a crucial step toward a better understanding and realizing designed NWs based on VS or VLS for a particular application.

5.1.3. Patterned growth and direct integration of NW on contact electrodes

Up to now the "pick and place" approach is the most widely used method for nanomaterial integration. However, it is not considered to be suitable for large-scale manufacturing because of the inherent drawbacks such as random placement, contaminations, and a general incompatibility with Si processing. Another approach is the deposition of the electrode directly on top of the NWs by e-beam evaporation. However, because of the multiple and slow processes involved, this method is also not practical on a large scale. Hence, a more elegant approach would be to selectively grow nanostructures directly onto desired areas of the substrate such as, for instance, directly on electrodes. This is particularly significant for sensor applications where a good ohmic contact between the metal electrode and the semiconductor NWs would ensure better performance.

We have demonstrated the catalyst-free and area-selective vapor solid (VS) growth of ZnO NW arrays on Si substrates using topographical confinement (Type I) and directly on Au electrode using preferential chemisorption on self-assembled monolayers (SAMs) (Type II) as shown in Figure 36. Self integration of ZnO NW bridges on Si substrates and Au electrodes has been achieved without severe defects over large area [101]. Obviously, the preformed ZnO seed layer with SAMs in Type II approach acts as a passivating layer against the deformation of underlying contact electrode. Electrodes are deposited at the ends of the pattern which measure the effective resistance arising from the NW bridges [102]. The initial resistance of Type I sensor is high in comparison with that of Type II. The difference between the effective resistances could be a result of the different ZnO seed layer formation process. The uniform loading, the underneath Au layer and the NW bridges contribute equally to the effective resistance. In Type I sensor the ZA species distribution is dependent on the rate of evaporation and surface tension of the solution and hence does not promise a uniform and dense packing of seed layer. On the other hand, in case of Type II sensor, the chemical anchoring of ZA species onto the SAM modified Au ensures a uniform and dense packing of the seed layer. The presence of the gold underneath the ZnO seeds will definately influence the conductance and transport properties of the system. In case of Type I, the system is a kind of a Metal-Semiconductor-Metal (MSM) system, where the semiconductor in the middle is represented by millions of ZnO nanobridges in series as well as parallel and the metal contacts are the two outer electrodes. In contrast, Type II device represents a sequence of Metal-Semiconductor-Metal (MSM) devices connected in series. Also in this case the inner semiconductor part consists of millions of ZnO nanobridges in parallel. Further the non-linearity in the I-V measurements for Type II sensor is negligible, making the sample nearly ohmic [101]. It could indicate that the preformed ZnO seeds layer on Au electrode efficiently acts as a passivating or protecting layer against a severe deformation of the underlying Au electrode at high deposition temperatures. Consequently, the presence of Au underneath the ZnO nuclei seems to help realizing a good contact between the adjacent NWs and also the bridging NWs.

Sensor response towards different concentration of carbon monoxide (CO) at RT was also investigated. Interestingly, Type II sensor exhibited a better response with a high sensitivity (S) of about ~ 900 compared to that of Type I sensor which showed S = 50 to 40 of CO. The sensitivity increased with the gas concentration with the lowest detectable concentration for type I and type II sensors of ~ 1.2 ppm and 120 ppb, respectively. Importantly, even at relatively low CO concentration, the type II sensor exhibited a superior response variation (S ~9 for 120 ppb).

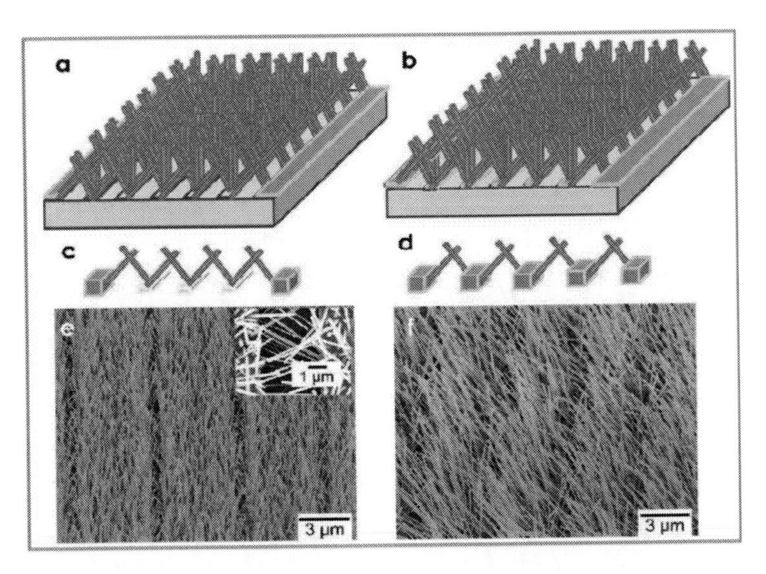

Figure 36. Schematic representation of type I and type II ZnO NW sensors (a, b) and the nanobridges formed between the adjacent lines (c, d). The corresponding SEM images of the NWs taken at a 45° tilt are shown in (e, f). The inset (e) shows an enlarged image of the ZnO nanobridges.

We demonstrated two new approaches for surface patterning of ZnO NWs based on selective deposition of ZA-derived seeds. The two approaches, namely, topographical confinement using a photoresist pattern and selective chemisorption on SAM modified gold electrodes, show high potential toward fabricating NW-based sensor materials with controlled location, size, and distribution. The NWs were grown selectively in the regions defined by designed pattern and features. The realized ZnO seed layers stabilized by strong gold-thiol and carboxylate-zinc species interactions efficiently protect the underneath Au electrodes from disintegration or collapse during high-temperature NW growth. Similar to this method, a number of specific and selective substrate-adsorbate interactions can be envisioned for the generation of nanostructures derived from a wide range of seed layer materials in the future. The SAM-modified Au electrode-based sensor type II showed an enhanced sensitivity in comparison with that of the photoresist-based sensor type I, which can be attributed to a series connection of the multi MSM devices based on millions of nanobridging ZnO NWs and the underneath Au electrodes. These results clearly demonstrate that the direct integration of

NWs on top of the electrodes is possible, ensures better conductance properties, and is thereby advantageous for fabricating sensor configuration.

CONCLUSIONS AND FUTURE SCOPE

NWs of different semiconducting oxides like ZnO, SnO_2, CuO and heterostructures have been grown easily using both physical as well as chemical methods. Techniques like thermal evaporation, vapor phase deposition and hydrothermal growth were used effectively to realize these nanostructures. Proper electrode contacts were realized using dielectrophoresis technique. Sensing properties of the nanostructures were investigated in detail employing various configurations.

ZnO NWs:

- Drop casted Mat-type thin films detected H_2S and NO selectively at room temperature with fast response and recovery times. For example, typical response and recovery times towards 10 ppm of H_2S were 250 and 700 s and towards NO were 250 and 150 s, respectively.
- Hydrothermally grown NW film detected H_2S with faster response and recovery time of 13 and 78 s, respectively towards 20 ppm at an operating temperature of 350°C.
- Cu incorporation over NW surface and surface modification using chemical self organization of Cu species has been demonstrated to improve both the sensitivity and selectivity towards H_2S.
- ZnO NW and polymer composites resulted in an improved selectivity towards gases. For example, the composite films of ZnO: P3HT with 1:1 composition exhibited a highly selective response towards NO_2, while the films of ZnO: PPy containing 50 wt.% NWs exhibited a highly selective response towards Cl_2 at room temperature.

SnO_2 NWs:

- Single NWs could detect H_2S reversibly with response and recovery time of 170 and 550 s, respectively.
- Mat-type sensors could detect 100 ppb of Cl_2 and 1 ppm of H_2S at an operating temperature of 150°C.

CuO NWs:

- Single NW sensors exhibited good sensitivity towards both H_2S (~200% for 10 ppm) and NO (~200% for 200 ppm).
- Mat-type NW films were sensitive to H_2S concentrations, as low as 200 ppb, with maximum sensitivity of 150% and fast response (~1 min.) and recovery (~4 min.) times at 200°C.

Our results clearly illustrate that NWs are the potential candidates for the realization of next generation of sensors. However, to realize an ultimate sensor that could be deployed to the commercial market is still a daunting challenge. Various problems are needed to be address before their successful utility. These include; complete control over growth processes and mechanism (crystal quality and aspect ratio), patterning of NWs over a large area (position control), good ohmic contact between NWs and electrode, addressing individual NWs in an array. Attempts were made to answer some of these using ZnO as a model system and the success achieved clearly emphasizes the possibility of the same. Hopes are high that significant practical devices will soon arise from the integration of these structures with conventional microelectronics.

ACKNOWLEDGMENTS

This work is supported by "DAE-SRC Outstanding Research Investigator Award" (2008/21/05-BRNS) and "Prospective Research Funds" (2008/38/02-BRNS) granted to D.K.A. Authors acknowledge Prof. M. Zacharias, Dr. S. Kittitat, Mrs. S. Kailasa Ganapati, Dr. Shovit Bhattacharya, Dr. S. Sen, Mr. S. K. Mishra, Mr. V. Rikka, Ms. C. Jain and Mr. V. Mukund for their support and help with characterization.

REFERENCES

[1] Gross, R.; Tagirov, L.; Editor Nanoscale Devices – Fundamental and Applications, Springer: Dordrecht. (2006).

[2] D.K. Aswal, S.K. Gupta (eds), Science and Technology of Chemiresistive Gas Sensors, Nova Science Publisher, NY, USA, (2007).

[3] Henzie, J.; Barton, J. E.; Stender, C. L.; Odom, T. W. Acc. Chem. Res. 2006, 39, 249.

[4] Wang, Z. L. ACS Nano 2008, 2, 1987. (b) Ramgir, N. S.; Mulla, I. S.; Pillai, V. K. J. Phys. Chem. B 2006, 110, 3995. (c) Ramgir, N. S.; Late, D. J.; Bhise, A. B.; More, M. A.; Mulla, I. S.; Joag, D. S.; Vijayamohanan, K. J. Phys. Chem. B 2006, 110, 18236. (d) Yan, X.; Li, Z.; Zou, C.; Li, S.; Yang, J.; Chen, R.; Han, J.; Gao, W. J. Phys. Chem. C 2010, 114, 1436. (e) Yang, J.; Qiu, Y.; Yang, S. Cry. Grow. Des. 2007, 7, 2562.

[5] Joshi, R. K.; Hu, Q.; Alvi, F.; Joshi, N.; Kumar, A. J. Phys. Chem. C 2009, 113, 16199. (b) Hong, W.-K.; Sohn, J. I.; Hwang, D.-K.; Kwon, S.-S.; Jo, G.; Song, S.; Kim, S.-M.; Ko, H.-J.; Park, S.-Ju.; Welland, M. E.; Lee, T. Nano Lett. 2008, 8, 950.

[6] Ramgir, N. S.; Yang, Y.; Zacharias, M. Small 2010, 6, 1705-1722.

[7] Tomchenko, A. A.; Harmer, G. P.; Marquis, B.T.; Allen, J.W. Sens. Actuators B 2003, 93, 126. (b) Mitzner, K. D.; Sternhagen, J.; Galipeau, D. W. Sens. Actuators B 2003, 93, 92. (c) Wollenstein, J.; Plaza, J.A.; Cane, C.; Min, Y.; Botttner, H.; Tuller, H.L. Sens. Actuators B 2003, 93 350.

[8] Hwang, J. K.; Cho, S.; Seo, E. K.; Myoung, J. M.; Sung, M. M. ACS Appl. Mater. Interfaces 2009, 1, 2843. (b) Cui, J.; Gibson, U. J. J. Phys. Chem. B 2005, 109, 22074.

[9] Wang, Z. L. Appl. Phys. A 2007, 88, 7–15. (b) Satyanarayana, V.N.T.; Kuchibhaltla, A.S.; Bera, K. D.; Seal, S. 2007, 52, 699. (c) Rao, C.N.R.; Deepak, F.L.; Gundiah, G.; Govindaraj, A. Prog. Solid State Chem. 2003, 31, 5. (d) Greene, E. L.; Yuhas, B. D.; Law, M.; Zitoun, D.; Yang, P. Inorg. Chem. 2006, 45, 7535. (e) Law, M.; Goldberger, J.; Yang, P. Ann. Rev. Mater. Res. 2004, 34, 83. (f) Gao, P.-X.; Ding, Y.; Wang, Z. L. Nano Letters 2009, 9, 137.

[10] Kong, X. Y.; Ding, Y.; Yang, R.; Wang, Z. L. Science 2004, 303, 1348. (b) Pan, Z. W.; Dai, Z. R.; Wang, Z. L. Science 2001, 291, 1947.
[11] Chang, Y.-C.; Yang, W.-C.; Chang, C.-M.; Hsu, P.-C.; Chen, L.-J.; Cry. Grow. Des. 2009, 9, 3161. (b) Liu, B.; Zeng, H. C.; Langmuir 2004, 20, 4196.
[12] Sysoev, V. V.; Goschnick, J. ; Schneider, T.; Strelcov, E.; Kolmakov, A. Nano Lett. 2007, 7, 3182. (b) Gao, F.; Mukherjee, S.; Cui, Q.; Gu, Z. J. Phys. Chem. C 2009, 113, 9546.
[13] Wu, X.; Kulkarni, J. S.; Collins, G.; Petkov, N.; Alm ́ecija, D.; Boland, J. J.; Erts, D.; Holmes, J. D. Chem. Mater. 2008, 20, 5954.
[14] Haghiri-Gosnet, A. M.; Vieu, C.; Simon, G.; Mejıas, M.; Carcenac, F.; Launois, H. J. Phys. 1999, 9, 133. (b) Marrian, C. R. K.; Tennant, D. M. J. Vac. Sci. Technol. A 2003, 21, 15.
[15] Chen, J.; Wiley, B. J. ; Xia, Y. Langmuir 2007, 23, 4120. (b) Schmidt, V.; Wittemann, J. V.; Gosele, U.; Chem. Rev. 2010, 110, 361. (c) Hochbaum, A. I.; Yang, P. Chem. Rev., 2010, 110, 527.
[16] Dang, H.Y.; Wang, J.; Fan, S.S.; Nanotechnology 2003, 14, 738.
[17] Kaur, M.; Bhattacharya, S.; Roy, M.; Deshpande, S. K.; Sharma, P.; Gupta, S. K.; Yakhmi, J. V. Appl. Phys. A 2007, 87, 91-96.
[18] Ramgir, N. S.; Ghosh, M.; Veerender, P.; Datta, N.; Kaur, M.; Aswal, D. K.; Gupta, S. K. Sens. Actuators B 2011, in press.
[19] Pacholski, C.; Kornowski, A.; Weller, H. Angew. Chem .Int. Ed. 2002, 41, 1188-1191.
[20] Greene, L. E.; Yuhas, B. D.; Law, M.; Zitoun, D.; Yang, P. Inorg. Chem. 2006, 45, 7535-7543.
[21] Vayssieres, L. Adv. Mater. 2003, 15, 464-466.
[22] Govender, K.; Boyle, D. S.; Kenway, P. B.; O'Brien, P. J. Mater. Chem. 2004, 14, 2575-2591.
[23] Wang, Z.; Qian, X. F.; Yin, J.; Zhu, Z. K. Langmuir 2004, 20, 3441-3448.
[24] Kaur, M.; Chauhan, S. V. S.; Sinha, S.; Bharti, M.; Mohan, R.; Gupta, S. K.; Yakhmi, J. V. J. Nanoscience Nanotechnology 2009, 9, 5293.
[25] Lao, C. S.; Liu, J.; Gao, P.; Zhang, L.; Davidovic, D.; Tummala, R. Nano. Lett. 2006, 6, 263.
[26] Suehiro, J.; Nakagawa, N.; Hidaka, S.-I.; Ueda, M.; Imasaka, K.; Higashihata, M. Nanotechnology 2006, 17, 2567.
[27] Wang, D.; Zhu, R.; Zhou, Z.; Ye, X. Appl. Phys. Lett. 2007, 90, 103110.
[28] Sun, Z. -P.; Liu, L.; Zhang, L.; Jia, D.-Z. Nanotechnology 2006, 17, 2266. (b) Zhang, H.; Wu, J.; Zhai, C.; Du, N.; Ma, X.; Yang, D. Nanotechnology 18 455604 2007. (c) Jing, Z.; Zhan, J. Adv. Mater. 2008, 20, 4547. (d) Kar, S.; Pal, B. N.; Chaudhuri, S.; Chakravorty, D.J. Phys. Chem. B 2006, 110, 4605.
[29] Baratto, C.; Sberveglieri, G.; Onischuk, A.; Caruso, B.; di Stasio, S. Sens. Actuators B 2004, 100, 261. (b) Zhang, J.; Wang, S.; Wang, Y.; Xu, M.; Xia, H.; Zhang, S.; Huang, W.; Guo, X.; Wu, S. Sens. Actuators B 2009, 139, 411. (c) Ahn, M.-W.; Park, K.-S.; Heo, J.-H.; Park, J.-G.; Kim, D.-W.; Choi, K. J.; Lee, J.-H.; Hong, S.-H. Appl. Phys. Lett. 2008, 93, 263103.
[30] Liao, L.; Lu, H. B.; Li, J. C.; He, H.; Wang, D. F.; Fu, D. J.; Liu, C.; Zhang, W.F. J. Phys. Chem. C 2007, 111, 1900.

[31] Ramgir, N. S.; Rikka, V.; Kaur, M.; Kailasa Ganapathi, S.; Mishra, S. K.; Datta, N.; Aswal, D. K.; Gupta, S. K.; Yakhmi, J. V. CP1313, International Conference on Physics of Emerging Functional Materials (PEFM-2010) Ed. D. K. Aswal, A. K. Debnath, 2010, 322-324.
[32] Wang, C.; Chu, X.; Wu, M. Sens. Actuators B 2006, 113, 320–323.
[33] Ahn, M.-W.; Park, K.-S.; Heo, J.-H.; Park, J.-G.; Kim, D.-W.; Choi, K. J.; Lee, J.-H.; Hong, S.-H. Appl. Phys. Lett. 2008, 93, 263103.
[34] Sen, S.; Bhandarkar, V.; Muthe, K. P.; Roy, M.; Deshpande, S. K.; Aiyer, R. C.; Gupta, S. K.; Yakhmi, J. V.; Sahni, V. C. Sens. Actuators B 2006, 115, 270-275.
[35] Yamazoe, N.; Shimanoe, K. Sens. Actuators B 2008, 128, 566–573.
[36] Datta, N.; Ramgir, N. S.; Kaur, M.; Aswal, D. K.; Gupta, S. K. 2011, manuscript under preparation.
[37] Saxena, V.; Aswal, D. K.; Kaur, M.; Koiry, S. P.; Gupta, S. K.; Yakhmi, J. V.; Kshirsagar, R. K.; Deshpande, S. K. Appl. Phys. Lett. 2007, 90, 043516.
[38] Turgeman, R.; Gershevitz, O.; Deutsch, M.; Ocko, B. M.; Gedanken, A.; Sukenik, C. N. Chem. Mater. 2005, 17, 5048.
[39] Lyu, S. C.; Zhang, Y.; Ruh, H.; Lee, H. J.; Shim, H. W.; Suh, E. K.; Lee, C. J. Chem. Phys. Lett. 2002, 363, 134.
[40] Briggs, D.; Seah, M. P. Editors, *Practical Surface Analysis by Auger and X-ray Photoelectron Spectroscopy,* Wiley, New York 359, 1983.
[41] Wang, Z. L.; Kong, X. Y.; Ding, Y.; Gao, P.; Hughes, W. L.; Yang, R.; Zhang, Y. Adv. Funct. Mater. 2004, 14, 943.
[42] Cotton, F.A.; Wilkinson, G.; Murillo, C.; Bochman, M. Adv. Inorg. Chem. 6th ed., John Wiley and Sons (Asia) Pvt. Ltd., 2003, p. 550.
[43] Dawson, D. H.; Williams, D. E. J. Mater. Chem. 1996, 6, 409. (b) C.F.; Gurnham, Edior, H. W. Gehm, in: *Industrial Waste Water Control*, Academic Press, New York, p 357 1965.
[44] Joshi, A.; Aswal, D. K.; Gupta, S. K.; Yakhmi, J. V.; Gangal, S. A. Appl. Phys. Lett. 2009, 94, 103115.
[45] Singh, A., Joshi, A.; Samanta, S.; Debnath, A. K.; Aswal, D. K.; Gupta, S. K.; Yakhmi, J. V.; Appl. Phys. Lett. 2009, 95, 202106.
[46] Baughman, R.H.; Shacklette, L.W. Phys. Rev B 1989, 39, 5872.
[47] Hakansson, E.; Lin, T.; Wang, H.; Kaynak, A. Synth. Met. 2006, 156, 1194.
[48] Vasquez, M.; Cruz, G.J.; Olayo, M.G.; Timoshita, T.; Morales, J.; Olayo, R. Polymer 2006, 47, 7864.
[49] Kang, E.T.; Neoh, K.G.; Tan, K.L. Adv. Polm. Sci. 1993, 106, 135.
[50] Niranjan R.S.; Sainkar, S. R.; Vijayamohanan, K.; Mulla, I. S. Sens. Actuators B 2002, 82, 82-88.
[51] Ramgir, N. S.; Mulla, I. S.; Vijayamohanan, K. P. J. Phys. Chem. B 2005, 109, 12297-12303.
[52] Ramgir, N.; Sen, S.; Kaur, M.; Mishra, S. K.; Rikka, V.; Choukikar, R.; Muthe, K. P. Asian J. Phys. 2010, 19, 25-30.
[53] Sen, S.; Bhandarkar, V.; Muthe, K.P.; Roy, M.; Deshpande, S.K.; Aiyer, R.C.; Gupta, S.K.; Yakhmi, J.V.; Sahni, V.C. *Sens. Actuators B* 2006, 115, 270-275.
[54] Kumar, V.; Sen, S.; Muthe, K.P.; Gaur, N. K.; Gupta, S.K.; Yakhmi, J.V. Sensors and Actuators B 2009, 138, 587–590.

[55] Sen, S.; Kanitkar, P.; Sharma, A.; Muthe, K.P.; Rath, A.; Deshpande, S.K.; Kaur, M.; Aiyer, R.C.; Gupta,S.K.; Yakhmi, J.V. Sens. Actuators B 2010, 147, 453.
[56] Rakhshni, A. E. Solid State Electron. 1986, 29, 7.
[57] Morales, J.; Sanchez, L.; Martın, F.; Ramos-Barrado, J.R.; Sanchez, M.; Thin Solid Films 2005, 474, 133.
[58] Ramgir, N. S.; Kailasa Ganapathi, S.; Kaur, M.; Datta, N.; Muthe, K. P.; Aswal, D. K.; Gupta, S. K.; Yakhmi, J. V. Sens. Actuators B 2010, 151, 90-96.
[59] Cruccolini, A.; Narducci, R.; Palombari, R.; Sens. Actuators B 2004, 98, 227.
[60] Fan, H.; Yang, L.; Hua, W.; Wu, X.; Wu, Z.; Xie, S.; Zou, B. Nanotechnology 2004, 15, 37.
[61] Ponce, A.A.; Klabunde, K.J. J. Mol. Catal. A 2005, 225, 1.
[62] Hsieh, C.-T.; Chen, J.-M.; Lin, H.-H.; Shih, H.-C. Appl. Phys. Lett. 2003, 83, 3383.
[63] Xu, C.; Liu, Y.; Xu, G.; Wang, G. Mater. Res. Bull. 2002, 37, 2365.
[64] Chen, D.; Shen, G.; Tang, K.; Qian, Y. J. Crystal Growth 2003, 254, 225.
[65] Cao, M.; Hsieh, C.-T.; Chen, J.-M.; Lin, H.-H.; Shih, H.-C. Appl. Phys. Lett. 2003, 82, 3316.
[66] Hu, C.; Wang, Y.; Guo, Y.; Guo, C.; Wang, E. Chem. Comm. 2003, 15, 1884.
[67] Zhu, Y.W.; Yu, T.; Cheong, F.C.; Xu, X.J.; Lim, C.T.; Tan, V.B.C.; Thong, J.T.L.; Sow, C.H. Nanotechnology 2005, 16, 88.
[68] Kumar, A.; Srivastava, A.K.; Tiwari, P.; Nandedkar, R.V. J. Phys.: Condens. Matter 2003, 16, 8531.
[69] Kaur, M.; Muthe, K.P.; Despande, S.K.; Choudhury, S.; Singh, J.B.; Verma, N.; Gupta, S.K.; Yakhmi, J.V. J. Cry. Grow. 2006, 289, 670–675.
[70] Kaur, M.; Kailasaganapati, S.; Datta, N.; Muthe, K.P.; Gupta, S.K. Int. J. Nanoscience, 2007, 1, 1–4.
[71] Eshelby, J. D. Phys. Rev. 1953, 91, 755.
[72] Arnold, S.M.; Eloise Koonce S. J. Appl. Phys. 1956, 27, 964.
[73] Kumar, A.; Srivastava, A.K.; Tiwari, P.; Nandedkar, R.V. J. Phys.: Condens. Matter 2003, 16, 8531.
[74] Datta, N.; Kaur, M.; Ramgir, N. S.; Aswal, D.K.; Gupta, S.K. International Symposium of Materials Chemistry (ISMA-2010), 2010, Bhabha Atomic Research Centre, Mumbai.
[75] Sen, S.; Kanitkar, P.; Sharma, A.; Muthe, K.P.; Rath, A.; Deshpande, S.K.; Kaur, M.; Aiyer, R.C.; Gupta, S.K.; Yakhmi, J.V. Sens. Actuators B 2010, 147, 453–460.
[76] Jain, C.; Mukund, V.; Kaur, M.; Kailasa Ganapathi, S.; Ramgir, N. S.; Datta, N. Aswal, D.K.; Gupta, S. K. CP1313, International Conference on Physics of Emerging Functional Materials (PEFM-2010) Ed. D. K. Aswal, A. K. Debnath, 2010, 132-134.
[77] Madhukar, A.; Lu, S.; Konkar, A.; Zhang, Y.; Ho, M.; Hughes, S. M.; Alivisatos, A. P. Nano Lett. (2005)
[78] Rajeshwar, K.; de Tacconi, N. R.; Chenthamarakshan, C. R. Chem. Mater. 2001, 13, 2765.
[79] Zhang, R.-Q.; Liftshitz, Y.; Lee, S. T. Adv. Mat. 2003, 15, 635.
[80] Subannajui, K.; Ramgir, N. S.; Grimm, R.; Yang, Y.; Zacharias, M.; Crys. Grow. Des. 2010, 10, 1585.
[81] Zhang, Y.; Wang, L.; Liu, X.; Yan, Y.; Chen, C.; Zhu, J. J. Phys. Chem. B 2005, 109, 13091.

[82] Wagner, R. S.; Ellis, W. C. Appl. Phys. Lett. 1964, 4, 89.
[83] Levitt, A. P.; Editor Wagner, R. S.; VLS Mechanism of Crystal Growth. In Whisker Technology Wiley: NewYork, 1970.
[84] Burda, C.; Chen, X.; Narayanan, R.; El-Sayed, M. A. Chem. Rev. 2005, 105, 1025.
[85] Zhu, Z.; Chen, T.-L.; Gu, Y.; Warren, J.; Osgood, R. M. Chem. Mater. 2005, 17, 4227.
[86] Hannon, J. B.; Kodambaka, S.; Ross, F. M.; Tromp, R. M. Nature 2006, 440, 69.
[87] Ho, S.-T.; Wang, C.-Y.; Liu, H.-L.; Lin. H.-N. Chem. Phys. Lett. 2008, 463, 141.
[88] Morris, R. J. H.; Dowsett, M. G.; Dalal, S. H.; Baptista, D. L.; Teo, K. B. K.; Milne, W. I. Surf. Interface Anal. 2007, 39, 898.
[89] Fan, H. J.; Bertram, R.; Dadgar, A.; Christen, J.; Krost, A.; Zacharias, M.; Nanotechnology 2004, 15, 1401.
[90] Ramgir, N. S.; Subannajui, K.; Grimm, R.; Yang, Y.; Micheiles, R.; Zacharias, M. J. Phys. Chem. C 2010, 114, 10323.
[91] Osborne, J. M.; Rankin, W. J.; Mccarthy, D. J.; Swinbourne, D. R. Metull. Mater. Trans. B 2001, 32B, 37.
[92] Fan, H. J.; Lee, W.; Hauschild, R.; Alexe, M.; Rhun, G. L.; Scholz, R.; Dadgar, A.; Nielsch, K.; Kalt, H.; Krost, A.; Zacharias, M.; Gösele, U. Small 2006, 2, 561.
[93] Ohtani, H.; Ishimura, S.; Kumai, M. Anal. Sci. 2008, 24, 1335.
[94] Yanagida, T.; Marcu, A.; Matsui, H.; Nagashima, K.; Oka, K.; Yokota, K.; Taniguchi, M.; Kawai, T. J. Phys. Chem. C, 2008, 112, 18923.
[95] Banerjee, D.; Jo, S. H.; Rem, Z. F. Adv. Mater. 2002, 22, 2028.
[96] Cheng, C.; Lei, M.; Feng, L.; Wong, T. L.; Ho, K. M.; Fung, K. K.; Loy, M. M. T.; Yu, D.; Wang, N. ACS Nano, 2009, 3, 53.
[97] Okamoto, H.; Massalski, B. T. Bull. Alloy Phase Diagrams, 1989, 10, 59.
[98] Bae, S. Y.; Seo, H. W.; Park, J. J. Phys. Chem. B 2004, 108, 5206.
[99] Zhang, Y.; Lu, F.; Wang, Z.; Zhang, L. J. Phys. Chem. C 2007, 111, 4519.
[100] Kuo, T.-J.; Lin, C.-N.; Kuo, C.-L.; Huang, M. H. Chem. Mater. 2007, 19, 5143.
[101] Yuon, S. K.; Ramgir, N. S.; Wang, C.; Subannajui, K.; Cimalla, V.; Zacharias, M.; J. Phys. Chem. C 2010, 114, 10092.
[102] Ahn, M.W.; Park, K.S.; Heo, J. H.; Kim, D.W.; Choi, K.J.; Park, J.G. Sens. Actuators B., 2009, 138, 168.

In: Nanowires: Properties, Synthesis and Applications
Editor: Vincent Lefevre
ISBN: 978-1-61470-129-3

Chapter 3

TRANSPORT PROPERTIES OF NANOSTRUCTURED MATERIALS

K. K. Choudhary [†]

Department of Physics, Shri Vaishnav Institute of Technology and Science, Baroli, Indore, India

Abstract

This review focuses on the transport properties of nanostructured materials. The complete theoretical understanding of transport properties such as electrical resistivity, thermal conductivity and thermoelectric power is presented in this article. Resistivity in metallic phase of Zn composite nanowires is analyzed within the framework of Bloch-Gruneisen (BG) model of resistivity as well the effect of electron-electron scattering is evaluated. The resistivity in Semiconducting phase of Zn composite nanowires is discussed with small polaron conduction (SPC) model. Mott expression is used to generate the electron diffusive thermoelectric power ($S_c^{diff.}$) and phonon drag thermoelectric power (S_{ph}^{drag}) is calculated within the relaxation time approximation when thermoelectric power is limited by the scattering of phonons with impurities, grain boundaries, charge careers and phonons in the nanowires. The thermal conductivity (κ) and S_{ph}^{drag} shows anomalous temperature dependent behaviour, which is an artifact of various operating scattering mechanisms. The anomalies are well accounted in terms of interaction among the phonons-impurity, phonon-grain boundaries, phonon-electron and the umklapp scattering. Furthermore, the effect of embedding nanoparticles on thermal conductivity of crystalline semiconductors is presented under phonon scattering mechanism. Numerical results obtained from the theoretical analysis are also compared with experimental results.

Keywords: nanocomposite materials, thermoelectric power, Thermal Conductivity; electrical resistivity.

[†] E-mail address: kkchoudhary1@yahoo.com

1. INTRODUCTION

In recent years, much attention has been paid to the preparation and characterization of one-dimensional (1D) nanomaterials because of their great potential to test fundamental quantum mechanic concepts [1, 2]. In many applications, the sensitivity or efficiency is proportional to the surface area, nanorods or nanowires offer a larger surface area per unit mass compared to that of films or the bulk material. Electrical properties of nanowires strongly depend on temperature and the diameter of the nanowires [3-6]. Zn nanowires also show metal to semiconductor transition when its diameter reduced from 15 nm to 4 nm [7]. The localization effects have been observed in Bi and Sb nanowire systems [8-10] and are expected to increase the electrical resistance. Enhancement of the thermoelectric figure of merit by embedding nanoparticles in the crystal [8, 11] opens the possibility of novel efficient thermoelectric materials, in many applications, the sensitivity or efficiency is proportional to the surface area. Nanorods or nanowires offer a larger surface area per unit mass compared to that of films or the bulk material.

J.P. Heremans *et al.* reported that the resistance of the 15 nm Zn/SiO2 composite shows the metallic face, it is linear with temperature above 40 K and saturates below 10 K, with a residual resistivity ratio of 13. However, the resistance of the 4 nm Zn wires follows essentially a $T^{-1/2}$ law i. e. the semiconducting phase [7]. This diameter dependent resistivity transition from semiconducting phase to metallic phase is presented in figure 1 [7]. Similarly the nature of resistivity also strongly depends on temperature, temperature-dependent resistivity $\rho(T)$ of ZnO nanostructures shows semiconducting phase in low temperature range (140 K $<$ T $<$ 180 K), shows an absolute minimum near 180K and increases linearly with T at high temperatures (200 K $<$ T $<$ 300 K) [12]. The metallic conduction at higher temperature is an indication of electron-electron and electron-phonon scattering mechanism of resistivity where the semiconducting behavior at low temperature shows strong possibility of small polaron conduction in these materials.

The resistivity of single-wall carbon nanotubes have been reported by Kane *et al.* they found it exhibit metallic behavior with an intrinsic resistivity which increases approximately linearly with temperature over a wide temperature range [13]. This behaviour of resistivity of nanotubes is well explained by a one-dimensional theory of the scattering of electrons by twistons which predicts an intrinsic resistivity proportional to the absolute temperature. Temperature dependence of the resistance of metallic nanowires of diameter ≥15 nm is also explained successfully using the Bloch-Grüneisen theorem by Bid et al. [14]. It is well known that electron–phonon scattering plays a key role in the electrical transport [15].

Nanostructures also offer the opportunity to study the physical properties of one-dimensional structures. Zinc oxide with hexagonal structure belongs to C^4_{6v} space group. The Raman peaks at 378 and 437 cm^{-1} are attributed to the A1 mode and the E2 mode of ZnO, respectively. The broadening asymmetry Raman peak at 437 cm^{-1} is typical of ZnO Raman active branches, which is also one of the characteristics of ZnO nanoparticles [16].

The electrical and field emission measurements were carried out [17] to study the correlation between resistance and field emission performance of individual one-dimensional ZnO nanostructures. Their results indicate that, besides the uniformity in the geometrical structure, the uniformity in electrical resistance of the emitters in an array should be ensured, in order to meet the requirement of device application. The conductivity of the ZnO nanowires is extremely sensitive to ultraviolet light exposure. The light-induced conductivity increase that allows us to produce the switches using the nanowires [18-20], above phenomena directly related with electrical resistivity behavior of ZnO nanowires.

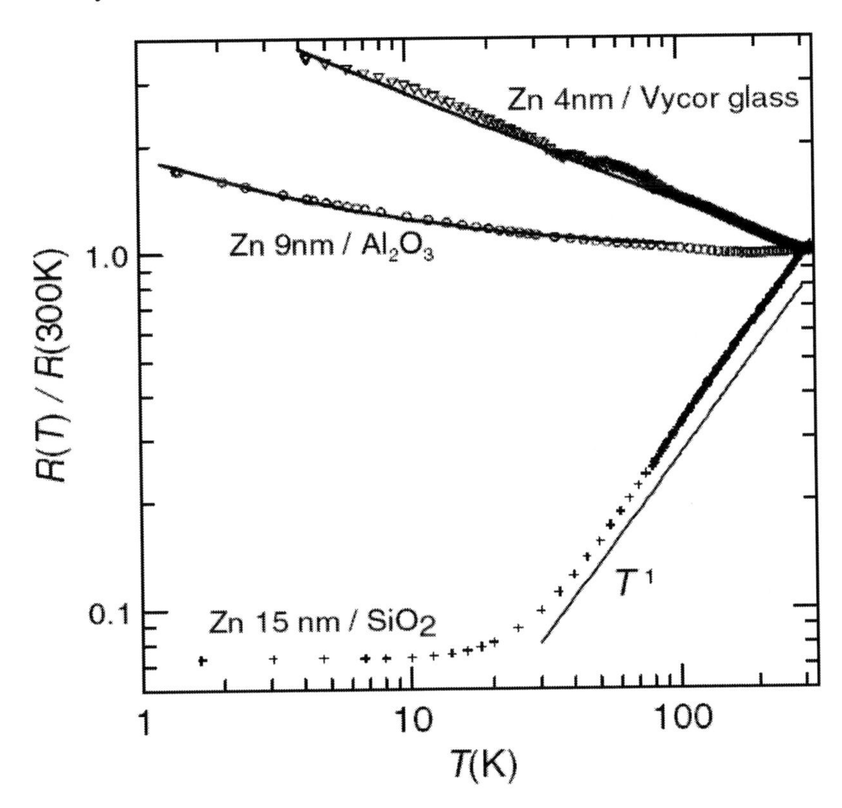

Figure 1. Temperature dependence of the resistance of Zn nanowire composites, normalized to the resistance at 300 K. A typical metallic behavior is seen in the Zn/SiO2 composite, with average wire diameters of 15 nm, while semiconducting behavior is seen for 4 nm and 9 nm wire diameters. Reprinted figure with permission from J. P. Heremans, C. M. Thrush, D. T. Morelli, and M. C. Wu, *Phys. Rev. Lett.* **91,** 076804 (2003). Copyright (©2003) by the American Physical Society.

The one-dimensional structure, characteristic electrical properties and its effect on technological applications of ZnO nanowires provides a good candidate for future nano-electronic devices [21]. For nanowires with diameters approaching molecular dimensions, the wavelength of the de-Broglie wave associated with the conducting electron becomes grater then size of nanowire, thus, the transport is likely to be quantum in nature. A proper understanding of the resistivity in this regime is needed because it allows one to get a quantitative estimate of the resistance of the wire.

The electrical transport is especially interesting because of their potential applications in future nano-electronic devices. Despite its technological importance, the electrical property of ZnO nanowires is not well understood and the resistivity behavior needs to be investigated systematically. It is required to have a comprehensive theoretical mechanism based on electron–phonon, electron-electron and small polaron interactions to understand the physical properties of nano-structures. The above facts triggered the theoretical analysis on the temperature dependent behaviours of resistivity in ZnO nanostructures. This review focuses on the issue of the electron-phonon scattering, electron-electron scattering and small polaron conduction to the resistivity of ZnO nanowires that explain the resistivity behavior in wide temperature range.

For metals to be good thermoelectric materials, they require a room temperature Seebeck coefficient exceeding 100 to 150 μV/K. A very large increase of the thermoelectric power of 4 nm Zn nanowires is reported which saturates at 130 μV/K [7], this effect is ascribed to one-dimensional (1D) size quantization. Figure 2 shows the temperature dependent Seebeck coefficient data for 4 nm Zn/Vycor glass and 9 nm Zn/Al_2O_3 along with the bulk material which clearly indicates the dimeter dependence of the thermoelectric power [7]. The peaks in the electronic density of states that are the signature of a 1D system result in an increase in thermopower. Experimental data for thermoelectric power *S*(*T*) of Zn nanowire composites shoes anomalous temperature dependent behaviour. The thermoelectric power is a powerful probe to understand not only the electronic structure but also shed light on the electron-phonon interaction in nanostructures.

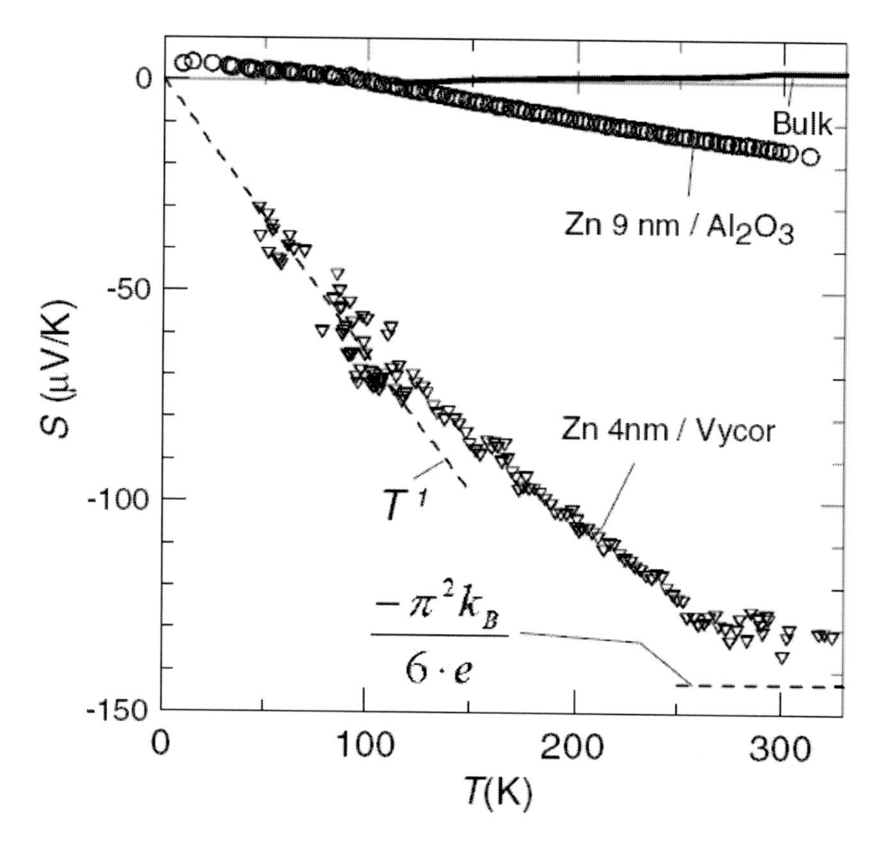

Figure 2. Temperature dependence of the thermoelectric power of the 9 nm Zn=Al2O3 and of the 4 nm Zn=Vycor glass nanocomposites along with bulk Zn. Reprinted figure with permission from J. P. Heremans, C. M. Thrush, D. T. Morelli, and M. C. Wu, *Phys. Rev. Lett.* **91,** 076804 (2003). Copyright (©2003) by the American Physical Society.

It has presented by various authors that the thermoelectric figure of merit *ZT* ($=S^2\sigma T/\kappa$) can be increased beyond unity by nanostructuring thermoelectric materials and the key reason for increase in *ZT* was the reduction of thermal conductivity and increase in thermoelectric power [22-24]. The fundamental reasons for how and why nanostructuring reduces thermal conductivity and increase the thermoelectric power in crystalline materials was not fully understood.

Kim *et al.* have reported that the atomic substitution in alloys can efficiently scatter phonons, they demonstrate that the thermal conductivity reduction by almost a factor of 2 below the alloy limit and a corresponding increase in the thermoelectric figure of merit by a factor of 2, by embedding ErAs nanoparticles in $In_{0.53}Ga_{0.47}As$ crystal [25]. Alloying is known to be the best solution for tuning thermal transport in a crystalline solid. For example, thermal conductivity of silicon and germanium [26] at room temperature is around 140 $Wm^{-1}K^{-1}$, and 60 $Wm^{-1}K^{-1}$, respectively. However, thermal conductivity of Si_xGe_{1-x} alloys [27] can be as low as 5-10 $Wm^{-1}K^{-1}$ at room temperature. In such alloys, atomic substitutions scatter phonons due to differences in mass and/or bond stiffness. The effect of ErAs deposition, monolayer (ML) thickness on the thermal conductivity of $ErAs/In_{0.53}Ga_{0.47}As$ superlattice is presented in figure 3 [25], Which clearly suggest that the thermal conductivity strongly vary with ErAs deposition, it is maximum for smallest ML thickness (0.05) and decrease with increase in ML thickness.

Various efforts to improve the thermoelectric performance of SiGe alloys have also been reported. Addition of GaP, to lower the thermal conductivity [28], is now believed to act as a superior dopant [29, 30]. Theoretical calculations indicating improved figure of merit values for hot-pressed SiGe alloys due to grain boundary scattering of phonons have also been verified [31, 32, 33]. Phonon propagation in the disordered nanostructures at about the Debye temperature or higher temperature is considered by Braginsky *et al.* Scattering at the grain boundaries is assumed to be the main mechanism restricting the thermal conductivity. Influence of the structure the grain size and its dispersion, the intergrain interface structure as well as temperature on the thermal conductivity is discussed [34]. In present article, we review the theoretical studies based on phonon scattering effects and interaction of phonon with increased interfaces on embedding nanoparticles in crystalline materials.

2. RESISTIVITY

The Zinc nanowire has a polycrystalline structure with grain size ranging from the wire diameter up to a few hundred nanometers [35]. Zinc is a direct-band gap semi conducting materials with an energy gap of 3.3ev at room temperature. It has higher exciton binding energy. In the test material, we anticipate that both the acoustic and the optical phonons participate in the process of electrical conduction.

The resistivity of metallic 15nm Zn/SiO_2 composite is calculated within the framework of electron–phonon scattering mechanism i. e. Bloch-Grüneisen model of resistivity [36]. The contributions to the resistivity by inherent acoustic phonons (ω_{ac}) as well as high frequency optical phonons (ω_{op}) were estimated. Estimated contribution to resistivity by considering both phonons i.e., ω_{ac} and ω_{op}, along with the zero limited resistivity is found, which is significantly lower then the experimental data. Thus the difference is obtained and the possibility of electron-electron scattering and scattering of electrons by twistons is assessed. The resistivity of Semiconducting 9nm Zn/Vycor glass nanowires is discussed with small polaron conduction (SPC) model [36].

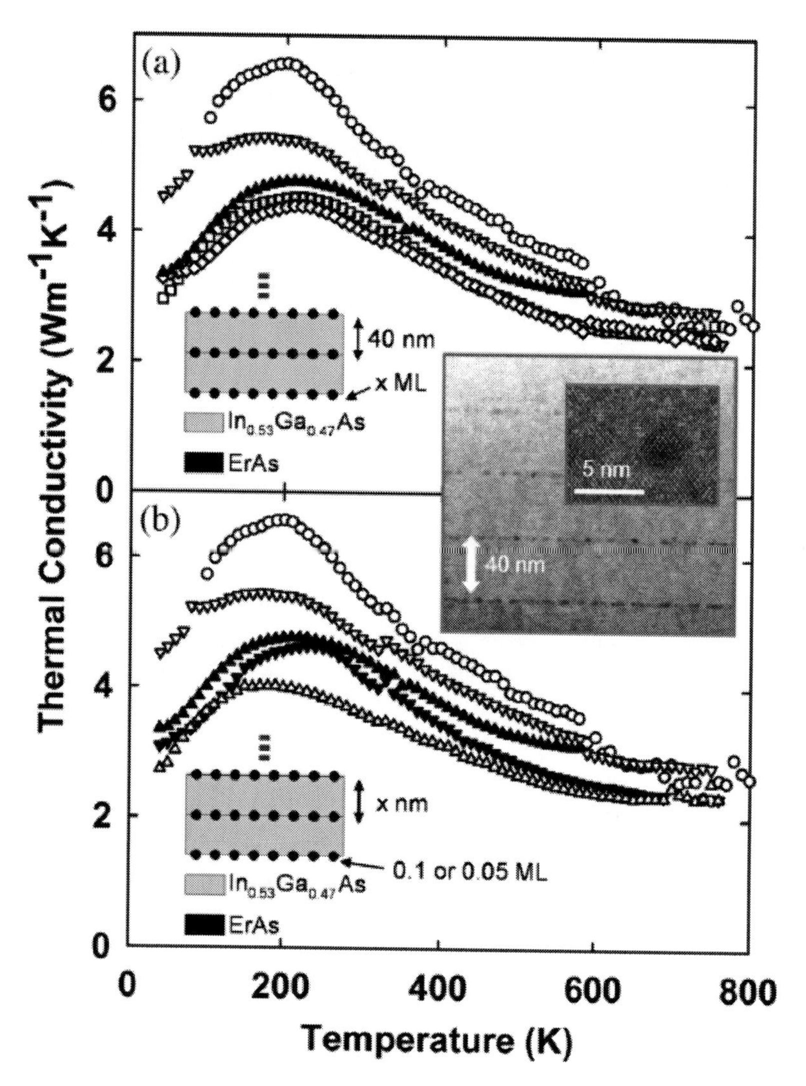

Figure 3. The effect of ML thickness on the thermal conductivity of ErAs/In0:53Ga0:47As superlattice. The thermal conductivity of $In_{0.53}Ga_{0.47}As$ (open circles) is shown as a comparison. ErAs deposition is varied from 0.05 ML (open downward triangles), 0.1 ML (solid upward triangles), 0.2 ML (open diamonds), to 0.4 ML (open squares). The inset shows TEM pictures of 0.4 ML with a 40 nm period thickness ErAs/$In_{0.53}Ga_{0.47}As$ superlattice. Reprinted figure with permission from Woochul Kim, Joshua Zide, Arthur Gossard, Dmitri Klenov, Susanne Stemmer, Ali Shakouri, and Arun Majumdar *Phys. Rev. Lett.* **96,** 045901 (2006). Copyright (©2006) by the American Physical Society.

2.1. Electron-phonon mechanism of resistivity for metallic nanowires.

To formulate a specific model, we start with the general expression for the temperature dependent part of the resistivity, given by

$$\rho = \frac{3\pi}{\hbar e^2 v_F^2} \int_0^{2k_f} |v(q)|^2 \left\langle |S(q)|^2 \right\rangle \left(\frac{1}{2k_F}\right)^4 q^3 dq. \quad (1)$$

v (q) is the Fourier transform of the potential associated with nanowire and S(q) being the structure factor and following the Debye model it takes the following form

$$|S(q))|^2 \approx \frac{k_B T}{M v_s^2} f(\hbar\omega / k_B T) \quad (2)$$

$$f(x) = x^2 [e^x - 1]^{-1} [1 - e^{-x}]^{-1} \quad (3)$$

f(x) represents the statistical factor. Thus the resistivity expression leads to

$$\rho \approx \left(\frac{3}{\hbar e^2 v_F^2}\right) \frac{k_B T}{M v_s^2} \int_0^{2k_F} |v(q)|^2 \left[\frac{(\hbar\omega / k_B T)^2 q^3 dq}{(\exp(\hbar\omega / k_B T) - 1)(1 - \exp(-\hbar\omega / k_B T))}\right]. \quad (4)$$

v_s being the sound velocity. Acoustic phonon contribution yields the Bloch-Gruneisen function of temperature dependence resistivity:

$$\rho_{ac}(T, \theta_D) = 4A_{ac}(T / \theta_D)^4 \times T \int_0^{\theta_D / T} x^5 (e^x - 1)^{-1} (1 - e^{-x})^{-1} dx. \quad (5)$$

Where, $x = \hbar\omega / k_B T$. A_{ac} is being a constant of proportionality defined as

$$A_{ac} \cong \frac{3\pi^2 e^2 k_B}{k_F^2 v_s^2 L \hbar v_F^2 M} \quad (6)$$

The phonon spectrum can be conveniently separated into two parts of phonon density of states. Therefore it is natural to choose a model phonon spectrum consisting of two parts: an acoustic Debye branch characterized by the Debye temperature Θ_D and an optical mode of vibration defined by the Einstein temperature Θ_E. In case of the Einstein type of phonon spectrum (an optical mode) ρ_{op} (T) may be described as follows

$$\rho_{op}(T, \theta_D) = A_{op} \theta_E^2 T^{-1} [\exp(\theta_E / T) - 1]^{-1} \times [1 - \exp(-\theta_E / T)]^{-1}. \quad (7)$$

A_{op} is defined analogously to equation (6). Finally, the phonon resistivity can be conveniently modelled as

$$\rho_{e-ph}(T) = \rho_{ac}(T, \theta_D) + \rho_{op}(T, \theta_E) \quad (8)$$

If the Matthiessen rule is obeyed, the total resistivity may be represented as a sum $\rho(T) = \rho_0 + \rho_{e\text{-}ph}(T)$, Finally, the total resistivity is now rewritten as

$$\rho(T,\theta_D,\theta_E) = \rho_0 + \rho_{ac}(T,\theta_D) + \rho_{op}(T,\theta_E)$$
$$= \rho_0 + 4A_{ac}(T/\theta_D)^4 T \times \int_0^{\theta_D/T} x^5 (e^x - 1)^{-1}(1 - e^{-x})^{-1} dx$$
$$+ A_{op}\theta_E^2 T^{-1}[\exp(\theta_E/T) - 1]^{-1}[1 - \exp(-\theta_E/T)]^{-1}. \quad (9)$$

where ρ_0 is the residual resistivity that does not depend on temperature. The zero temperature elastic scattering rate and plasma frequency will allow us to have an independent estimation of zero temperature-limited resistivity. Zero temperature -limited resistivity is expressed as

$$\rho(0) = \frac{4\pi\tau^{*-1}}{\omega_p^2} \quad (10)$$

We estimate the temperature dependent restivity of 15nm Zn/SiO_2 composite nanowire using well known electron-phonon scattering mechanism for metallic behaviour of the nanowires. Use of electron-phonon model of resistivity is justified for the nanowires, the applicability of the Bloch-Grüneisen theorem based on electron-phonon scattering has also been verified by Bid et al. [14] for interpretation of temperature dependence resistance of metallic nanowires of diameter ≥15 nm. The importance of phonons scattering mechanism in nanostructures has also been verified by thermal conductivity analysis in earlier reporting [11]. For numerical calculation of resistivity we use the Debye temperature θ_D = 543 K, Einstein temperature θ_E = 628 K and zero temperature-limited resistivity $\rho_0/\rho(300K) = 0.72$ [7] for 15nm Zn/SiO_2 composite nanowire. Figure 4 shows the results of temperature dependence of resistivity (normalized by resistivity at 300K) via the ordinary electron-phonon interaction from equation (9). The contributions of acoustic and optical phonons towards resistivity are shown separately along with the total resistivity in the inset of Figure 4. It is inferred from the curve that ρ_{ac} increases linearly, while to that ρ_{op} increase exponentially with the increase in temperature. Both the contributions are clubbed and the resultant resistivity is exponential at low temperatures, and nearly linear at high temperatures till room temperature.

It is noticed from the plot that the estimated ρ is lower than the reported data. The difference in between the measured ρ and calculated, $\rho_{diff.}$ [$= \rho_{exp.} - \{\rho_0 + \rho_{e\text{-}ph} (= \rho_{ac} + \rho_{op})\}$] is plotted in Figure 5. A linear temperature dependence of $\rho_{diff.}$ is inferred in a wide temperature range with change in slope at around 200 K. The linear temperature contribution for resistivity is an indication of conventional electron-electron scattering and the scattering of electrons by twistons in the nanowires. The feature of linear temperature dependence of $\rho_{diff.}$ is also consistent with a one-dimensional theory of the scattering of electrons by twistons which predicts an intrinsic resistivity proportional to the absolute temperature [7]. The additional term due to electron-electron and electrons by twistons contribution was required in understanding the resistivity behaviour, as extensive attempts to fit the data with residual resistivity and phonon resistivity were unsuccessful. Here it is required to add some more terms in the model to explain the electron-electron scattering.

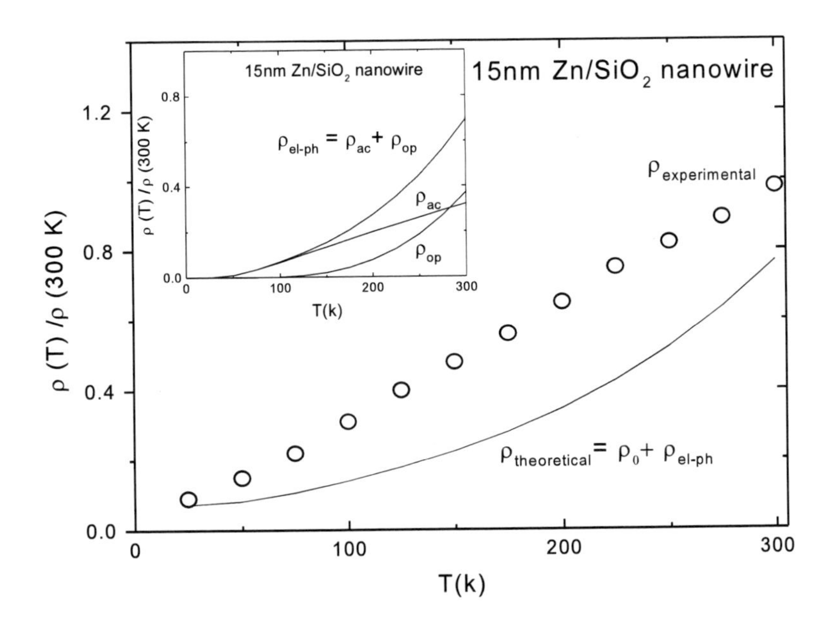

Figure 4. Variation of normalized resistivity [ρ(T)/ ρ(300K)] of 15nm Zn/SiO_2 nanowire with temperature. Open circles are the experimental data taken from Heremans *et al.* 2003, Solid line is the fitting by electron-phonon scattering mechanism. Inset shows the contribution of acoustic phonon ρ_{ac} as well of optical phonon ρ_{op} to the resistivity.

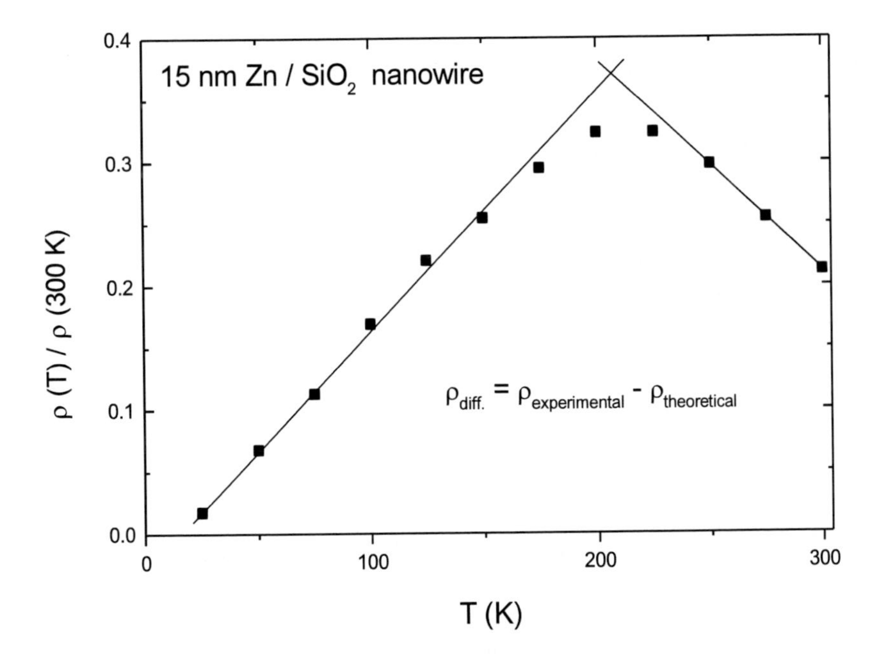

Figure 5. Variation of difference of experimental and theoretical (electron-phonon) resistivity of the 15nm Zn/SiO_2 nanowire with Temperature; Solid line is linear fit to data.

The size dependence on the phonon contribution to the resistivity of a nanowires come from the value of x in equation 5, where $x=\hbar\omega/kBT$. Here $\omega = c/\lambda$, c is the sound velocity averaged over all the acoustic modes and λ is the phonon wavelength. At a given temperature T the diameter dependence to the integral in equation 1 comes from the value of x for which the integrand has a maximum value. The dominant value of x depends on the diameter which decides the dominant value of ω and λ in the test material.

2.2. Small polaron conduction (SPC) model of resistivity for semiconducting nanowires

We shall now switch to a brief description of the temperature dependent resistivity, due to adiabatic Small Polaron Conduction (SPC) model. We must mention that the most rapid motion of a small Polaron occurs when the carrier hops each time the configuration of vibrating atoms in an adjacent site coincides with that in the occupied site. Henceforth, the charge carrier motion within the adiabatic regime is faster than the lattice vibrations and the resistivity for SPC follows:

$$\rho = \rho_{os} T \exp\left(\frac{E_p}{k_B T}\right), \tag{11}$$

E_p is being the Polaron formation energy, k_B is Boltzman constant. The resistivity coefficient ρ_{os} is given by

$$\rho_{os} = \frac{k_B}{ne^2 D}. \tag{12}$$

where n is the charge carrier density ($\sim 10^{18}$ cm^{-3}), e is the electronic charge, and D is the Polaron diffusion constant.

To estimate the semiconducting behavior of resistivity data of Zn/Vycor Glass composite nanowire we have used the adiabatic small Polaron conduction (SPC) model. Keeping in mind that the charge carrier motion is faster than the atomic vibrations in the adiabatic regime and hence the nearest-neighbour hopping of a small Polaron leads to mobility with a thermally activated form. We obtain the best-fitted value of polaron formation energy $E_p = 26$ meV using the diffusion coefficient as $D = 0.232$ cm^2 s^{-1}. The plots of normalized resistivity ρ/ρ(300K) versus T have been fitted with SPC model shown in Figure 6. The SPC model with realistic physical parameters successfully retraces the reported experimental behaviour at temperature T > 100K.

3. Thermoelectric Power

The role of scattering of phonon with defects, grain boundaries, phonon and electrons for phonon drag thermoelectric power (S_{ph}^{drag}) as well the carrier diffusion contribution (S_c^{diff}) is investigated [37]. This study improves our understanding of the interplay of scattering processes between the heat carriers themselves and between the carriers and the impurities for the explanation of the reported behaviour of the thermoelectric power $S(T)$. Also it is important to look for the relative magnitudes of these scattering processes, which lead to the anomalous behavior, and this is our motivation for the present investigation.

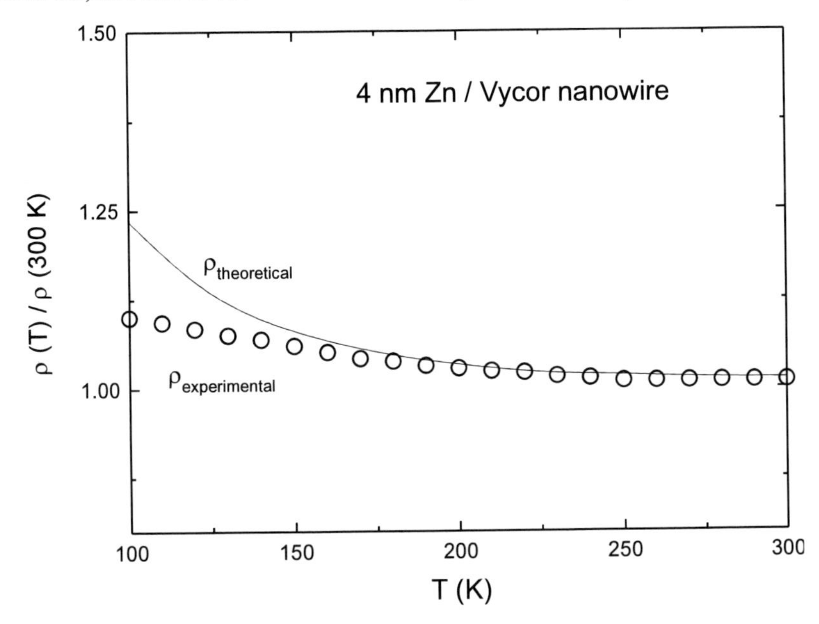

Figure 6. Variation of normalized resistivity [ρ(T)/ρ(300K)] of 4nm Zn/Vycor composite nanowire with temperature. Open circles are the experimental data taken from Heremans *et al.* 2003. Solid line is the fitting by Small Polaron conduction model.

3.1 Carrier Diffusion thermoelectric power

Let us begin with estimation of the carrier diffusion thermoelectric power. We use the well-known Mott formula to estimate the contribution of electrons towards thermoelectric power. The low temperature carrier diffusion thermoelectric power [38] is

$$S_c^{diff.}(T) = -\frac{\pi^2 k_B^2 T}{3|e|}\left[\frac{\partial \ln \sigma(\varepsilon)}{\partial \varepsilon}\right]_{\varepsilon=\varepsilon_F} \tag{13}$$

with σ (ω) [= $ne^2\tau(\varepsilon)/m$] is the energy dependence of electrical conductivity in the relaxation time approximation. Here, $n(m)$ is the density (mass) of carriers and τ the relaxation time. Moreover this expression is true if also the free electron approximation is assumed and not the relaxation time approximation alone. For the sake of simplicity, it is sufficient to neglect the energy dependence in τ (ε) taking τ (ε) = τ (ε_F), for three-dimensional free electron model the equation (13) becomes,

$$S_c^{diff.}(T) = -\frac{\pi^2 k_B^2 T}{6|e|\varepsilon_F} \tag{14}$$

Keeping in mind that $\tau = \ell/v_F$, the method point to the scattering of carriers by impurities is dominant for constant relaxation times. Such a procedure has found success in explaining previous data on nanocomposites on a wide temperature range [39].

The carrier diffusion thermoelectric power ($S_c^{diff.}$) is estimated [37] for 4 nm Zn/Vycor composite where Zn nanowires imbedded in porous host materials, which is given by the expressions (13 and 14). We take the Fermi energy ε_F of Zn nanowires is about 30.7 meV, which is used for the computation of S_c^{dif}. The carrier diffusion contribution to the thermoelectric power S_c^{diff} is documented in figure 7 as a function of temperature. It is evident from the plot that $S_c^{diff.}$ (T) increases linearly for increasing temperature. Estimated $S_c^{diff.}$ has been subtracted from the experimental data [7] and the difference ($S_{difference} = S_{\text{experimental}}$ - S_c^{diff}) has obtained. The difference $S_{difference}$ is characterized as phonon drag thermoelectric power and fitted within the relaxation time approximation in next step.

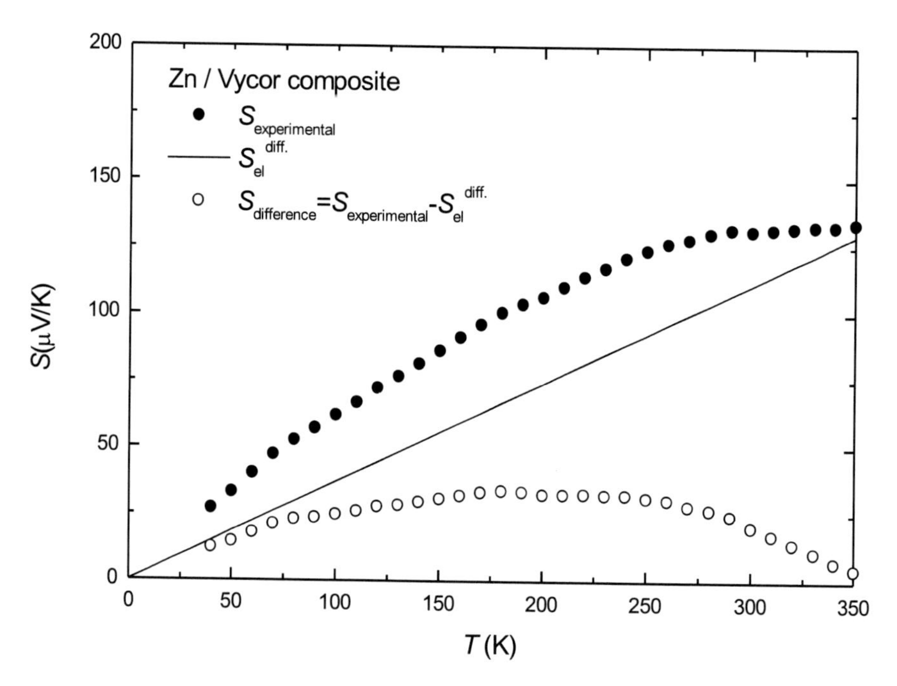

Figure 7. Variation of carrier diffusion thermoelectric power (S_c^{diff}) with temperature. Compared with experimental data, taken from Heremans *et al.* 2003 [5]. Reprinted from *Journal of Physics and Chemistry of Solids*, Vol **71** Issue 1, by K.K. Choudhary, D. Prasad, N. Kaurav, Dinesh Varshney, Entitled: Interpretation of thermoelectric power of Zinc Nanowire Composites: Phonon scattering mechanism, Pages No. 47-50, Copyright (©2010), with permission from Elsevier.

3.2. Phonon Drag thermoelectric power

In order to assess the role of phonon drag thermoelectric power, we use a model Hamiltonian where the low energy interatomic vibrations (phonons) in Zn nanowire are described in the Debye model. The charge carriers are quasiparticles in a periodic arrangement, and hence have a well-defined dispersion relation $\varepsilon_{\mathbf{k}}$. The phonon inelastic scattering events are assumed to be independent and scattering of phonon with various scattering sources are additive. We shall use the isotropic BCS-like model with interatomic acoustic phonon to derive qualitative results for thermoelectric power. The model Hamiltonian is [40]

$$H = \sum_p \varepsilon_p a_p^+ a_p + \sum_q \omega_q b_q^+ b_q + \sum_{p_1, p_2} \varphi(p_1, p_2) a_{p_1}^+ a_{p_2}$$
$$+ D_p \sum_{p,q} q \left[\frac{\hbar}{2\rho\omega_q}\right]^{1/2} a_{p+q}^+ a_p (b_p + b_{-q}^+) \tag{15}$$
$$+ \frac{R}{2n} \sum_{q_1, q_2} e^{i(q_1+q_2)R_i} \left[\frac{\hbar\omega_{q_1}\hbar\omega_{q_2}}{4}\right]^{1/2} (b_{q_1} - b_{-q_1}^+)(b_{q_2} - b_{-q_2}^+) + H_{p-p}.$$

The initial two terms are carrier (electron), and phonon excitation, the third and fourth terms represent carrier-impurity and carrier-phonon interactions, respectively. The fifth term is phonon-impurity interaction and last term stands for the phonon-phonon interaction. The symbols appeared in equation (15) are: ε_p the carrier free energy, a (a^+) and b (b^+) are the creation (annihilation) operators for phonon and electron, φ is coupling parameter of electron and impurity potential, D_p is deformation-potential constant, ρ is ionic mass density, ω_q is acoustic phonon frequency of a wave vector **q**, R is relative ionic-mass difference $[(M''-M)/M'']$, M (M''), number of cells is n and R_i stands for the position of defects.

To estimate the phonon drag thermoelectric power, we shall use the Kubo formula [41] following model Hamiltonian. It has contributions from both the phonons and the carriers. In the continuum approximation the lattice part follows [42]

$$S_{ph}^{drag}(T) = -\frac{k_B}{|e|}\left[\frac{T}{\theta_D}\right]^3 \int_0^{\omega_D} d\omega (\beta\omega)^4 A(\omega) (\beta\omega)^4 \frac{e^{\beta\omega}}{(e^{\beta\omega}-1)^2}. \tag{16}$$

k_B is being the Boltzman constant, e is the charge of carriers, ω_D is the Debye frequency and $\beta = \hbar/k_B T$.

The relaxation time is inhibited in A (ω), the total relaxation rate is proportional to the imaginary part of the phonon self-energy P, where ω is the phonon frequency. The relaxation times ratio can be calculated to the lowest order of the various interactions in the weak interaction case. The phonon drag thermoelectric power relaxation times ratio A (ω) follows

$$A(\omega) = \left[1/\tau_{ph-d} + 1/\tau_{ph-gb} + 1/\tau_{ph-ph}\right]^{-1} \times \left[1/\tau_{ph-d} + 1/\tau_{ph-gb} + 1/\tau_{ph-ph} + 1/\tau_{ph-c}\right], \tag{17}$$

The various relaxation times are defined in terms of transport coefficients as

$$\tau_{ph-d}^{-1}(\omega) = D_{phd} / k_B^3 \omega^4 \hbar^3, \tag{18}$$

$$\tau_{ph-gb}^{-1}(\omega) = D_{phgb} v_s / L, \qquad (19)$$

$$\tau_{ph-ph}^{-1}(\omega) = D_{phph}(T\omega\hbar / k_B)^3. \qquad (20)$$

and

$$\tau_{ph-c}^{-1}(\omega) = D_{phe}\,\omega n_F(\Delta), \qquad (21)$$

v_s being the velocity of sound, L is the length of nanowire, n_F is the Fermi-Dirac distribution function and Δ is the energy gap parameter. The notation $\tau_{ph\text{-}d}$, $\tau_{ph\text{-}gb}$, $\tau_{ph\text{-}ph}$ and $\tau_{ph\text{-}c}$ are the phonon scattering relaxation time due to substitutional defects, grain boundaries, phonon and phonon-carrier interactions respectively. We note that to this order Mathiessen's rule holds namely, that the inverse of the total relaxation time is the sum of the various contributions for the different scattering channels and the relaxation rate due to different scattering processes have been normalized by total relaxation rate. Here the transport parameters $D_{ph\text{-}d}$, $D_{ph\text{-}gb}$, $D_{ph\text{-}ph}$ and $D_{ph\text{-}c}$, shows the strength of phonon-defects, phonon-grain boundaries, phonon-phonon and phonon-carrier scatterings respectively.

The transport coefficients appearing in equations (18 – 21) are defined as

$$D_{phd} = \left[\frac{3n_i R^2}{4\theta_D^3}\right] \qquad (22)$$

with n_i is the density of impurities or point defects, θ_D the Debye temperature. The strength of the electron-phonon scattering in terms of deformation potential D_p, carrier mass m, ionic mass M and the Fermi energy (ε_F) is

$$D_{phe} = \frac{9\pi}{4}\left[\frac{m}{3M}\right]^{1/2} \frac{D_p^{\,2}}{\varepsilon_F^2} \qquad (23)$$

Herein, the Thomas-Fermi approximation defines electron-acoustic phonon coupling strength $D_p = -\varepsilon_F/2$.

The phonon drag thermoelectric power is a powerful probe to study the nature of carriers and scattering process between them in the phonon system with either absence or presence of a magnetic field. For the actual calculation of the phonon drag thermoelectric power in 4 nm diameter Zn nanowire (Zn/Vycor composite), more realistic values of some physical parameters obtained from numerical analysis are as follows: The transport parameters appeared in equations (18-21) [41] have been evaluated from fitting method as $D_{ph\text{-}d} = 2.6 \times 10^{-9}\ K^{-3}$, $D_{ph\text{-}gb} = 2.14 \times 10^{-2}$, $D_{ph\text{-}ph} = 0.8 \times 10^{-2}\ K^{-6}\ sec^{-1}$ and $D_{ph\text{-}c} = 3.4$ respectively. Values of these parameters have optimized for the system under investigation to match the theoretically estimated results with experimental data. These are material dependent parameters characterize the strengths of the phonon-defects, phonon-grain boundary, phonon-phonon and phonon-carrier scattering process for phonon drag thermoelectric power in the present model. It is instructive to mention that the electron-phonon interaction is limited to the coupling of acoustic phonons.

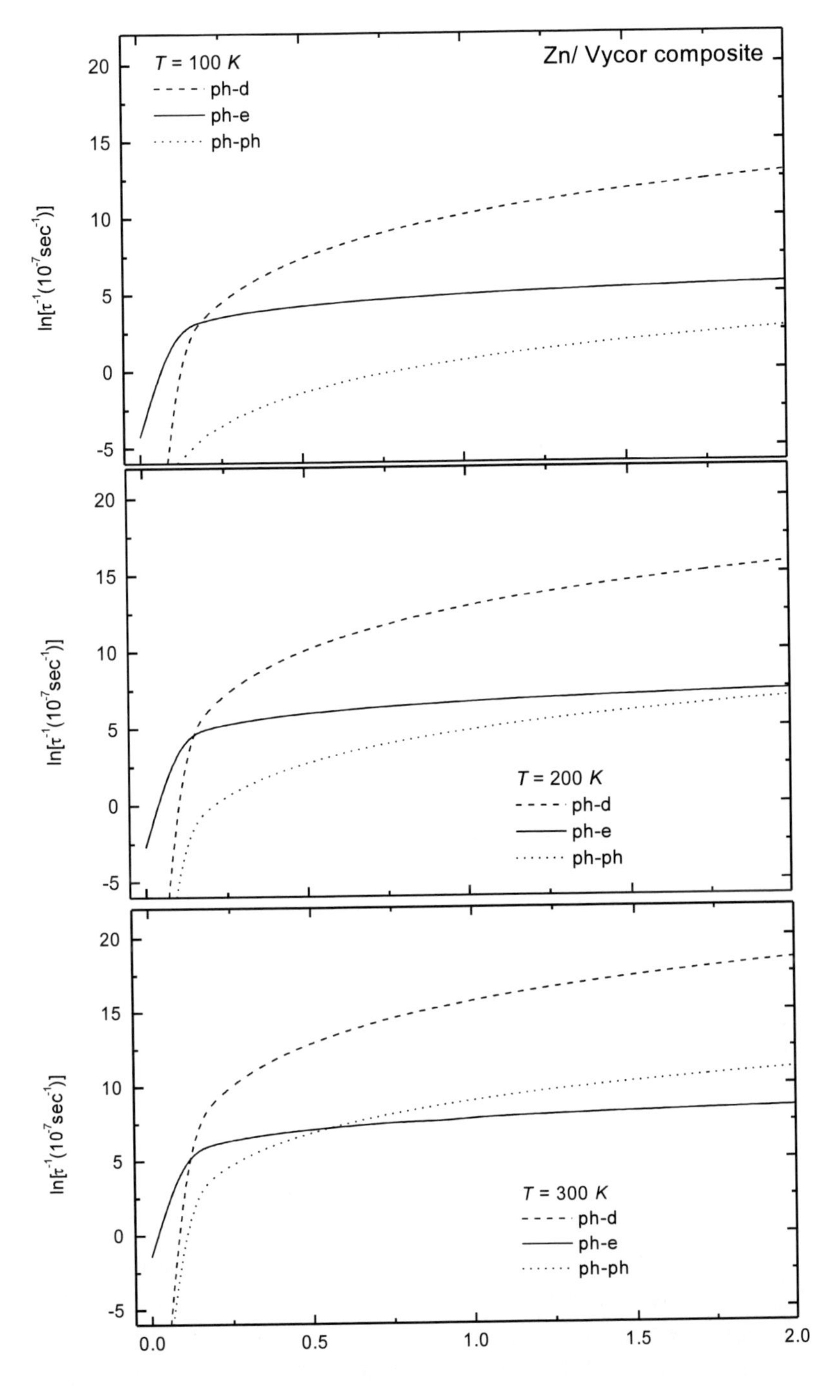

Figure 8. Variation of various phonon relaxation rates as a function of ξ (= $\hbar\omega/k_BT$) for T = 100, 200 and 300 K. Reprinted with permission from K. K. Choudhary, D. Prasad, K. Jayakumar, Dinesh Varshney, *International Journal of Nanoscience,* **9**, 453-459 (2010), Copyright © World Scientific Publishing Company.

Now we discuss the relative magnitudes of the various scattering rates at different temperatures. A plot of various phonon scattering relaxation rates as a function of ξ (= $\hbar\omega/k_BT$) in terms of frequency at T = 100, 200 and 300 K from equations (18 - 21) are shown in figure 8. It can be seen that, for low value of ξ, the various scattering rates increases abruptly and become almost saturate at higher values of ξ. At low temperatures (T = 100 K), phonon-electron scattering dominates over phonon-phonon scatterings while, at higher temperature (T = 3000 K) where phonons are exited in large number phonon-phonon scattering dominate over phonon-electron scattering. In general all the scatterings improve as temperature increases because of increase in energy of queasy particles. The temperature dependent variation of phonon scattering relaxation rates is plotted in figure 9 for a constant value of ξ = *1.0*.

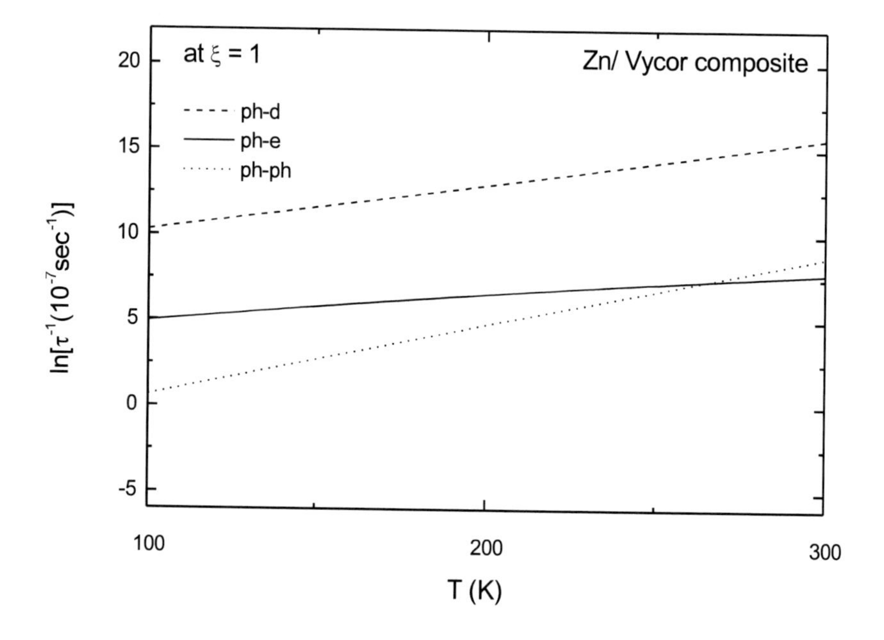

Figure 9. Variation of various phonon relaxation rates as a function of temperature at ξ = 1. Reprinted with permission from K. K. Choudhary, D. Prasad, K. Jayakumar, Dinesh Varshney, *International Journal of Nanoscience,* **9**, 453-459 (2010), Copyright © World Scientific Publishing Company.

We now qualitatively discuss the phonon drag thermoelectric power (S_{ph}^{drag}) in the presence of various scattering mechanisms (please see figure 10). At low temperatures, the quasi particle excitations as acoustic phonon are freezed and they can not scatter phonons. Due to which the phonons mean free pat increases, more heat is then carried by the phonons thus. S_{ph}^{drag}, thus, increases exponentially with temperature in the low temperature regime (T < 50 K). S_{ph}^{drag} develops a broad peak at moderate temperature ($T \approx$ 200 K) because of competition among several operating scattering mechanisms and falls when temperature is further increased because, at higher temperature more and more phonons get excited and get scattered by various irregularities, interfaces and by themselves in the Zn nanowire. Scattering of phonons decrease the heat carrying capacity (thermal conductivity) of the phonons in the Zn nanowire results in increase in thermoelectric power. In figure 10, we can also compare the effect of various operating scatterings on thermo power. Initially when only phonon-grain boundaries (ph-gb) scattering is available, thermoelectric power is minimum, as phonon-defect (ph-d) and phonon-electron (ph-e) scatterings are also included they improves the thermo power almost by a factor of two. Further, when phonon-phonon (ph-ph) scattering is also included, since phonons are more active at higher temperatures, it increases the thermo power mainly at higher temperatures. The results shows that the effect of scatterings on thermoelectric power is additive and the final variation depend on the relative magnitude of different scattering processes available. It is worth to mention that S is zero ($T \rightarrow 0$ K) and that every non-zero value is an artifact due to the model. Four scattering mechanism are considered to check their importance and are not compared due to lack of scattering information's and shall be considered as predictions of the proposed model for the thermoelectric power [37].

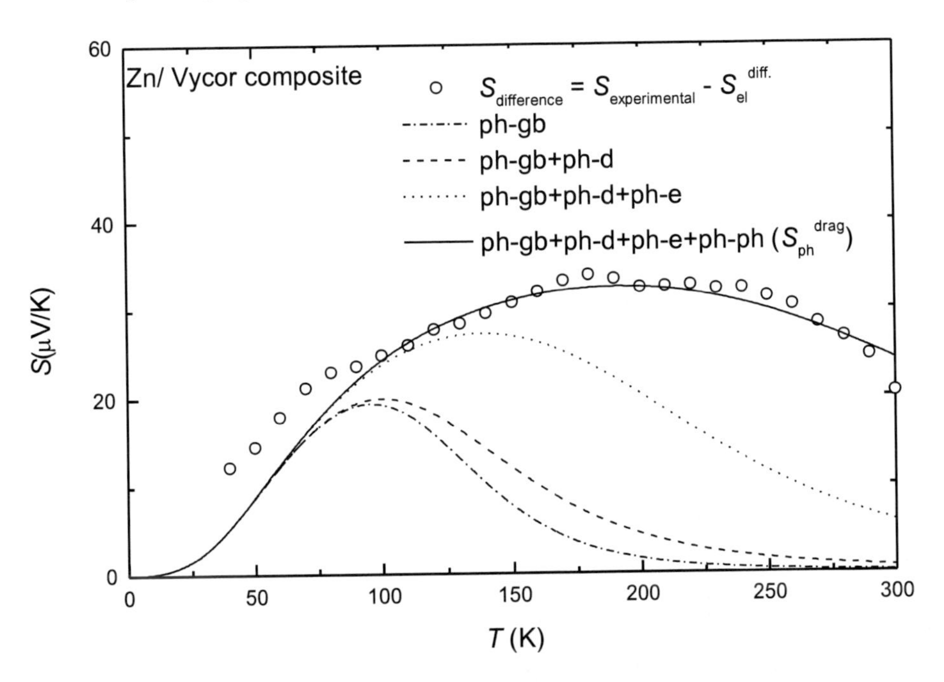

Figure 10. Variation of phonon drag thermoelectric power as function of temperature in presence of various phonons scattering mechanism. Reprinted from *Journal of Physics and Chemistry of Solids*, Vol **71** Issue 1, by K.K. Choudhary, D. Prasad, N. Kaurav, Dinesh Varshney, Entitled: Interpretation of thermoelectric power of Zinc Nanowire Composites: Phonon scattering mechanism, Pages No. 47-50, Copyright (©2010), with permission from Elsevier.

To ascertain the physical significance of the density of impurities, we evaluate the transport coefficients $D_{ph\text{-}d}$ appeared in equations (18 and 22). We estimate the product of density of impurities and square of relative ionic mass difference, $n_iR^2 = 0.316$ from the value of coefficient $D_{ph\text{-}d}$. Due to the fact that the transport parameter $D_{ph\text{-}d}$ is determined by the magnitude of the phonon-impurity interaction, we are able to roughly estimate the density of impurity scatterers which may point to the fact that the quasi particles in the metallic nanowires are essentially localized. By this way, one can set a limit to the concentration of impurities if the impurities as scatterers are of isotope in origin. Herein, we believe that the density of impurity is constant with respect to temperature.

The S (T) behavior depends on the competition among the various scattering mechanisms for the heat carriers and balance between the electron and phonon competition. It is worth stressing that the S (T) is well reproduced from the present theoretical model, this phenomenon is attributed to shortened phonon mean free path. It may be seen that the slope change in S_{ph}^{drag} is much more pronounced than that in S_c^{diff}. The reason for being this change is due to the competition in between phonon-impurity and phonon-electron scattering mechanism leads to faster change of slope in S_{ph}^{drag}. It is customary to mention that a change in slope of $S_{ph}^{drag.}$ is obtained at about $T \approx \theta_D$ as in the normal metals [43]. Indeed below $T \approx \theta_D$ the temperature dependence of the phonon drag contribution is attributed to be a competitive process. Deduced results on temperature dependence of thermoelectric power from the present model are consistent qualitatively with the experimental data.

4. Thermal Conductivity

Theoretical model to explain the effect of embedding ErAs nanoparticles on thermal conductivity of $In_{0.53}Ga_{0.47}As$ crystalline semiconductor is presented [11]. The role of scattering of phonon with defects, grain boundaries, phonon and electrons is studied for thermal conductivity. We will calculate the strength of various scattering mechanisms leads to anomalous behaviour of thermal conductivity. We have also quantitatively estimated the strength of various phonon scattering mechanisms before and after embedding nanoparticles, based on transport parameters appear in the theoretical model. A significant change in the strength of phonon-grain boundaries and phonon-defect scattering and their relaxation rates have been observed due to increased interfaces and irregularities in the crystal.

We follow a simple model where the phonons are described in the Debye model. The use of the Debye model is reasonable since the temperature region of interest lies around the Debye temperature. As the simplest approximation to the problem at hand, the isotropic Debye model approach can be used to derive qualitative results, as we will demonstrate later. The model Hamiltonian [equation (15)] that presented in the previous section has been used for quantitative estimation of thermal conductivity.

The thermal conductivity following model Hamiltonian [equation (15)] can be calculated from the Kubo formula [41]. The lattice thermal conductivity in continuum approximation follows

$$\kappa_{ph} = \frac{k_B \hbar^2}{2\pi^2 v_s} \int_0^{\omega_D} d\omega \omega^2 \tau(\omega)(\beta\omega)^2 \frac{e^{\beta\omega}}{(e^{\beta\omega}-1)^2}. \tag{24}$$

with k_B is the Boltzman constant, v_s is the sound velocity, ω_D is the Debye frequency and $\beta = \hbar/k_B T$. The relaxation time is proportional to the imaginary part of the phonon self-energy. In the weak interaction case, it has been calculated to the lowest order of the various interactions. The relaxation time is expressed as

$$1/\tau(\omega) = 2\left|\mathrm{Im}\, P(\omega/v_s, \omega)\right| \tag{25}$$

$$= 1/\tau_{ph-d} + 1/\tau_{ph-e} + 1/\tau_{ph-gb} + 1/\tau_{ph-ph}, \tag{26}$$

where the various relaxation times are similar as presented in previous section by equations (18 - 21). the temperature dependent thermal conduction of $In_{0.53}Ga_{0.47}As$ crystalline semiconductors, before and after embedding ErAs nanoparticles is numerically calculated [11]. For pure $In_{0.53}Ga_{0.47}As$ crystalline semiconductors we obtain the following physical parameters using the curve fitting method, which characterize the strengths of the phonon-defect, phonon-electron, phonon-grain boundaries and phonon-phonon scattering process: $D_{ph\text{-}d} = 0.2 \times 10^{-8}$ K^{-3}, $D_{ph\text{-}c} = 0.3 \times 10^{-3}$, $D_{ph\text{-}gb} = 1.3 \times 10^{3}$ and $D_{ph\text{-}ph} = 0.9 \times 10^{-2}$ K^{-6} sec^{-1}, respectively. These are indeed free parameters for thermal conductivity in the present model and we shall look for their physical interpretation later on. The velocity of sound is used as $v_s = 3 \times 10^{5}$ cm sec^{-1} to evaluate the model equations.

We first qualitatively discuss the variation of thermal conductivity due to various phonons scattering processes. Figure 11 show our numerical results for phonon thermal conductivity of pure $In_{0.53}Ga_{0.47}As$ (do not contain the nanoparticles). As the temperature is lowered around 100 K, the quasi-particle excitations condense into the ground state and indeed they cannot scatter phonons. The phonon thermal conductivity thus increases exponentially as the temperature decreases in the absence of the other scattering mechanisms. Although the phonon thermal conductivity experiences an exponential increase at low temperatures, the presence of the grain boundaries, and defects set a limit on its growth, as a consequence the phonon thermal conductivity diminishes as the temperature decreases. Further, the phonon-electron scattering reduces the mean free path of phonon and limited the thermal conductivity as temperature increases. As we go on higher temperature (above 200 K) the phonon excitation increases, due to that phonon-phonon scattering become more prominent and is responsible for reducing thermal conductivity at higher temperatures.

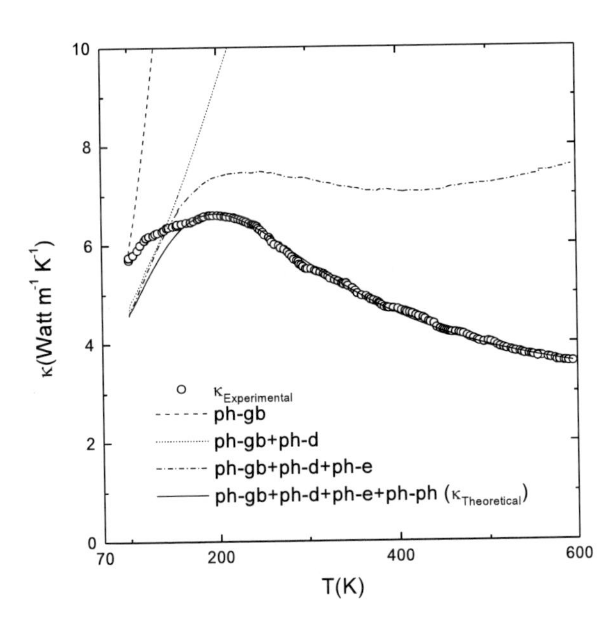

Figure 11. Variation of thermal conductivity as function of temperature for pure $In_{0.53}Ga_{0.47}As$ in the presence of various phonon scattering mechanisms. Open circles are experimental data, taken from Woochul Kim *et al.* 2006, compared with theoretical fit as discussed in the text (Solid line). Reprinted with permission from K. K. Choudhary, D. Prasad, K. Jayakumar, Dinesh Varshney, *International Journal of Nanoscience,* **8**, 551-556 (2009), Copyright © World Scientific Publishing Company.

The maximum position depends on the relative magnitudes of the phonon-electron and phonon-defect interactions below 100 *K*, the defects become the effective phonon scatterers and the thermal conductivity exhibits a typical quadratic temperature behavior at even lower temperatures, grain-boundary scattering dominates and the usual Debye T^3 behaviour appears. Furthermore, κ develops a broad peak at about 200 K, before falling off at lower temperatures. This brings us to point out that the phonon peak originates from the competition between the increase in the phonon population and decrease in phonon mean free path due to phonon-phonon scattering with increasing temperature. These are the features contained in equations (18 - 21) and are consistent with the experimental observations [25].

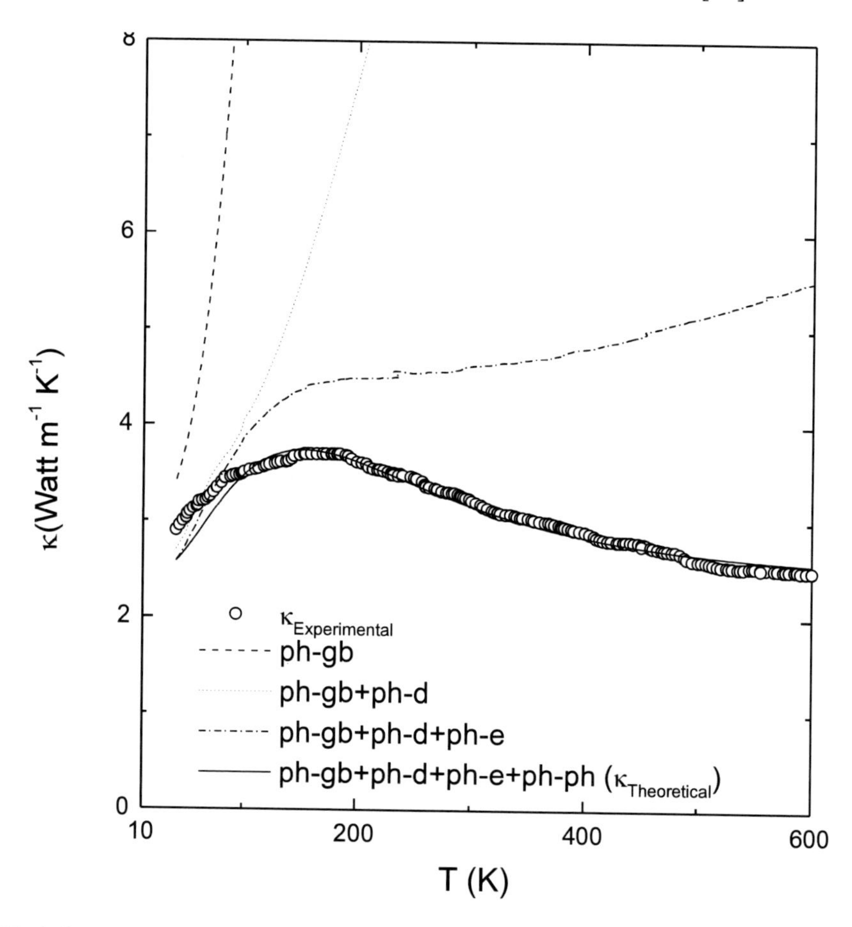

Figure 12. Variation of thermal conductivity as function of temperature for $In_{0.53}Ga_{0.47}As$ containing randomly distributed ErAs nanoparticles, in the presence of various phonon scattering mechanism. Open circles are experimental data, taken from Woochul Kim *et al.* 2006, compared with theoretical fit as discussed in the text (Solid line). Reprinted with permission from K. K. Choudhary, D. Prasad, K. Jayakumar, Dinesh Varshney, *International Journal of Nanoscience,* **8**, 551-556 (2009), Copyright © World Scientific Publishing Company.

Figure 12 shows the temperature dependent thermal conductivity of $In_{0.53}Ga_{0.47}As$ after embedding ErAs nanoparticles where ErAs nanoparticles are randomly distributed in the crystalline $In_{0.53}Ga_{0.47}As$. The characteristics of the curves are as similar as in figure 11, except the magnitude of the thermal conductivity. Embedded ErAs nanoparticles provide an additional source of scattering to phonons, which cause in decrease in magnitude of thermal conductivity. Because of embedded nanoparticles the amount of interfaces increases in the crystal, which increases manly the phonon-defect and phonon-grain boundaries scatterings. The characteristic parameters obtain from the present model are: $D_{ph\text{-}d} = 1.4 \times 10^{-8}$ K^{-3}, $D_{ph\text{-}c} = 0.18 \times 10^{-3}$, $D_{ph\text{-}gb} = 1.5 \times 10^{5}$ and $D_{ph\text{-}ph} = 3.5 \times 10^{-2}$ K^{-6} sec^{-1}. If we compare these characteristic parameters with the parameters for pure $In_{0.53}Ga_{0.47}As$, then it can be clearly seen that, inserted nanoparticles increase the strength of Phonon-grain boundaries and phonon-defect scattering processes.

To ascertain the physical significance of the density of impurities and the ratio of the deformation potential constant to the Fermi energy, we evaluate the transport coefficients $D_{ph\text{-}d}$ and $D_{ph\text{-}c}$ appeared in Equation (20 and 21). We estimate $n_i R^2 = 0.17(1.19)$ from the value of $D_{ph\text{-}d}$ before(after) inserting nanoparticles. These values reflect that the density of defects increased by seven times on inserting nanoparticles. Due to the fact that the parameter $D_{ph\text{-}d}$ is determined by the magnitude of the phonon-impurity interaction, we are able to roughly estimate the density of impurity scatterers, which may point to the fact that the nanoparticles are essentially localized and scatter the phonons. We next consider the parameter $D_{ph\text{-}c}$ that essentially points to the interaction between electrons as carriers and phonons. The estimated ratio of the deformation potential constant to the Fermi energy is about one half from the $D_{ph\text{-}c}$ value. We found here that no significant change occur in the ratio on embedding the nanoparticles.

Quite generally the heat conduction is adequately depends on way of energy transport via the scattering of quasi-particles with impurity as scatterers, interaction of the heat carriers among themselves and a balance between various scattering processes. Deduced results on temperature dependent thermal conductivity of $In_{0.53}Ga_{0.47}As$ before and after embedding ErAs nanoparticles from the present model and with a suitable choice of free parameters, we are able to reproduce the experimental result [25].

5. Conclusion

A complete review of transport properties viz. electrical resistivity, thermal conductivity and thermoelectric power of nanostructured materials are presented in this article. A theoretical analysis of the electrical resistivity for Zn nanowires of different diameters is presented. The electrical properties of Zn nanowires changes with change in diameter. Our results on the resistivity analysis show that the electron-phonon scattering mechanism is successful to explain the metallic behaviour of resistivity of 15nm Zn/SiO_2 composite nanowire. The linear temperature dependence of $\rho_{diff.}$ obtained in present analysis indicates the strong possibility of scattering of electron with electron and twistons in the nanowires. The additional term due to electron-electron and electron-twistons contribution was required in understanding the resistivity behavior, as extensive attempts to fit the data with residual resistivity and phonon resistivity were unsuccessful. Further, we have presented the analysis of electrical resistivity behavior of semiconducting Zn/Vycor composite nanowire based on small polaron conduction by using the diffusion coefficient and the polaron formation energy. The developed approach consistently explains the reported behaviour in wide temperature region for both matellic as well semiconducting Zn nanowires.

The thermoelectric power $S(T)$ behavior which is an instructive probe to reveal the phonon drag and carrier diffusion effects is well described as well the interaction of these excitations with one another, with impurities, grain boundaries and defects are studied. Among various transport probes it brings information about available subsystems and the measurements are fairly simple. The present study intends to contribute towards a thorough understanding of the scattering processes taking place in these fascinating nanowires. The thermoelectric power behavior exhibits exponential temperature dependence at low temperatures, develops a broad peak at moderate temperature ($T \approx 200$ K) and decreases when temperature is further increased, this anomalous behavior can be well explained considering two channels for $S(T)$: carrier diffusion (S_c^{diff}) and phonon drag (S_{ph}^{drag}) and the competition between various scattering mechanisms used.

The phonon drag thermoelectric power (S_{ph}^{drag}) is discussed within the Debye-type relaxation rate approximation in terms of the acoustic phonon frequency, a relaxation time τ and the sound velocity. The total phonon relaxation rate is the sum of terms corresponding to independent scattering mechanism, like defects, grain boundaries, phonons and electrons dominating in different temperature intervals. We have made a careful analysis taking into account of several different processes that can exist in this material whose interaction can yield the observed dependence.

To investigate the temperature dependent thermal transport i.e., thermal conductivity κ(T) behavior we consider the phonon relaxation time approximation. The presented investigation deals with a quantitative description of temperature dependent behavior of κ and the effect of embedding nanoparticles on the thermal conductivity. The thermal conductivity behavior is an artifact of strong phonon-defects, phonon-phonon, phonon-electron and phonon grain boundaries scattering mechanism. However, we have made a careful analysis taking into account of several scattering processes that can exist in this material whose interaction can yield the observed dependence. The fitting parameters in the present scheme that characterize the strengths of the phonon-defect, phonon-electron, phonon-grain boundaries and phonon-phonon scattering process: $P_d = 0.2 \times 10^{-8}$ (1.4×10^{-8}) K^{-3}, $P_e = 0.3 \times 10^{-3}$ (0.18×10^{-3}), $P_{gb} = 1.3 \times 10^3$ (1.5×10^5) and $P_p = 0.9 \times 10^{-2}$ (3.5×10^{-2}) K^{-6} sec^{-1}, respectively before(after) embedding ErAs nanoparticles in $In_{0.53}Ga_{0.47}As$, leads to a result that successfully retraces the experimental curve. Above parameters show that the nanoparticles provide an additional source for phonon scattering, because of enhanced interfaces and irregularities in the crystal.

It is stressed that the phonon-phonon scattering dominates and their contribution to thermal conduction is significant. In brief, behavior of κ is determined by competition among the several operating scattering rates for the heat carriers. In principle, the acoustic phonons are major source of thermal transport. The presented model calculations successfully explain the effect of embedding ErAs nanoparticles in $In_{0.53}Ga_{0.47}As$ crystalline semiconductors on the temperature dependent thermal conductivity κ. Despite of the limitations and use of material parameters for the estimation of transport coupling strengths, the presented theoretical models on the resistivity, thermoelectric power and thermal conductivity consistently reveals the interesting behavior reported experimentally. Although we have provided a simple phenomenological explanation of this effect, there is clearly a need for a good theoretical understanding of the transport properties.

ACKNOWLEDGMENTS

Financial assistance from M. P. Council of Science and Technology, Bhopal, India is gratefully acknowledged. Author is also thankful to American Physical Society, Elsevier and World Scientific Publishing Company for permission to reprint their published materials.

REFERENCES

[1] S. M. Prokes and K. L Wang, *Mater.Res. Sci. Bill.* **24** (1999) 13.

[2] J. Hu, T. W. Odom, and C. M. Leiber, *Acc. Chem. Res.* **32** (1999) 435.

[3] C. Durkan and M. E. Welland, *Phys. Rev. B* **61** (2000) 14215.

[4] M. Venkata Kamalakar and A. K. Raychaudhuri *Phys. Rev. B* **79** (2009) 205417.

[5] Werner Steinhögl, G. Schindler, Gernot Steinlesberger, and Manfred Engelhardt, *Phys. Rev. B* **66** (2002) 075414.

[6] W. Steinhogl, G. Schindler, G. Steinlesberger, M. Traving, and M. Engelhardt, J. *Appl. Phys.* **97** (2005) 023706.

[7] J. P. Heremans, C. M. Thrush, D. T. Morelli, and M. C. Wu, *Phys. Rev. Lett.* **91** (2003) 076804.

[8] L. D. Hicks, and M. S. Dresselhaus, *Phys. Rev. B* **47** (1993) 16631.

[9] J. Heremans, C. M. Thrush, Z. Zang, X. Sun, M. S. Dresshaus, J. Y. Ying, and D. T. Morelli, *Phys. Rev. B* **58** (1998) R10 091.

[10] J. Heremans, C. M. Thrush, Y.-M. Lin, S. B. Cronin, and M. S Dresselhaus *Phys. Rev. B* **63** (2001) 085406.

[11] K. K. Choudhary, D. Prasad, K Jayakumar and Dinesh Varshney *Int. J. Nanoscience* **8** (2009) 551.

[12] Edward M. Likovich, Kasey J. Russell, Eric W. Petersen, and Venkatesh Narayanamurti, *Phys. Rev. B* **80** (2009) 245318.

[13] C. L. Kane, E. J. Mele, R. S. Lee, J. E. Fischer, P. Petit, H. Dai, A. Thesh, R. E. Smalley, A. R. M. Verschueren, S. J. Tans and C. Dekker, Euro*phys. Lett.* **40**(6) (1998) 683.

[14] A. Bid, A. Bora and A. K. Raychaudhary, *Phys. Rev. B* **74** (2006) 035426.

[15] Z.Yao, C.L.Kane and C.Dekker, *Phys. Rev. Lett.* **84** (2000) 2941.

[16] Li Zhengquan, Yujie Xiong, and Xie Yi, *Inorganic Chemistry* **42** (2003) 8105.

[17] Juncong She, Zhiming Xiao, Yuhua Yang, Shaozhi Deng, Jun Chen, Guowei Yang, and Ningsheng Xu, *ACC nano* **2** (2008) 2015.

[18] S. J. Tans, R. M. Verschueren and C. Dekker, *Nature* **393** (1998) 49.

[19] R. Martel, T. Schmidt, H. R. Shea, T. Hertel and P. Avouris, *Appl. Phys. Lett.* **73** (1998) 2447.

[20] J. Kong and H. Dai, *J. Phys. Chem. B* **105** (2001) 2890.

[21] S. Frank, Philippe Poncharal, Z. L. Wang and Walt A. de Heer *Science* **280**(1998) 1744.

[22] T. C. Harman, P. J. Taylor, M. P. Walsh and B. E. LaForge, Science **297**, (2002) 2229.

[23] K. F. Hsu, S. Loo, F. Guo, W. Chen, J. S. Dyck, C. Uher, T. Hogan, E. K. Polychroniadis, and M. G. Kanatzidis, Science **303**, (2004) 818.

[24] R. Venkatasubramanian, E. Siivola, T. Colpitts and B. O'Quinn, Nature (London) **413**, (2001) 597.

[25] Woochul Kim, Joshua Zide, Arthur Gossard, Dmitri Klenov, Susanne Stemmer, Ali Shakouri, and Arun Majumdar Phys. Rev. Lett. **96,** (2006) 045901.

[26] Y. S. Touloukian *Thermophysical Properties of Matter* (1970)

[27] J. P. Dismukes, E. Ekstrom, D. S. Beers, E. F. Steigmeier and I. Kudman J. Appl. Phys. **35,** (1964) 2899.

[28] P. K. Piiharody and L. P. Garvey, Proceedings of the 7th Intersociety Energy Conwrstbn Engineering Conference (IEEE), New York, (1978) 1963.

[29] J. W. Vandersande, C. Wood and S. Draper, Mater. Res. Sot. Symp. Proc. **97,** (1987) 347.

[30] J. W. Vandemande, A. Borshchevsky, J. Parker and C. Wood, Proceedings of the 7th International Conference on Thermwlectrk Energy Conwrsion, University of Texas, (1988), 76.

[31] H. J. Goldsmid and A. W. Penn, Phys. Lett. **27A**, (1968) 523.

[32] D. M. Rowe, J. Phys. D **7**, (1974) 1843.

[33] Cronin B. Vining, J. Appl. Phys. **69**, (1991) 331.

[34] L. Braginsky, N. Lukzen, V. Shklover and H. Hofmann, Phys. Rev. B **66**, (2002) 134203

[35] J. G. Wang, M. L. Tian, N. Kumar, and T. E. Mallouk, *Nano Lett.* **5** (2005)1247.

[36] K. K. Choudhary, D. Prasad, K Jayakumar and Dinesh Varshney *J.Compu. and Theo. Nanoscience* (to be published in 2011).

[37] K. K. Choudhary, D. Prasad, N. kaurav and Dinesh Varshney, *Journal of Phys. and Chem. of Solids* **71** (2010) 47.

[38] T. Inabe, H. Ogata, Y. Maruyama, Y. Achiba, S. Suzuki, K. Kikuchi and I. Ikemoto *Phys. Rev. Lett.* **69** (1992) 3797; Y Maruyama, T. Inabe, H. Ogata, Y. Achiba, S. Suzuki, K. Kikuchi and I. Ikmoto, *Chem. Lett.* **10** (1991) 1849.

[39] K. K. Choudhary, N. Kaurav, N. Gupta and Dinesh Varshney, *Journal of Phys.: Conference Series* **92** (2007) 012146

[40] Dinesh Varshney, K. K. Choudhary, and R. K. Singh, New Journal of Physics **5,** (2003) 72.1

[41] J. Callaway, Quantum theory of the Solid State, *Academic Press London* (1991).

[42] R. D. Barnard, Thermoelectricity in metals and alloys, *Taylor and Francis Ltd. London* (1972).

[43] N. F. Mott, E. A. Davis, Electronic processes in non-crystalline materials, *Clarendon, Oxford* (1979).

In: Nanowires: Properties, Synthesis and Applications
Editor: Vincent Lefevre
ISBN: 978-1-61470-129-3

Chapter 4

NANOWIRE ARRAY ELECTRODES IN BIOSENSOR APPLICATIONS

Kafil M. Razeeb, Mamun Jamal, Ju Xu and Maksudul Hasan
Tyndall National Institute, University College Cork, Lee Malting,
Prospect Row, Cork, Ireland

Abstract

This chapter investigates the area of nanowire based biosensors for different analyte detection. The electrochemical properties of two different platforms made of gold nanowire array (NAE) and Pt nanoparticle modified gold nanowire array (PtNP/NAE) are examined in details for the detection of hydrogen peroxide. Third-generation H_2O_2 biosensor is prepared by covalent immobilization of horseradish peroxidase (HRP) on the self-assembled monolayer modified NAEs. Also PtNP/NAEs are used to fabricate oxidase enzyme based biosensor. A comparative study between these two systems has been performed and a state of the art comparison with these two systems is reported. Moreover, a detailed electrochemical and physical characterization has been performed using cyclic voltammetry, amperometry, scanning electron microscopy (SEM) and transmission electron microscopy (TEM).

1. INTRODUCTION

Direct electron transfer between redox proteins and electrodes has attracted considerable attention due to its significance in both theoretical [1] and practical applications [2,3] in biosensors, bioelectronics and energy systems [4]. Studies based on direct electrochemistry of redox proteins have been employed to investigate the three dimensional (3-D) structures of enzymes and the relations between the structures and the catalytic properties, for a better understanding of the kinetics and thermodynamics of the biological redox process [1]. Direct electro-communication of biomolecules and electrodes established a foundation for fabricating mediator-free and sensitive biosensors [5]. Third-generation biosensors usually offer better selectivity by operating in a potential range close to the redox potential of the enzyme itself, thus being less exposure to interfering reactions [2]. These biosensors are also advantageous for in-vivo detection due to its simplicity and harmlessness [6,7].

To achieve direct electron transfer, enzyme immobilization is crucial as the enzymatic redox centres are usually embedded deeply into the low conductive amino acid chains, resulting in a slow electron transfer rate in most orientations [5]. Enzymes denature easily or become inactive when directly adsorbed on a conventional electrode. To overcome its drawback, different approaches such as covalent, cross-linking attachment or affinity interaction [8,9], entrapment or encapsulation in a polymer or a sol-gel/inorganic matrix [10] and a nanostructure combined method [11,12] have been utilized for enzyme immobilization [7]. The 2D immobilization methods [3], such as self-assembled monolayer (SAM) [8] and layer-by-layer (LBL) [9], allow the enzyme to anchor oriented on the electrode surface, thus achieving direct electron transfer by reducing the distance between the active site and electrode surface [2,13]. Entrapment/encapsulation immobilization [10] is widely applied owing to its ability to immobilize more electroactive enzymes on to the electrode [14]. In comparison to the 2D approach, entrapment/encapsulation provides a 3D porous structure which drastically increases the active sites for enzyme immobilization. Further improvements have been reported by combining nanostructures with either SAM and/or entrapment [5,15]. However, the distribution of enzyme on the gel or the polymer 3D matrix is usually not uniform, sometimes spatially hindered, resulting in relatively low electron transfer efficiency because only a very small percentage of the enzyme immobilized is electroactive. Thus, ordered 3D matrices, such as nanotube arrays [16,17], ordered porous materials [18] and nanowire/nanorod/nanopillar electrode ensembles [19-21], have been used for increasing the active surface area for enzyme immobilization. Nanowire electrode ensembles (NEEs) prepared from a polycarbonate (PC) membrane [22,23] and nanowire array electrodes (NAEs) prepared from an anodic alumina membrane (AAM) [24] have been reported for various enzyme immobilizations. The ordered 3D nanowire arrays with high aspect ratios improve the signal-to-noise ratio significantly, resulting in significantly higher sensitivity and a lower detection limit [25,26]. Direct electrochemistry of a H_2O_2 biosensor on NEEs fabricated from a PC membrane based on direct adsorption immobilization of haemoglobin (Hb) has been reported [23]. The results showed that the quantity of electroactive Hb varied with the changing of the electrode morphology and increased with the nanowire length.

Hydrogen peroxide (H_2O_2) is necessary for the metabolism of proteins, carbohydrates, fats, vitamins, and minerals. Besides its essential role for the production of estrogens, progesterone, and thyroxin in the body, H_2O_2 helps regulate blood sugar and cellular energy production. Oxidative damages in the body are caused by cellular H_2O_2 imbalance as this chemical plays an important role in cell signalling and communication [27]. Cisplatin used in testicular cancer treatment, affects an increase in the H_2O_2 production to kill the cells. Cancer cells that are resistant to cisplatin or other cancer therapies are capable of synthesizing larger amounts of peroxiredoxin to degrade H_2O_2 [28]. A significant amount of work in enzyme based biosensors is related to the detection of H_2O_2, a co-product of enzymatic oxidation using oxidases [29]. Therefore, the detection of H_2O_2 is of great interest and importance in analytical, environmental and biomedical chemistry as well as food and environmental monitoring. Horseradish peroxidase (HRP) is one of the most important peroxidase enzymes for H_2O_2 or other peroxides [2]. The enzyme has a heme catalytic centre and the surrounding protein. The electrochemically active centres in HRP are normally buried deeply in its extended 3D structure, which makes direct electron transfer between HRP and the electrode surface prohibitively slow [3]. Direct physical adsorption of HRP on a bare electrode, often with denaturation because of multiple contacts and interaction with the surface, can create an insulating layer that blocks electron transfer to the electrode [2,3]. Therefore, nanoparticle modified 3D electrode is getting popular for last couple of years due to their extraordinary catalytic activities [30-33]. Nanoparticles can act as tiny conduction centres and can facilitate the transfer of electrons. Furthermore, it has been reported that enzymes maintained their enzymatic activity when immobilised on nanoparticle modified electrodes.

In this chapter, NAEs with different nanowire lengths were prepared using the template of AAM. For combining the advantages of SAM modification such as oriented attachment of enzymes, biocompatibility and electron transfer properties of the gold nanowires and the ordered 3D nanowire matrix having a large active surface area, HRP was covalently immobilized on these self-assembled monolayer modified gold NAEs. The direct electrochemistry and biosensing performance of HRP/NAEs were presented and discussed in detail. This chapter will further discuss the fabrication of PtNP modified 3D Au NAEs and its catalytic behaviour with oxidase based enzyme glutmate oxidase (GlOx). A detailed electrochemical and microscopic characterization such as cyclic voltammetry, amperometry, SEM and TEM has been described in this chapter.

2. Fabrication Procedures

2.1. Fabrication of Gold Nanowire Array Electrodes (NAEs)

Vertically standing gold nanowire arrays with different lengths were fabricated using a template electrodeposition technique. Typically, a thin layer of Au (thickness of 300 nm, 5 mm in diameter) was first sputtered onto one side of the 3M transparent polymer film. The AAM template was attached on the gold-coated film by a specially designed sample holder. Electrodeposition of Au was performed at a constant current density at 1.0 mA cm^{-2} for 2-5 h in the Au bath (PURAMET 402). After deposition, the AAM template was etched out by submerging the sample in 6.0 M NaOH solution for one hour. The resulting NAEs were rinsed and kept in deionised water.

SEM images of the gold nanowire modified electrodes made from different deposition times (2, 4 and 5 h) are shown in Figure 1A, B, C. The average lengths of the nanowires were ~5μm, 15μm and 20μm for NAE_{2h}, NAE_{4h} and NAE_{5h} respectively. The diameters of the nanowires were ~200 nm, consistent with the pore size of the AAM template. Nanowires of NAEs were bended or bundled together on the top part due to the surface tension force exerted on them during evaporation of the washing solvent. The TEM image in Figure 1D showed that the Au nanowires have rough surfaces with small bumps of ~10 nm, possibly owing to the casting of the rough surface of the template pore wall. A further HRTEM micrograph in Figure 1E showed these bumps exhibit single crystalline structure nature with clearly visible lattice fringes, a useful feature for the biosensing behaviour.

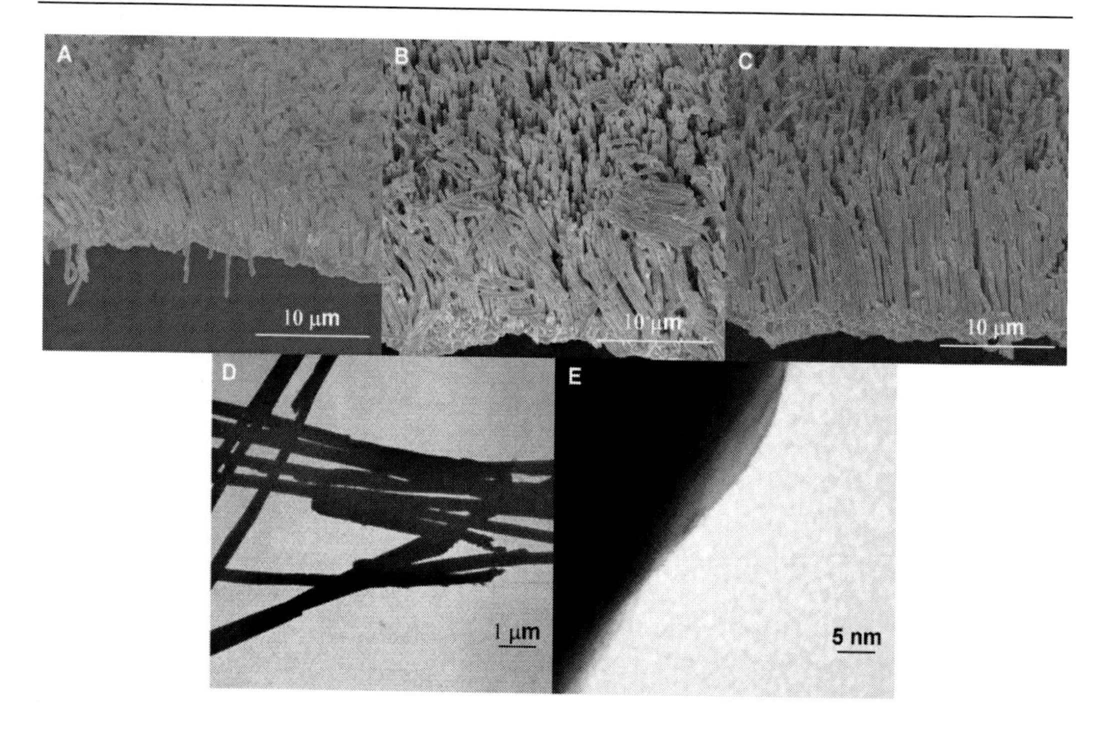

Figure 1. (A-C) SEM image of the NAE_{2h}, NAE_{4h} and NAE_{5h}; (D) TEM image of gold nanowires after AAM removed, most nanowires have rough surfaces; (E) HRTEM image of a single Au nanowire, bump on nanowire has a single crystalline nature. Reproduced with permission from [26].

2.2. Synthesis of Pt Nanoparticle Modified Gold Nanowire Array Electrodes (PtNP/NAEs)

The synthesised gold NAEs were rinsed in deionised water and dipped into the Pt solution made of 5 mM potassium hexachloro-platinate (Sigma-Aldrich, Ireland) and 10 mM HCl and electrodeposition was performed with a current density of 1.0 mA cm^{-2} in a two electrode system using platinised Ti mesh as the counter electrodes from 60 to 900 s to fabricate PtNP/NAEs.

Fig. 2 shows SEM images of gold NAE, PtNP modified gold NAEs and enzyme immobilised PtNP/NAE. The average length of the nanowires is approximately 5 μm with the diameter of ~200 nm (Fig. 2A). Fig. 2A shows the fabricated nanowires after removing the AAM template. Fig. 2B clearly shows that PtNPs were formed on the nanowire surface after 100 s of deposition at a current density of 5 mA cm^{-2}. With increasing deposition time, more PtNPs were deposited on the nanowire, covers more or less all over the area and concentrated on top of the nanowires like a flower (Fig. 2C).

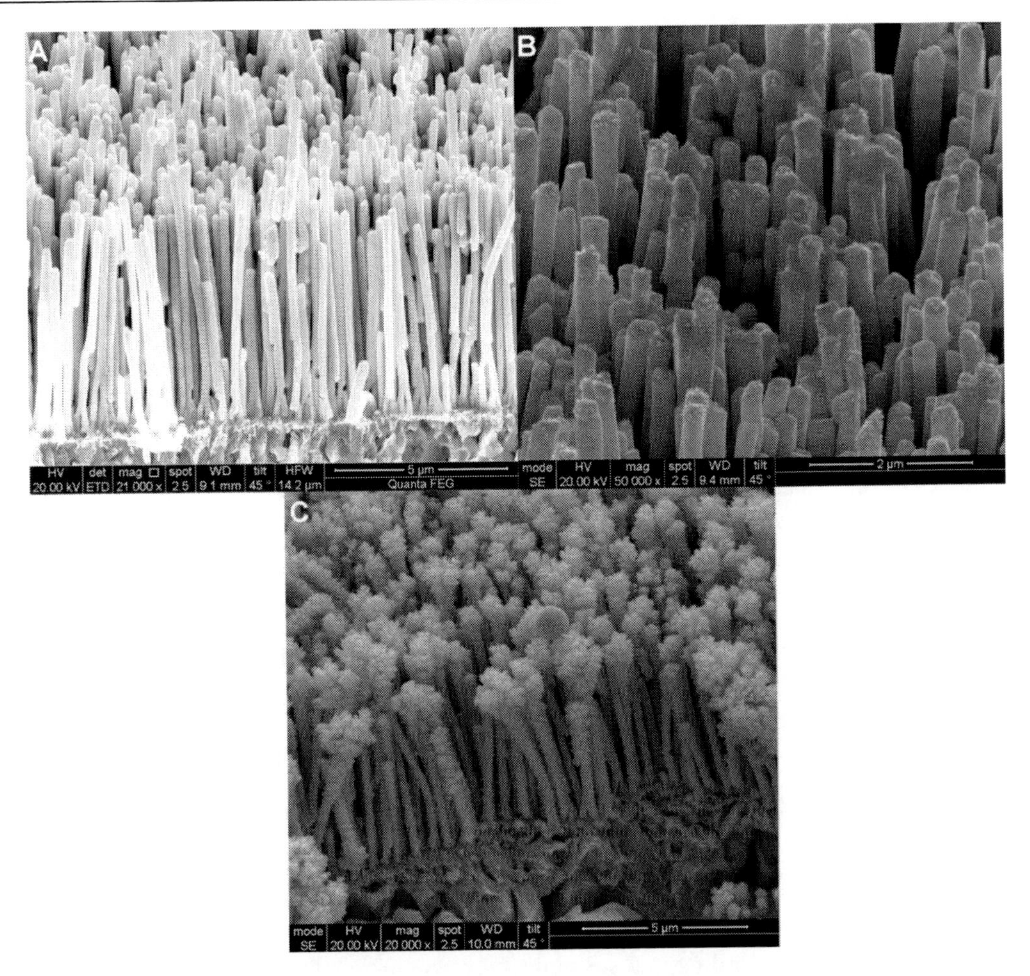

Figure 2. SEM images of the (A) NAE; (B) 100s PtNP deposited NAE; (C) 600s PtNP modified NAE. Reproduced with permission from [34].

3. ELECTOCHEMICAL CHARACTERISATION

3.1. Gold Nanowire Array Electrodes (NAEs)

Cyclic voltammetry of the flat gold electrode (circular Au coated polymer with same diameter of 5 mm) and the NAEs were performed in 0.5 M H_2SO_4 at a scan rate of 0.1 V s^{-1} connected to a potentiostat CHI 660C (Figure 3A). Roughness ratios of the NAEs are presented in Table 1, calculated as the ratio of the reduction peak area (integrating the voltammogram from +0.7 to +1.2 V) of each NAE relative to the flat electrode. The roughness ratios of the NAEs enlarged several tens of times compared to the flat gold electrode, thus, providing more active sites for enzyme immobilization. Of interest was the comparison of the nanowire electrodeposition time (nanowire length) and the roughness data. The roughness ratio of the NAE_{4h} was about twice of the NAE_{2h}, while the roughness ratio increased 1.3-fold from the NAE_{4h} to the NAE_{5h}.

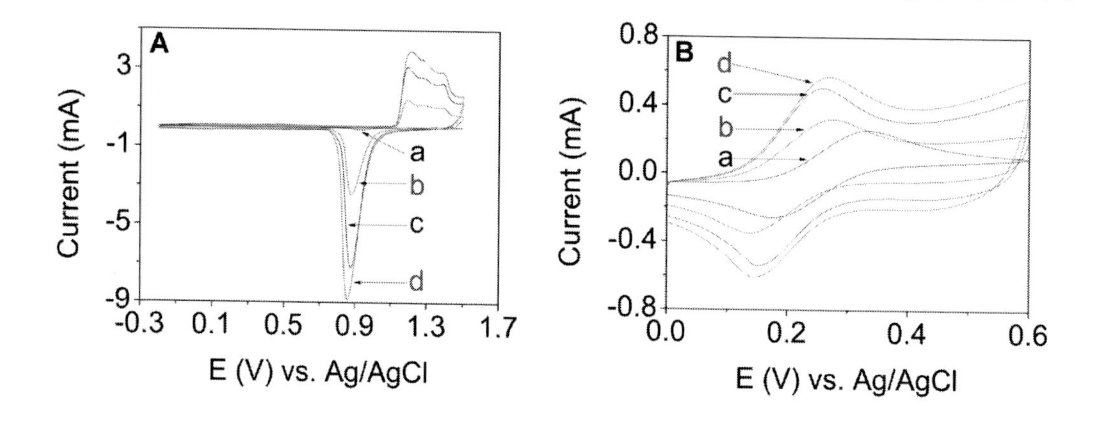

Figure 3. (A) Cyclic voltammograms of the flat gold electrode (a), NAE_{2h} (b), NAE_{4h} (c) and NAE_{5h} (d) in 0.5 M H_2SO_4 at 0.1 V s^{-1}. (B) Cyclic voltammograms of the flat gold electrode (a), NAE_{2h} (b), NAE_{4h} (c) and NAE_{5h} (d) in 0.1 M KCl containing 5 mM $K_3Fe(CN)_6$ at 0.1 V s^{-1}. Reproduced with permission from [26].

The effective surface areas of the NAEs were estimated using ferrocyanide as a redox probe. Figure 3B shows the cyclic voltammograms of the flat gold electrode and different NAEs in 0.1 M KCl containing 5 mM $K_3Fe(CN)_6$ at a scan rate of 0.1 V s^{-1}. The well defined redox peaks were observed, with ΔE_p = 155, 137, 107 and 66 mV for the flat gold electrode, NAE_{2h}, NAE_{4h} and NAE_{5h}, respectively, indicating a quasi-reversible reaction at each electrode [35]. The CV features of the high-density NAEs were similar to the flat gold electrode rather than a sigmoidal steady-state curve, due to the overlapping of the diffusion layers from individual nanowires [36]. The electroactive surface areas (A_{eff}) of the flat electrode, NAE_{2h}, NAE_{4h} and NAE_{5h} were calculated as 0.22, 0.28, 0.40 and 0.44 cm^2, respectively (Table 1), according to the Randles-Sevcik equation [16]:

$$I_p = 2.69 \times 10^5 A n^{3/2} D_0^{1/2} v^{1/2} C_0$$

where I_p, A, n, D_0, v and C_0 represent the redox peak current (A), the effective surface area of the electrode (cm^2), electrons per molecule oxidized or reduced (n = 1), the diffusion coefficient of $K_3Fe(CN)_6$ in 0.1 M KCl (0.673×10^{-5} cm^2 s^{-1}) [37], the scan rate (V s^{-1}) and the concentration of redox species (mol cm^{-3}). Similar to the roughness ratios, the NAE with longer deposition time showed larger electroactive surface area although the enlargement of the effective surface area was moderate compared to the increase of the roughness ratio (e.g. A_{eff} of NAE_{5h} was only slightly larger than that of NAE_{4h}, whereas the roughness ratio of NAE_{5h} and NAE_{4h} was ~1.3). This may be attributed to the fast reaction rate of $K_3Fe(CN)_6$, resulting in only the top part of the nanowires contributing to the increase of the electroactive surface for electron transfer [25,38].

3.2. Pt Nanoparticle Modified Gold Nanowire Array Electrodes (PtNP/NAE)

Fig. 4B shows the cyclic voltammetry of the flat gold electrode (AuE) (circular Au coated polymer with 5 mm diameter) and PtNPs modified flat gold electrode (PtNP/AuE). Fig. 4C shows the same for gold nanowire array electrode (NAE) and PtNPs modified NAEs (PtNP/NAE). All these experiments were performed in 0.5 M H_2SO_4 at a scan rate of 0.1 V s^{-1}

connected to a potentiostat CHI 660C. The electrochemically active surface area of PtNP was determined from the charge consumed during the hydrogen adsorption/ desorption [39]. The loading of nanoparticle was optimised on the surface of NAE, where 600 to 800 s found to be the optimum deposition time at a current density of 5 mA cm^{-2}. Longer deposition time of PtNP on NAE has negative effect on the active surface area, which may be attributed to the blocking effect by the nanoparticles, due to larger amount of deposition at the top of the NAEs as evident from Fig. 3C.

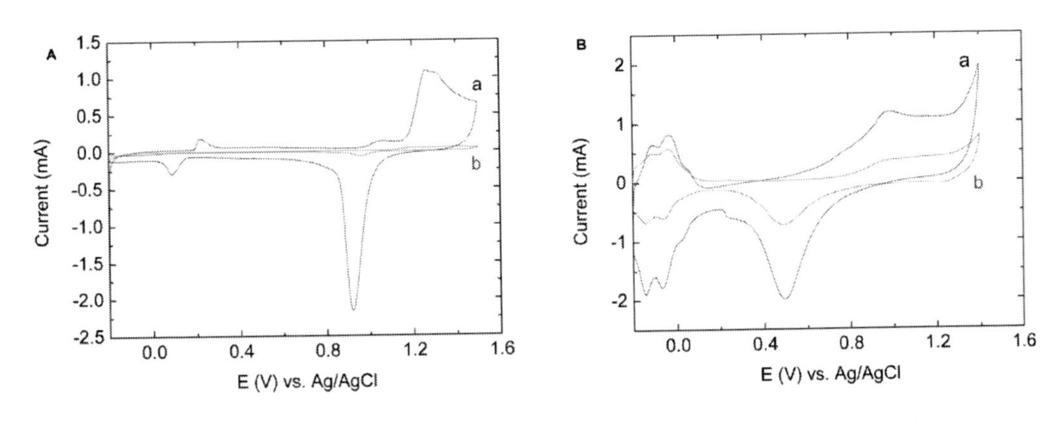

Figure 4. (A) Cyclic voltammograms of the Au NAE (a) and flat Au electrode (b). (B) Cyclic voltammograms of the PtNP/NAE (a) and PtNP on flat Au electrode (b) in 0.5 M H_2SO_4 at 0.1 Vs^{-1}. (Active surface area of blank Au 0.2 cm^2, NAE 1.3 cm^2, PtNP/Au 6.7 cm^2 and PtNP/NAE 10 cm^2, surface areas were calculated using Randles-Sevcik equation using 4 mM Ferricyanide solution). Reproduced with permission from [34].

Moreover, large amount of PtNP reduce the inter particle distance, which effectively reduce the active surface area [31]. Roughness ratio of the flat Au, NAE and PtNP/NAE is calculated by integrating the voltammogram from +0.72 to +1.2 V for flat Au and NAE (Fig. 4B); from +0.20 to +0.80 V for PtNP/AuE and PtNP/NAE (Fig. 4C) electrodes [26]. Roughness is increased by 26.6 times for NAE electrodes compared to the flat Au electrode and increased by 4.3 times for PtNP/NAE compared to flat Au/PtNP electrode.

4. Electrochemical Study of Immobilisied HRP on Au NAEs

4.1. Direct electron transfer of HRP on the Au NAE

The gold nanowire array electrode was cleaned with Piranha (3:1 v/v mixture of concentrated H_2SO_4 and H_2O_2), followed by rinsing with ethanol and deionised water. The electrode was attached to an electrochemical cell containing 0.5 M H_2SO_4 for cleaning by cyclic voltammetry between -0.2 and +1.5 V at 0.1 V/s until a stable CV profile was obtained. The resulting electrode was immersed in a freshly prepared and deaerated 10 mM cysteine (Sigma-Aldrich) for 30 min. After rising with water, the cysteine-modified gold nanowire array electrode was incubated into 5 mg HRP (EC 1.11.1.7, Type II, 150-250 units/mg, Sigma-Aldrich, Ireland) in 1.0 mL of 50 mM pH 7.0 phosphate buffer containing 2% glutaraldehyde (GA) (Sigma-Aldrich, Ireland, Grade I, 25% w/v) for 12 h at 4 °C to effect enzyme immobilisation via glutaraldehyde cross-linking.

The performance of electron transfer depends strongly on the immobilization procedure. Self-assembled monolayer (SAM) exhibits high organisation and homogeneity, making it very attractive for surfaces tailored with desired properties [2]. Figure 5A illustrates the stepwise fabrication process of the HRP/NAEs. L-cysteine was self-assembled onto the gold nanowires, leaving the amine groups for oriented enzyme attachment [40]. Covalent immobilization of the redox protein on the SAM avoided the influence of the mass transfer and enabled a diffusion-less electron-transfer process [3]. As L-cysteine and glutaraldehyde are both small molecules, the distance between the electroactive group of HRP and the electrode was close to facilitate direct electron transfer.

Figure 5B compares the cyclic voltammograms of different HRP/NAEs and the HRP/flat Au electrode in 50 mM deoxygenated phosphate buffer (pH 7.0). Well-defined redox peaks were observed on the HRP/NAEs, with the peak-to-peak separation of 72 mV and the formal potential ($E^{\circ\prime} = (E_{pa} + E_{pc})/2$) of 18 mV at 0.1 V s^{-1} versus 3.0 M Ag/AgCl. The value of $E^{\circ\prime}$ was smaller than the HRP/Nafion-cysteine/Au electrode (60 mV vs. Ag/AgCl or 257 mV vs. NHE) [41], suggesting better electron transfer efficiency.

The scan rate effect on the response of the HRP/NAE_{2h} was investigated to determine kinetic parameters of HRP (Figure 5C). As shown in Figure 5C inset, the anodic or cathodic peak current was proportional to the scan rate, as I_{pa} (μA) = 1.0488 + 113.7v (V s^{-1}) (R = 0.9987) and I_{pc} (μA) = -1.0063 - 113.84v (V $^{s-1}$) (R = 0.9984). This further suggested the electrochemical behaviour of HRP on the NAEs was typical of a surface-controlled quasi-reversible process [15], verifying that immobilized HRP was stable on the NAEs.

The relationship between the peak potential (E_p) and the natural logarithm of scan rate (ln v) for HRP/NAEs in 50 mM PB, pH 7.0 was studied. The cathodic peak potential (E_{pc}) changed proportional to ln v with a linear regression equation of E_{pc} (V) = -0.0494 ln v (V s^{-1}) - 0.2802 (R = 0.9958) in the range from 0.35 to 0.5 V s^{-1}. According to the Laviron equation [42]:

$$E_{\mathrm{p}} = E^{0'} + \frac{RT}{\alpha nF} - \frac{RT}{\alpha nF} \ln v$$

where α is the cathodic electron transfer coefficient, n is the electron transfer number, R is the gas constant (R = 8.314 J mol^{-1} K^{-1}), T is the temperature in Kelvin (T = 298 K) and F is the Faraday constant (F = 96493 C mol^{-1}). The αn was calculated to be 0.52. Given $0.3 < \alpha < 0.7$ in general [43], it could be concluded that $n = 1$ and $\alpha = 0.52$. So, the redox reaction between HRP and the NAE was a single electron transfer process. When $n\Delta E_p < 200$ mV, the electron transfer rate k_s can be estimated with the Laviron's equation $k_s = \alpha nFv/RT$ [42]. The heterogeneous electron transfer rate of each NAE was shown in Table 1. The HRP/NAEs exhibited much higher ks values compared to the HRP/flat Au electrode, due to the high electrocatalytic reaction with elevated amounts of HRP on the NAEs and the excellent electron transfer accelerating property of the well-aligned gold nanowires. The electron transfer rate of HRP/NAEs was also much higher than the ks values of HRP immobilized on zinc oxide nanorods (1.15 s^{-1}) [20] and a Nafion-cysteine/Au electrode (1.13 s^{-1}) [41], suggesting superior electron-transfer kinetics of HRP on the NAEs.

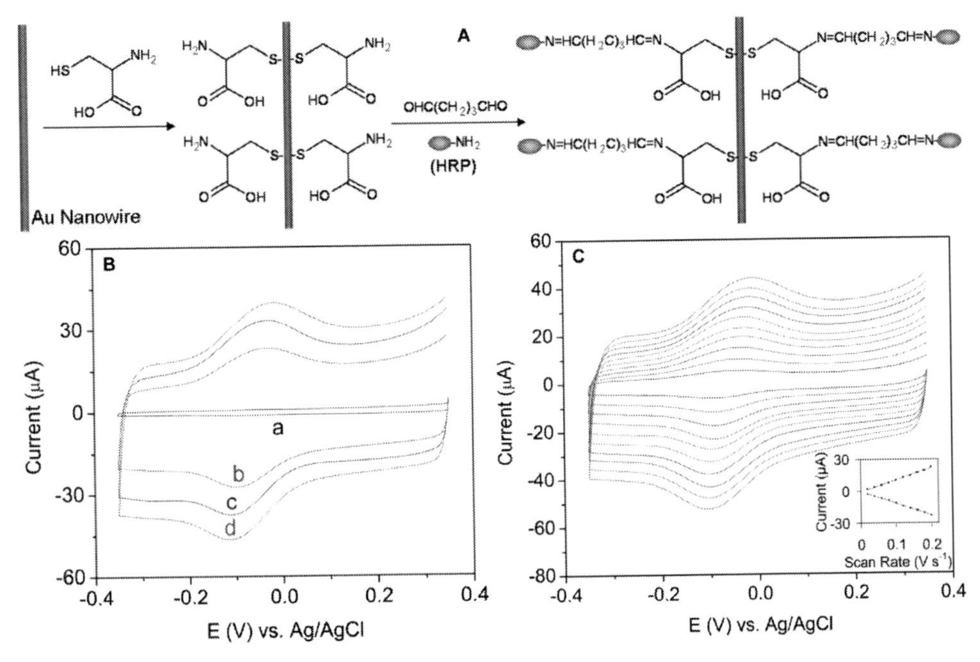

Figure 5. (A) Schematic illustration of step wise fabrication of the HRP/NAEs electrodes. (B) Cyclic voltammograms of the HRP/flat Au electrode (a), NAE_{2h} (b), NAE_{4h} (c) and NAE_{5h} (d) in 50 mM deoxygenated PB (pH 7.0) at 0.1 V s-1. (C) Cyclic voltammograms of the HRP/NAE_{2h} in 50 mM deoxygenated PB (pH 7.0) at different scan rates (20 – 200 mV s^{-1} with an increase of 20 mV s^{-1} each step). Inset: anodic and cathodic current plotted against the scan rate. Reproduced with permission from [26].

According to Faraday's law:

$$I_p = \frac{nFQv}{4RT} = \frac{n^2F^2A\Gamma v}{4RT}$$

the surface average concentration of electroactive HRP (Γ^*, mol cm^{-2}) on the SAM-modified NAEs was estimated as follows:

$$\Gamma^* = \frac{Q}{nFA}$$

where I_p is the reduction current (A), v is scan rate (V s^{-1}), Q is the charge (C) by integrating of the anodic peak of the HRP / NAE, n is the electron transfer number, F is the Faraday constant and A is the electrode area (cm^2). The surface average concentration of active HRP on the NAE_{2h}, NAE_{4h} and NAE_{5h} were 1.42×10^{-10}, 1.94×10^{-10} and 2.48×10^{-10} mol cm^{-2}, respectively (Table 1). The value of Γ^* increased with the length of the nanowires, which was the reason that the direct electron transfer peak current increased towards the enhancement of the roughness ratio. The surface concentration of electroactive HRP was larger than the saturated monolayer concentration of HRP (5.0×10^{-11} mol cm^{-2}) [11] because HRP was coated over the entire surface topology of the nanowire electrode. Comparing to those reported for HRP surface concentration in a multilayer modified ZnO nanorod electrode [20], cysteine/AuNWs are more efficient for HRP immobilization.

Table 1. Key parameters of the HRP modified flat gold and gold nanowire array electrodes (NAEs)

	Roughness ratio[a]	A_{eff} (cm^2)	$\Gamma^* \times 10^{11}$ (mol cm^{-2})	k_s (s^{-1})	Linearity (mM)	Detection limit[b] (µM)	Sensitivity[b] (µA $mM^{-1}cm^{-2}$)	K_M^{app} (mM)
Flat AuE	1	0.22	4.2	0.62	1.8	3.90	0.07	2.06
NAE_{2h}	43.3	0.28	14.2	2.03	15	0.97	10.78	1.87
NAE_{4h}	91.6	0.40	19.8	2.14	15	0.64	26.65	1.54
NAE_{5h}	125.1	0.44	24.8	2.22	15	0.42	45.86	0.64

[a] Calculated as the ratio of the reduction peak area (integrating the voltammogram from +0.7 to +1.2 V) of each NAE relative to the flat electrode. [b] Detection potential: -0.1 V.

4.2. Electrocatalytic reduction of H_2O_2

The enzyme immobilised gold NAEs obtained via glutaraldehyde cross-linking during HRP immobilisation was also used to measure the electrocatalytic reduction of H_2O_2. Electrocatalytic reduction of hydrogen peroxide by the HRP/NAEs was a single electron transfer process, which could be expressed as [41]:

$$\text{HRP[heme(Fe}^{III}\text{)]} + e^- + H^+ \leftrightarrow \text{HRP[heme(H}^+\text{-Fe}^{III}\text{)]}$$

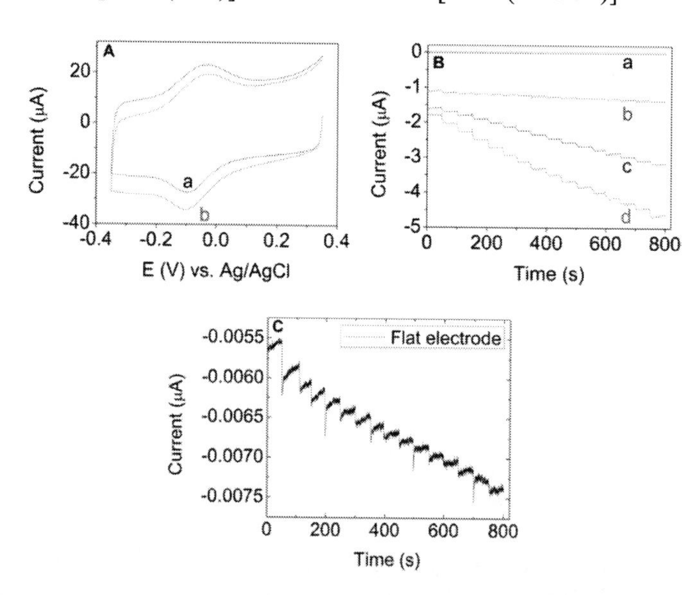

Figure 6. (A) Cyclic voltammograms of the HRP/NAE_{2h} in 50 mM PB, pH 7.0 in the absence (a) and presence (b) of 100 µM H_2O_2. (B) The typical current-time responses of the HRP/flat Au electrode (a), NAE_{2h} (b), NAE_{4h} (c) and NAE_{5h} (d) at -0.1 V with successive addition of 10 µM of H_2O_2, (C) Amplified current-time response of the HRP/flat Au electrode at -0.1 V with successive addition of 10 µM of H_2O_2. Reproduced with permission from [26].

Figure 6A shows the cyclic voltammograms of the HRP/NAE$_{2h}$ in 50 mM PB, pH 7.0 in the absence and presence of 100 μM H_2O_2. The cathodic peak increased greatly upon the addition of H_2O_2, indicating an enzymatic catalytic reaction on the HRP/NAE$_{2h}$. The reduction peak observed at -0.1 V was selected as the detection potential throughout this study.

The typical current-time responses of different HRP/NAEs and the HRP/Au electrode at -0.1 V with successive addition of 10 μM of H_2O_2 were illustrated in Figure 6B. The signal response of H_2O_2 at the HRP/NAEs was large compared to that of the flat electrode, reflecting superior performances of the gold nanowire electrodes. Nevertheless, the amplified current response of the HRP/flat electrode towards H_2O_2 was shown in Figure 6C. The response of the HRP/NAEs towards H_2O_2 was fast (3-4 s), with linearity up to 15 mM, suggesting a fast catalytic process and high catalytic activity of the immobilized enzyme. The response time was faster than the reported results of HRP immobilized on different types of 3D electrodes as shown in Table 2. This could be attributed to three aspects: First, HRP was exposed to the surface of the gold nanowires, therefore, H_2O_2 could diffuse freely to the enzyme. Second, favourable orientation of HRP on cysteine/AuNWs enabled high biocatalytic efficiency. Third, nano-gold on the nanowire surface may accelerate electron transfer between the enzyme and the electrode [14]. The HRP/NAE$_{2h}$, HRP/NAE$_{4h}$ and HRP/NAE$_{5h}$ exhibited high sensitivity of 10.78, 26.65 and 45.86 μA mM^{-1} cm^{-2} and low detection limits of 0.97, 0.64 and 0.42 μM (S/N = 3), respectively, compared to 0.07 μA mM^{-1} cm^{-2} and 3.90 μM at the HRP immobilised on the flat gold electrode. The sensitivity of the HRP/NAEs was much greater than that of HRP entrapped in 3D sol-gel or immobilized on nanotube or nanorod electrodes (see Table 2), evincing that well-aligned nanowires provided larger and more uniform electroactive sites for immobilizing biomolecules and facilitated electron transfer.

Table 2. Performance of the biosensors based-on direct electron transfer of HRP in 3-D matrices

	Response time (s)	**Detection limit** (μM)	**Linear range**	K_M^{app} (mM)	**References**
HRP-SAM-AuNAE$_{5h}$	< 4	0.42	0.74 μM – 15 mM	0.64	26
HRP-AuNPs–SF/GCE	< 8	5	10 μM – 1.8 mM	1.22	15
HRP-CNT-Chitsan-sol-gel/GCE	5	1.4	4.2 μM – 5 mM	6.51	5
HRP-LBL-ZnO nanorod	< 5	1.9	5 μM – 1.7 mM	10.72	20
HRP-TiO_2 nanotube arrays	NA	0.1	50 μM – 1 mM	1.9	44
HRP-AuNP-TiO_2 nanotube arrays	< 5	2.0	5 μM – 0.4 mM	NA	45
HRP-flower ZnO-AuNP-Nafion/GCE	< 5	9.0	15 μM – 1.1 mM	1.76	46
HRP-MSHS-Nafion/ GCE	< 5	1.2	3.9 μM – 0.14 mM	0.22	47

The apparent Michaelis-Menten constant (K_M^{app}), a reflection of the enzymatic affinity and the ratio of microscopic kinetic constants [41], can be obtained from the electrochemical version of the Lineweaver-Burk equation [48]:

$$\frac{1}{I_{ss}} = \frac{1}{I_{max}} + \frac{K_M^{app}}{I_{max} C}$$

where I_{ss} is the steady-state current after the addition of substrate, Imax is the maximum steady-state current measured under saturated substrate condition and C is the bulk concentration of the substrate. The values of HRP/NAE_{2h}, HRP/NAE_{4h} and HRP/NAE_{5h} were 1.87, 1.54 and 0.64 mM, respectively, significantly lower than reported values in Table 2. These results indicated that the immobilized HRP possessed high enzymatic activity and the HRP/NAEs exhibited high affinity for H_2O_2 [15].

When stored in pH 7 phosphate buffers at 4 °C, the biosensors based on HRP covalently immobilized on the NAEs were stable and retained 96% of its initial activity for H_2O_2 after four weeks. Among five electrodes prepared under the same condition, their response to hydrogen peroxide only varied 4.6% (RSD). Excellent stability and reproducibility could be attributed to the firm and consistent attachment of HRP on the SAM-modified NAEs and good biocompatibility of the gold nanowires.

5. Electrochemical Study Of Immobilisied GlOx on PtNP/ Au NAEs

Fig. 7A illustrates the stepwise fabrication process of the PtNP/NAEs and enzyme immobilized PtNP/NAEs. Prior to enzyme immobilisation, PtNP modified NAE was rinsed with acetone and deionised water, followed by running a cyclic voltammetry in 0.5 M H_2SO_4 with a potential range of -0.2 to +1.5 V at the scan rate of 0.1 V s^{-1} until a stable CV is obtained. 20 µl of glutamate oxidase (GlutOx) (Yamasa Corp., Japan, 25 units/vial in 250 µl of 0.01 M phosphate buffer saline (PBS) at pH 7.4) was mixed with 2 mg of bovine serum albumin (BSA) (Lennox, Ireland), 20 µl of glutaraldehyde (GA) and 10 µl of Nafion (perflurinated ion exchange resin, 0.5% w/v, Fluka, Ireland). 10 µl of this mixture was then dispensed onto the surface of a PtNP/NAE and allowed to dry for 60 min under ambient conditions. The electrodes were then stored at 4 ºC overnight in 0.01 M PBS buffer prior to use. Electrode stability was evaluated using amperometry at E_{app} = 0.65 V vs. Ag/AgCl by addition of 200 µM of L-glutamate at room temperature (22 °C) using a potentiostat Versastat 3F (AMETEK PAR, Oak Ridge, TN). Testing was performed every 3 days for a period of two weeks. In between measurement, electrode is stored in PBS at 4 °C.

Fig. 7B showed the amperogram of direct electro-oxidation of H_2O_2 on PtNP/NAE electrode at E_{app} = 0.65 V vs. Ag/AgCl using a potentiostat Versastat. To obtain the amperogram, 200 µM of H_2O_2 was added at a regular interval (Fig. 7B inset). Reproducible response was obtained upon repeated addition of H_2O_2, with a sensitivity of 194.6 µA mM^{-1} cm^{-2} at 20 ºC and a linear range of up to 20 mM. It confirms that electrode does not undergo deactivation during the experiments. Without using any redox mediator or enzyme, PtNP/NAE can detect H_2O_2 as low as 1 µM at ambient. Moreover, PtNP/NAE showed 24 times higher in sensitivity toward H_2O_2 compared to Au NAE and 82 times compare to flat Au electrode (Fig. 7C).

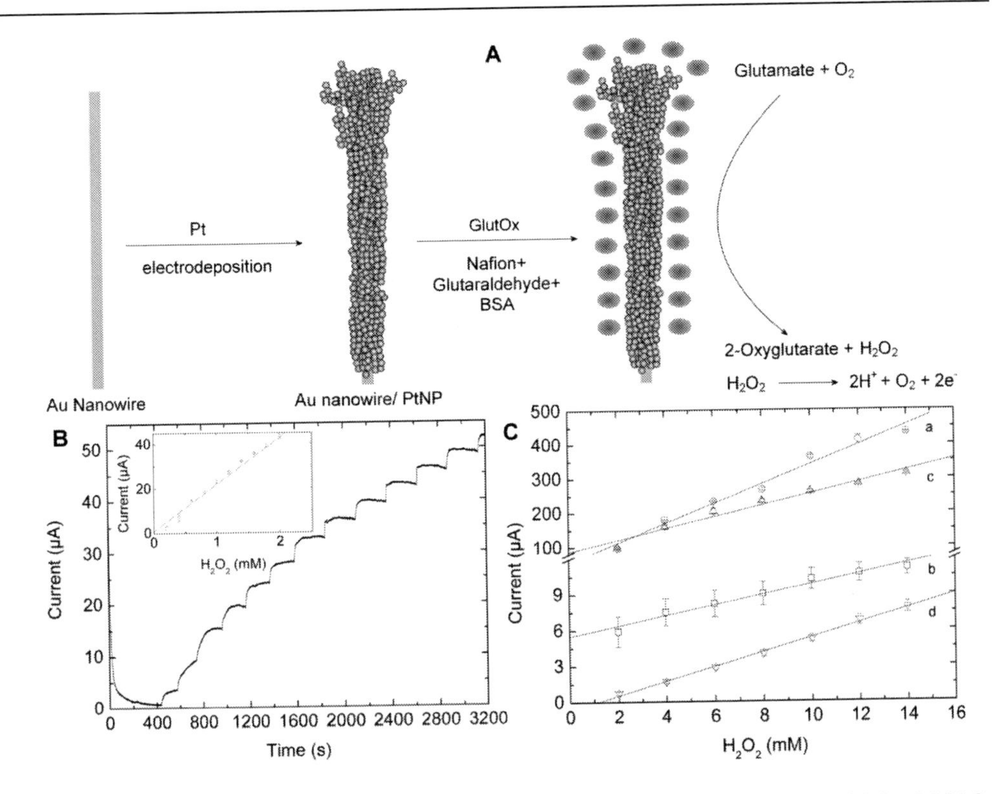

Figure 7. (A) Amperometric response of PtNP/NAE electrode for successive addition of 0.2 mM H_2O_2 (E_{app} = 0.65 V vs. Ag/AgCl) and corresponding concentration plot (inset). Background electrolyte is 0.01 M PBS (pH 7.4), (B) Plot of concentration of H_2O_2 vs. current for amperometric response of PtNP/NAE (a), NAE (b), nanoparticle modified flat Au (c) and flat Au (d) electrode. (Conditions: Background electrolyte is PBS, pH 7.4, E_{app} = 0.65 V). Reproduced with permission from [34].

Table 3. Analytical performance parameters of AuE, PtNP/AuE, NAE and PtNP/NAE using no enzyme for detection of H_2O_2

Electrodes	r^2	t_{90} (s)	Linearity (mM)	Detection limit (μM)	Sensitivity (μA mM^{-1} cm^{-2})
AuE	0.9972	55	2.0 – 20	47	2.35 ± 0.11
PtNP/AuE	0.9627	18	0.02 – 18	23	166.4 ± 7.9
NAE	0.9628	16	1.0 – 16	5	8.0 ± 0.38
PtNP/NAE	0.9847	11	0.02 - 20	1	194.6 ± 9.2

However, sensitivity of PtNP modified flat Au electrode found to be 166.4 μA mM^{-1} cm^{-2} closer to PtNP/NAE 194.6 μA mM^{-1} cm^{-2}, suggest that nanoparticles play a major role in H_2O_2 oxidation. The intra-electrode relative standard deviation (RSD) was found to be 5.96% (n = 3), while inter electrode RSD was 9.45% (n = 3) for 200 μM of H_2O_2 additions. The overall superiority of PtNP/NAE compared to other electrode platform prepared in this work is based on sensitivity, detection limit, t_{90} and r^2 value (Table 3). A comparison of the performance of some efficient sensor platforms using PtNPs on nano array electrodes for H_2O_2 detection are presented in Table 4. It can be seen that the proposed PtNP/NAE sensor template shows a good performance in terms of sensitivity, linear range and limit of detection

(LOD). However, sensitivity which is tabulated in Table 3 was calculated by considering the surface area of all the electrodes are same as all the electrodes are made on gold electrode with a surface area of 0.2 cm^2.

Table 4. A comparison of the performance of some sensor platforms using PtNPs on nano array electrodes for H_2O_2 detection

Electrodes	Sensitivity ($\mu A\ mM^{-1}\ cm^{-2}$)	Linear range (mM)	LOD (μM)	References
PtNP/NAE	194.6	0.02 – 20	1.0	49
PtNP-CNT arrays	140	5×10^{-3} – 25	1.5	50
PtNP-TiO_2 nanotube arrays	1.68	4×10^{-3} – 1.25	4.0	51
PtNP-CNT TiO_2 nanotube arrays	0.134	1×10^{-3} – 2	1.0	52
Pt-AuNP TiO_2 nanotube arrays	2.92	1×10^{-2} – 8×10^{-2}	10	53

The analytical performance of the PtNP/NAE platform was examined to detect glutamate. Immobilisation of glutamate oxidase (GlutOx) was achieved by cross-linking with glutaraldehyde (1% v/v) and BSA (2% w/v) onto the surface of PtNP/NAEs. The optimum enzyme loading was determined by examining the activity levels over the range of 0.02-0.8 U/electrode glutamate oxidase. An enzyme loading of 0.58 U GlutOx per electrode was found to be optimal where higher loadings resulted in poor substrate response. On the other hand, lower enzyme loadings resulting in poor response sensitivity. To obtain the best cross linking, the ratio of Nafion, glutaraldehyde and BSA were used following the article published by Jamal et al. [54].

The potential of the GlutOx immobilised electrode was held at 0.65V and aliquots of glutamate was injected into oxygen saturated PBS at a regular intervals (Fig. 8). This electrode could successfully detect enzymatically generated H_2O_2. The sensitivity of the electrode towards glutamate obtained from the calibration plot was 10.76 $\mu A\ mM^{-1}\ cm^{-2}$, which is higher than a large number of reported glutamate oxidase based sensor [54-59], where the sensitivity varies between 0.001 to 1.43 $\mu A\ mM^{-1}\ cm^{-2}$. This would be attributed to the uniform distribution of nano-particle and a higher amount of enzyme deposited onto the ordered 3D NAE surface. Fig. 8 (inset) also shows that the GlutOx sensor is linear up to 0.8 mM and starts attaining to a saturation level at higher concentrations as expected by Michaelis-Menten type enzyme kinetics. The limit of detection was found to be 14 μM with t90 of 4.6 s, where t_{90} suggest that the electrode responds rapidly to the change of the analyte concentration compared to the reported work. The intra-electrode relative standard deviation (RSD) was found to be 3.23% (n = 3), while inter electrode RSD was 6.65% (n = 3) for 200 μM of glutamate additions.

The stability of the GlutOx modified PtNP/NAEs was determined over a period of two weeks, with a analysis carried out every 3 days, 6 assay each day with 200 μM glutamate addition. In between the testing, the electrode is stored in 0.01 M PBS at 4°C. The enzyme modified electrode was found to retain 98% of its initial activity after two weeks, which compared well with Chang et al., 2007 (approx. 100% after two weeks).

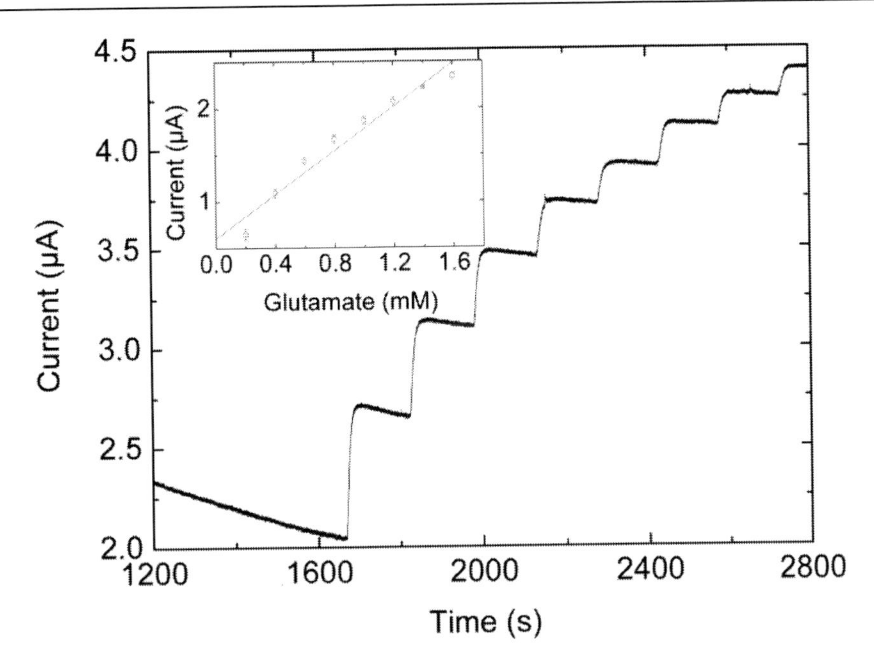

Figure 8. Amperometric response of GlutOx modified PtNP/NAE electrode for successive addition of 0.2 mM glutamate (E_{app} = 0.65 V vs. Ag/AgCl) and corresponding concentration plot (inset). Background electrolyte is 0.01 M PBS (pH 7.4). Reproduced with permission from [34].

6. Conclusions

Vertically aligned NAEs and PtNP/NAEs were prepared by electro-deposition of gold into an anodic aluminium oxide membrane (AAM), providing an ordered three dimensional (3D) matrix for immobilization of redox proteins. The direct electron transfer between HRP and the monolayer-modified NAEs was achieved by combining the superiorities of the ordered 3D nanowires and the self-assembled monolayer. The resulting HRP/NAEs displayed prominent improvement regarding electroactivity, sensitivity and stability, comparing with the HRP/flat gold electrodes. We also have reported a novel platform based on PtNP modified 3D gold nanowire array electrode to detect H_2O_2 without using any mediator or enzyme. PtNP/NAE modified electrode showed 24 times higher in sensitivity toward H_2O_2 compared to Au NAE electrode and 82 times compared to flat Au electrode. PtNP/NAE electrode platform is applied successfully to detect glutamate by immobilising GlutOx using cross linking method, where 14 μM of glutamate can be detected without using any mediator. Moreover, fast response time (<5 s) with high sensitivity towards glutamate (10.76 μA mM^{-1} cm^{-2}) was obtained with a linear range up to 0.8 mM. Glutamate sensitivity of PtNP/NAE electrode platform found to be higher compare to a large number of reported glutamate oxidase based sensor, where the sensitivity varied between 0.001 to 1.43 μA mM^{-1} cm^{-2}. This attractive and versatile procedure, with great clinical and environmental significance, opens up new opportunities for the construction of various mediator-free biosensors with high sensitivity, low detection limit and rapid response.

REFERENCES

[1] Leger, C.; Bertrand, P. *Chem. Rev.* 2008, *108*, 2379.
[2] Freire, R. S.; Pessoa, C. A.; Mello, L. D.; Kubota, L. T. *Journal of the Brazilian Chemical Society* 2003, *14*, 230.
[3] Wu, Y. H.; Hu, S. S. *Microchimica Acta* 2007, *159*, 1.
[4] Guo, C. X.; Hu, F. P.; Li, C. M.; Shen, P. K. *Biosens. Bioelectron.* 2008, *24*, 819.
[5] Kang, X. H.; Wang, J.; Tang, Z. W.; Wu, H.; Lin, Y. H. *Talanta* 2009, *78*, 120.
[6] Wang, J. *Chem. Rev.* 2008, *108*, 814.
[7] Wang, Y.; Xu, H.; Zhang, J. M.; Li, G. *Sensors* 2008, *8*, 2043.
[8] Chaki, N. K.; Vijayamohanan, K. *Biosens. Bioelectron.* 2002, *17*, 1.
[9] Zheng, L. Z.; Yao, X.; Li, J. H. *Current Analytical Chemistry* 2006, *2*, 279.
[10] Gupta, R.; Chaudhury, N. K. *Biosens. Bioelectron.* 2007, *22*, 2387.
[11] Xiao, Y.; Ju, H. X.; Chen, H. Y. *Analytical Biochemistry* 2000, *278*, 22.
[12] Zhao, Y. D.; Zhang, W. D.; Chen, H.; Luo, Q. M.; Li, S. F. Y. *Sens. Actuat. B* 2002, *87*, 168.
[13] Zimmermann, H.; Lindgren, A.; Schuhmann, W.; Gorton, L. *Chemistry-a European Journal* 2000, *6*, 592.
[14] Jia, J. B.; Wang, B. Q.; Wu, A. G.; Cheng, G. J.; Li, Z.; Dong, S. J. *Anal. Chem.* 2002, *74*, 2217.
[15] Yin, H. S.; Ai, S. Y.; Shi, W. J.; Zhu, L. S. *Sens. Actuat. B* 2009, *137*, 747.
[16] Cui, H. F.; Ye, J. S.; Zhang, W. D.; Li, C. M.; Luong, J. H. T.; Sheu, F. S. *Anal. Chim. Acta* 2007, *594*, 175.
[17] Lan, D.; Li, B. X.; Zhang, Z. J. *Biosens. Bioelectron.* 2008, *24*, 934.
[18] Zhu, Y. H.; Cao, H. M.; Tang, L. H.; Yang, X. L.; Li, C. Z. *Electrochim. Acta* 2009, *54*, 2823.
[19] Anandan, V.; Rao, Y. L.; Zhang, G. G. *Inter. J. Nanomedicin.* 2006, *1*, 73.
[20] Gu, B. X.; Xu, C. X.; Zhu, G. P.; Liu, S. Q.; Chen, L. Y.; Wang, M. L.; Zhu, J. J. *J. Phys. Chem. B* 2009, *113*, 6553.
[21] Yu, S. F.; Li, N. C.; Wharton, J.; Martin, C. R. *Nano Lett.* 2003, *3*, 815.
[22] Roberts, M. A.; Kelley, S. O. *J Am Chem Soc* 2007, *129*, 11356.
[23] Yang, M.; Qu, F.; Li, Y.; He, Y.; Shen, G.; Yu, R. *Biosens. Bioelectron.* 2007, *23*, 414.
[24] Lee, S. J.; Anandan, V.; Zhang, G. G. *Biosens. Bioelectron.* 2008, *23*, 1117.
[25] De Leo, M.; Kuhn, A.; Ugo, P. *Electroanal.* 2007, *19*, 227.
[26] Xu, J.; Shang, F.; Luong, J. H. T.; Razeeb, K. M.; Glennon, J. D. *Biosens. Bioelectron.* 2010, *25*, 1313.
[27] Dröge, W. *Physiological Reviews* 2002, *81*, 47.
[28] Wood, Z. A.; Poole, L. B.; Karplus, P. A. *Science* 2003, *300*, 650.
[29] Bakker, E. *Anal. Chem.* 2004, *76*, 3285.
[30] Hrapovic, S.; Liu, Y. L.; Male, K. B.; Luong, J. H. T. *Anal. Chem.* 2004, *76*, 1083.
[31] Karam, P.; Halaoui, L. I. *Anal. Chem.* 2008, *80*, 5441.
[32] Male, K. B.; Hrapovic, S.; Luong, J. H. T. *Analyst* 2007, *132*, 1254.
[33] You, T. Y.; Niwa, O.; Tomita, M.; Hirono, S. *Anal. Chem.* 2003, *75*, 2080.
[34] Jamal, M.; Xu, J.; Razeeb, K. M. *Biosens. Bioelectron.* 2010, *26*, 1420.
[35] Wang, J. *Analytical electrochemistry*; third ed.; John Wiley & Son Inc, 2006.

[36] Li, Z.; Hu, N. F. *J. Electroanal. Chem.* 2003, *558*, 155.
[37] Chailapakul, O.; Popa, E.; Tai, H.; Sarada, B. V.; Tryk, D. A.; Fujishima, A. *Electrochemistry Communications* 2000, *2*, 422.
[38] Anandan, V.; Yang, X.; Kim, E.; Rao, Y., L;; Zhang, G. *Journal of Biological Engineering* 2007, *1*, 5.
[39] Chakraborty, S.; Raj, C. R. *Biosens. Bioelectron.* 2009, *24*, 3264.
[40] Schaumburg, H. H.; Byck, R.; Gerst, R.; Mashman, J. H. *Science* 1969, *163*, 826.
[41] Hong, J.; Moosavi-Movahedi, A. A.; Ghourchian, H.; Rad, A. M.; Rezaei-Zarchi, S. *Electrochim. Acta* 2007, *52*, 6261.
[42] Laviron, E. *J. Electroanal. Chem.* 1979, *101*, 19.
[43] Ma, H. Y.; Hu, N. F.; Rusling, J. F. *Langmuir* 2000, *16*, 4969.
[44] Wu, F. H.; Xu, J. J.; Tian, Y.; Hu, Z. C.; Wang, L. W.; Xian, Y. Z.; Jin, L. T. *Biosens. Bioelectron.* 2008, *24*, 198.
[45] Kafi, A. K. M.; Wu, G.; Chen, A. *Biosens. Bioelectron.* 2008, *24*, 566.
[46] Xiang, C.; Zou, Y.; Sun, L. X.; Xu, F. *Sens. Actuat. B* 2009, *136*, 158.
[47] Cao, Z. X.; Zhang, J.; Zeng, J. L.; Sun, L. X.; Xu, F.; Cao, Z.; Zhang, L.; Yang, D. W. *Talanta* 2009, *77*, 943.
[48] Xiao, Y.; Ju, H. X.; Chen, H. Y. *Anal. Chim. Acta* 1999, *391*, 73.
[49] Jamal, M.; Xu, J.; Razeeb, K. M. *Biosensors and Bioelectronics* 2010, *26*, 1420.
[50] Wen, Z. H.; Ci, S. Q.; Li, J. H. *J. Phys. Chem. C* 2009, *113*, 13482.
[51] Cui, X. L.; Li, Z. Z.; Yang, Y. C.; Zhang, W.; Wang, Q. F. *Electroanal.* 2008, *20*, 970.
[52] Pang, X. Y.; He, D. M.; Luo, S. L.; Cai, Q. Y. *Sens. Actuat. B* 2009, *137*, 134.
[53] Kang, Q.; Yang, L. X.; Cai, Q. Y. *Bioelectrochem.* 2008, *74*, 62.
[54] Jamal, M.; Worsfold, O.; McCormac, T.; Dempsey, E. *Biosens. Bioelectron.* 2009, *24*, 2926.
[55] Cooper, J. M.; Foreman, P. L.; Glidle, A.; Ling, T. W.; Pritchard, D. J. *J. Electroanal. Chem.* 1995, *388*, 143.
[56] Kwon, N. H.; Won, M. S.; Choe, E. S.; Shim, Y. B. *Anal. Chem.* 2005, *77*, 4854.
[57] Kwong, A. W. K.; Grundig, B.; Hu, J.; Renneberg, R. *Biotechnol. Lett.* 2000, *22*, 267.
[58] Nakorn, P. N.; Suphantharika, M.; Udomsopagit, S.; Surareungchai, W. *World J. Microbiol. Biotechnol.*, 2003, *19*, 479.
[59] Oldenziel, W. H.; Beukema, W.; Westerink, B. H. C. *J. Neurosci. Methods* 2004, *140*, 117.

In: Nanowires: Properties, Synthesis and Applications
Editor: Vincent Lefevre
ISBN: 978-1-61470-129-3

Chapter 5

ANALOGIES BETWEEN METALLIC NANOWIRES AND CARBON NANOTUBES

M. A. Grado-Caffaro*[‡] *and M. Grado-Caffaro
C/ Julio Palacios 11, 9-B, 28029-Madrid, Spain

Abstract

In relation to quantum transport, the more remarkable analogies between metallic nanowires and carbon nanotubes are discussed from both qualitative and quantitative standpoints. In fact, we investigate in what aspects of electron conductance metallic nanowires are similar to metallic multiwalled carbon nanotubes. Within this framework, we establish a general mathematical relationship for the electrical conductance of both perfect and imperfect metallic nanowires and multiwalled carbon nanotubes.

Keywords: Metallic nanowires; Metallic multiwalled carbon nanotubes; Electrical conductance.

1. Introduction

Nanowires and carbon nanotubes are quantum electron waveguides which present certain significant analogies from the point of view of electron transport. Nanowires can be metallic as, for instance, made of gold or silver, and semiconducting

(semiconductor nanowires as, for example, GaAs or Si nanowires). On the other hand, carbon nanotubes can be either metallic or semiconducting. At this point, it is well-known that around $1/3$ of the carbon nanotubes are metallic so about $2/3$ are semiconducting. It is also well-known that the conductance through a single-walled carbon nanotube with zero bias is twice the value of the fundamental conductance quantum. However, the conductance of a multiwalled carbon nanotube is not the conductance of one of the layers of the tube multiplied

‡ E-mail address: ma.grado-caffaro@sapienzastudies.com; www.sapienzastudies.com

by the number of layers (walls) [1-4], each layer being, of course, a single-walled tube. The above fact was shown experimentally [1] and, in addition, important theoretical-analytical work on the subject has been carried out [2-4]. At any rate, the conductance through a perfect multiwalled carbon nanotube scales with the number of layers so it is quantized [1-5]. On the other hand, the conductance at resonance through a perfect metallic nanowire is proportional to the number of transverse conducting modes (channels) so it is also quantized (see, for example, refs.[6-8]). Then, one may envisage perfect metallic nanowires and perfect metallic multiwalled carbon nanotubes as similar quantum conductors from the point of view of conductance. This point of view will be analyzed in the following.

2. Theoretical Considerations

We shall establish a general mathematical relationship for the electrical conductance through a quantum wire. More precisely, we refer to both perfect and imperfect wires so that the latter are considered as disordered systems or wires with defects. Then, according to previous formulations as in refs.[2-4], the following expression for doubly quantized conductance (at resonance) represents the above-mentioned general relationship and is valid for metallic nanowires and multiwalled carbon nanotubes:

$$G_{mn} = \frac{aG_0 mn}{m+n} \tag{1}$$

where $m = 1,2,...$ and $n = 1,2,...$ are the two involved quantum numbers relative to their two respective spectra bands, G_0 is the fundamental conductance quantum, and a is a natural number. Formula (1) refers to resonance, that is, when the electronic energy equals the Fermi energy.

For $m = n$, relation (1) reduces to a simply quantized conductance G_n proportional to n. On the other hand, if a is even, the above linear law gives integer multiples of G_0 whereas it gives proper and improper rational fractions of G_0 when a is odd. Note that only one proper rational fraction of G_0 is found corresponding to $a = 1$ so that $G_n = G_0 n/2$. Proper or improper rational-fractional conductance quantization for $m = n$ and for $m \neq n$ occurs in metallic nanowires with disorder [7] while G_n as integer multiples of G_0 (a even) takes place in perfect metallic nanowires and also in perfect multiwalled carbon nanotubes; index n labels the number of walls of these tubes so conductance scales with the number of layers of a given multi-wall tube [1-5]. In multiwalled carbon nanotubes with defects, one has that $G_n = 2G_0/n$ ($a = 4$) or $G_n = 4G_0/n$ ($a = 8$) (see ref.[2]).

If $amn \leq m+n$, then the quantity (transmission factor) which multiplies by G_0 in eq.(1) means probability. This factor is the transmission probability at resonance so it is clear that $0 \leq T_{mn} \leq 1$ where T_{mn} is the above probability such that $G_{mn} = G_0 T_{mn}$. On the other hand, taking into account that $m \geq 1$ and $n \geq 1$, from the aforementioned inequality relative to a, m and n, it follows:

$$a \leq \frac{1}{n} + \frac{1}{m} \leq 2 \tag{2}$$

By inequality (2), it is evident that $a = 1$ or $a = 2$ for the transmission factor regarded as transmission probability.

Now we consider the ground state relative to one of the two involved bands for the doubly quantized conductance (with transmission probability) expressed in eq.(1) so we put, for instance, $m = 1$ in eq.(1). Therefore, one has that $G_{1n} = aG_0 n/(n+1)$ which, by formula (2) and given that $n/(n+1) < 1$, leads to $G_{1n} < 2G_0$. This result tells us that G_{1n} is strictly smaller than twice the fundamental conductance quantum. Hence, G_{1n} is a proper fraction (rational or not) of $2G_0$.

Next we will examine basic aspects of a perfect metallic single-walled carbon nanotube for extrapolating them adequately to a multiwalled carbon nanotube whose layers are metallic. Within this framework, first we recall that a graphene sheet is a zero-gap semiconductor with both bonding and antibonding $\pi -$bands which are degenerate at the corner (vertex) point of the hexagonal bidimensional first Brillouin zone. The periodic boundary conditions for a one-dimensional single-walled carbon nanotube of small diameter give rise to permitted wave-vectors in the circumferential direction of the tube. Then, metallic conduction takes place when one of these permitted wave-vectors passes through the aforementioned corner point (see, for example, ref.[9]). From another perspective, it is well-known that metallic conduction occurs when the following condition is satisfied:

$$2\eta + \mu = 3\alpha \tag{3}$$

where (η, μ) is the pair of indices which define the chirality and the diameter of the tube and α is a natural number.

Moreover, the diameter of the tube is given by:

$$d = \frac{a_0}{\pi}\sqrt{3\left(\eta^2 + \mu^2 + \eta\mu\right)} \tag{4}$$

where a_0 is the atomic lattice spacing (nearest-neighbour carbon atom-carbon atom distance).

On the other hand, the de Broglie wavelength of an electron in the tube is quantized and given by $\lambda_n = \pi d/n$ ($n = 1,2,...$). For the sake of simplicity, now we also employ index n as the same symbol used in the notation relative to the discussion of eq.(1). Since the quantized electron wavenumber reads $k_n = 2\pi/\lambda_n$, from the conjunction of these last two formulae with (3) and (4), one gets (eliminating, for example, μ):

$$k_n = \frac{2\pi n}{3a_0\sqrt{\eta^2 - 3\alpha\eta + 3\alpha^2}} \tag{5}$$

From relation (5) it follows that k_n is proportional to the number of transversal conducting modes. One of these modes, say N, corresponds to k_N as the modulus of the wave-vector that passes through the corner point of the hexagonal two-dimensional first Brillouin zone. Therefore, inserting $n = N$ into (5), one has the expression which gives k_N in terms of chirality.

Now we want to determine the tube with maximal k_n, that is, the values of η and μ which minimize the subradical quantity in formula (5). It is clear that η, μ and α must be natural numbers. For this minimization, one has that $\eta^2 - 3\eta - 4 = 0$; solving this equation, we have that $\eta = 4$ which, replaced into (3) together with the necessary optimum value $\alpha = 3$, yields $\mu = 1$. Consequently, we have found the $(4,1)$ metallic single-walled carbon nanotube. Substituting the above values of η, μ and α into (5), we get $k_{n(\max)} = 2\pi n/\left(3\sqrt{7}a_0\right)$. Assuming the $(4,1)$ tube in question as one (or more) of the layers of a certain multiwalled carbon nanotube, we may speak of an "optimum" perfect metallic multiwalled carbon nanotube.

3. Conclusion

In summary, we have established formulae (1) and (5) as main results of the present communication, relation (1) being a general formula for calculating the electron conductance of metallic nanowires and metallic multiwalled carbon nanotubes. In this context, unifying theoretical formulations valid to investigate electron transport through a wide class of quantum conductors is certainly necessary. On the other hand, relationship (5) provides useful information on electron wavenumber quantization with respect to metallic conduction. In this respect, examining the difference $k_{n(\max)} - k_N$ has a certain interest.

REFERENCES

[1] S. Frank, P. Poncharal, Z.L. Wang, W.A. de Heer: Carbon nanotube quantum resistors. Science 280 (1998) 1744-1746.

[2] M.A. Grado-Caffaro, M. Grado-Caffaro: Fractional conductance in multiwalled carbon nanotubes: a semiclassical theory. Mod. Phys. Lett. B 18 (2004) 761-767.

[3] M.A. Grado-Caffaro, M. Grado-Caffaro: Fractional conductance quantization in multiwalled carbon nanotubes with disorder. Optik 116 (2005) 409-410.

[4] M.A. Grado-Caffaro, M. Grado-Caffaro: A new theory for interpreting electron conductance quantization in multiwalled carbon nanotubes without defects. Physica B 404 (2009) 1544-1545.

[5] M.F. Lin, K.W.-K. Shung: Magnetoconductance of carbon nanotubes. Phys. Rev. B 51 (1995) 7592.

[6] A. Yazdani, D.M. Eigler, N.D. Lang: Off-resonance conduction through atomic wires. Science 272 (1996) 1921-1924.

[7] C. Shu, C.Z. Li, H.X. He, A. Bogozi, J.S. Bunch, N.J. Tao: Fractional conductance quantization in metallic nanoconstrictions under electrochemical potential control. Phys. Rev. Lett. 84 (2000) 5196-5199.

[8] M.A. Grado-Caffaro, M. Grado-Caffaro: On the probability distribution of conduction states in metallic nanowires. Optik 120 (2009) 797-798.

[9] M.S. Dresselhaus, P.C. Eklund: Phonons in carbon nanotubes. Adv. Phys. 49 (2000) 705-814.

In: Nanowires: Properties, Synthesis and Applications ISBN: 978-1-61470-129-3
Editors: Vincent Lefevre

Chapter 6

CHIRALITY DEPENDENT ELASTICITY OF SINGLE WALLED CARBON NANOTUBES

W. Mu and Z-C. Ou-Yang
Institute of Theoretical Physics, The Chinese Academy of Sciences,
Beijing 100190, China

Abstract

Carbon Nanotubes (CNTs), especially Single Walled Carbon Nanotubes (SWCNTs) are good candidates for the devices of nano-electromechanical systems (NEMS). Intuitively, the elasticity of SWCNT and electromechanical coupling of SWCNT should be chirality dependent. By carefully studying the structure and symmetry of SWCNT, an analytical approach to obtain SWCNT's chirality-curvature dependent anharmonic anisotropic elastic constants is developed, the harmonic elastic constants, $c_{11}, c_{12}, \ldots, c_{66}$ and anharmonic elastic constants $c_{111}, c_{112}, \ldots, c_{666}$ are expressed analytically as series expansion of curvature parameter (r_0/R), i.e., the ratio of carbon-carbon bond length to the radius of tube. We found constants reflecting the coupling between axial strain and circumferential strain, such as c_{16} and c_{116} have terms proportional to $(r_0/R)^2 \sin(6\theta_c)$, which imply the asymmetric axial-strain induced torsion (a-SIT) phenomenon. Base on our analytical method, we reproduced recently reported Molecular Dynamic (MD) simulations on asymmetric a-SIT accurately. Present method can be used to study various chirality-curvature dependent electromechanical coupling phenomena of SWCNTs.

Introduction

SWCNT was firstly synthesized in 1993 [1], which can be thought of as a graphene sheet rolled into a seamless cylinder [2]. SWCNT with the diameter in nanometer scale can be even several micrometers long, and shows quasi-one-dimensional electronic/ heat transport behaviors.

SWCNTs have many unique properties due to their special structures, for example, their electron energy band structure are chirality dependent: $1/3$ SWCNTs are metallic while other $2/3$ are semiconductive. Thus their electronic properties depend sensitively on their chirality [3–6].

CNTs/ SWCNTs also have extraordinary mechanical properties, such as high elastic modulus, exceptional directional stiffness, and low density, which make them ideal for the applications of NEMS devises [7–9]. Recent studies have demonstrated the possibilities of using CNT/ SWCNT as actuator [10], nano-tweezers [11], and nanorelay [12–15].

Many previous studies suggested that the elastic properties of SWCNT are insensitive to the tube diameter and chirality [16–21], which implies SWCNT can be modeled as an isotropic elastic thin shell. However, there is at least one counterexample of this over-simplified model: a-SIT response of chiral SWCNT [22], which is a convincing proof of chirality-dependent elasticity for SWCNT with small diameter [23–27]. In this chapter, we will theoretically discuss the chirality-dependent elasticity of SWCNT.

Our main work have two parts: one is to develop a method, the other is to study phenomena:

1) In continuum elastic theory, the elastic properties of SWCNT can be sufficiently described by a set of elastic constants. We determined all these constants from the fundamental carbon-carbon interatomic interaction potential.

2) Use these constants, we reproduced a chirality-dependent anharmonic elastic response: asymmetric a-SIT of SWCNT.

Structures of SWCNTS

To study the chirality dependent elasticity of SWCNT, we start with a brief discuss on lattice structure of SWCNT. The 2D lattice of a graphene is shown in Fig. 1, where several vectors are introduced to characterize the structure of a SWCNT.

The Fig.1 shows the commonly used coordinate system in SWCNT with basis

$$\begin{aligned} \vec{a}_1 &= \sqrt{3}r_0\,(\sqrt{3}/2,\ 1/2), \\ \vec{a}_2 &= \sqrt{3}r_0\,(\sqrt{3}/2,-1/2), \end{aligned}$$

and $r_0 = 1.42\,\text{Å}$ is the carbon-carbon bond length.

The chiral vector of a SWCNT is defined as $\mathbf{C}_h = n\vec{a}_1 + m\vec{a}_2$, a pair of integers (n,m) characterize the chirality of SWCNT [2]. When a graphene sheet is rolled into a cylinder, the points $(0,0)$ and (n,m) coincide, thus the radius of SWCNT is $R = |\mathbf{C}_h|/2\pi$.

The chiral angle θ_c of a SWCNT is defined as the angle between $\vec{a}_1$ and $\mathbf{C}_h$. For an (n,m) SWCNT, $\theta_c = \arccos\left[(2n+m)/2\sqrt{n^2+m^2+nm}\right]$. The translational vector $\mathbf{T}$ is pointing from origin to the first lattice point that the perpendicular line of chiral vector $\mathbf{C}_h$ passes. In 2D lattice, one unit cell of the SWCNT is a rectangle determined by vectors $\mathbf{C}_h$ and $\mathbf{T}$, and vectors $\vec{a}_1$ and $\vec{a}_2$ determine the area of a unit cell for 2D graphite. Therefore, the number of hexagons per unit cell of SWCNT is $N = |\mathbf{C}_h \times \mathbf{T}|/|\vec{a}_1 \times \vec{a}_2| = 2(n^2+m^2+nm)/d_R$, which can be very large for chiral SWCNTs. Here d_R is the greatest common divisor of $(2m+n)$ and $(2n+m)$. There are $2N$ carbon atoms in each unit cell of SWCNT because each hexagon contains two carbon atoms.

A SWCNT has cylindrical geometry, the basis of lattice vectors in cylindrical coordinate system can be written as, [**?**]

$$\vec{\alpha}_1 = (2\pi/N,\tau) \equiv (\psi,\tau), \quad \vec{\alpha}_2 = (0,T),$$

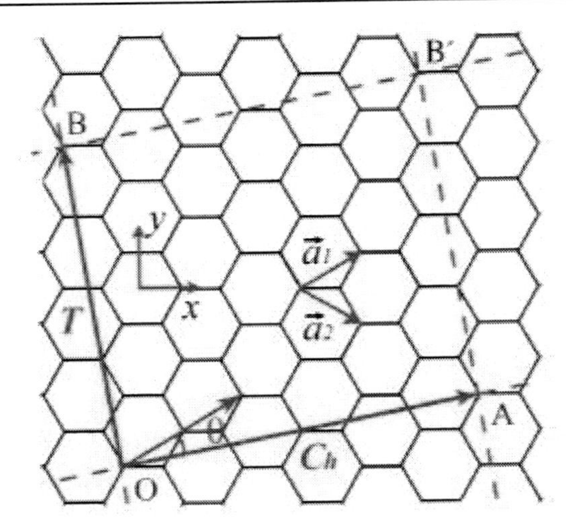

Figure 1. The 2D sketch map of single-wall carbon nanotubes. This image was made with VMD and is owned by the Theoretical and Computational Biophysics Group, an NIH Resource for Macromolecular Modeling and Bioinformatics, at the Beckman Institute, University of Illinois at Urbana-Champaign.

and the corresponding reciprocal lattice vectors give

$$\vec{\beta}_1 = (N,0), \quad \vec{\beta}_2 = (-\tau N/T, 2\pi/T),$$

with

$$T = \frac{3r_0\sqrt{n^2+nm+m^2}}{d_R}.$$

obviously, $\vec{\alpha}_i \cdot \vec{\beta}_j = 2\pi\delta_{ij}$. The electron states or phonon states of SWCNTs can be described by two quantum numbers (k,l). In the first Brillouin zone, $l = 0,1,2,\ldots,N-1$ and $-\pi/T \leq kr_0 \leq \pi/T$.

Continuum Description of the Elasticity of SWCNTs

In microscopic point of view, a small deformation of a SWCNT leads the stretching and bending of carbon carbon bonds, therefore changes both $\sigma-$ and $\pi-$ bonds in SWCNT, thus changes the total bond energy. This is the origin of elastic energy.

To describe the deformation energy of a single layered curved graphite, there are many types of carbon-carbon interaction potential in literatures. In this section, we adopt Lenosky's energy [28], which has a simple form and each energy term has clear physical meaning. In the next section, we will use so-called second generation potential (REBO-II potential) [29].

Lenosky's *et al.* energy is written as, [28]

$$\begin{aligned} E^g &= (\varepsilon_0/2)\sum_{(ij)}(r_{ij}-r_0)^2+\varepsilon_1\sum_i\left(\sum_{(j)}\mathbf{u}_{ij}\right)^2 \\ &+ \varepsilon_2\sum_{(ij)}(1-\mathbf{n}_i\cdot\mathbf{n}_j)+\varepsilon_3\sum_{(ij)}(\mathbf{n}_i\cdot\mathbf{u}_{ij})(\mathbf{n}_j\cdot\mathbf{u}_{ji}). \end{aligned} \tag{1}$$

The first two terms are bond stretching energy and bond bending energy. The last two terms come from the contribution of the π-electron resonance. The idea of π-electron resonance terms is natural: deforming a SWCNT changes the spatial orientation of a carbon atom's p_z orbit and affects the forming of delocalize $\pi-$ bonds in SWCNT. Lenosky *et al.* considered the orientation of a carbon atom's p_z orbital is the same as $\vec{n}_i$, the normal direction of the surface of SWCNT at the point where the atom located (the normal direction of the triangle determined by the atom's three nearest neighboring atoms) [28]. Thus, the last two terms of Eq. (1) describe the carbon-carbon interaction beyond the nearest neighboring interaction. Here, the reference state is planar graphite, i.e., $\vec{n}_i\cdot\vec{n}_j=1$.

In the first term, $r_0=1.42$ Å is the bond length of a graphene sheet without deformation, and r_{ij} is the length of the bond between atoms i and j with deformation. In the remaining terms, $\mathbf{u}_{ij}$ is a unit vector pointing from atom i to its neighbor atom j, and $\mathbf{n}_i$ is the unit vector normal to the plane determined by the atom i's three nearest neighbors. The summation $\Sigma_{(j)}$ is taken over the three nearest neighboring j atoms of i atom, and $\Sigma_{(ij)}$ is taken over all the nearest neighboring atom pairs (i,j). The parameters $(\varepsilon_1,\varepsilon_2,\varepsilon_3)=(0.96,1.29,0.05)$ eV were determined by Lenosky *et al.* [28] by the calculations based on local density approximation. The value of ε_0 was given as $\varepsilon_0=57\text{eV}/\text{Å}^2$ by Zhou *et al.* [30] using the force-constant method.

In 1997, Ou-Yang *et al.* [31] developed a method to find the continuum limit of Lenosky's energy without considering the bond length change and obtained the curvature elastic energy of a SWCNT

$$E^{(s)}=\int\left[\frac{1}{2}k_c(2H)^2+\bar{k}_1K\right]dA, \tag{2}$$

where the bending elastic constant

$$k_c=(18\varepsilon_1+24\varepsilon_2+9\varepsilon_3)r_0^2/(32\Omega)=1.17\text{eV} \tag{3}$$

with $\Omega=2.62$ Å^2 being the occupied area per atom, and

$$\bar{k}_1/k_c=-(8\varepsilon_2+3\varepsilon_3)/(6\varepsilon_1+8\varepsilon_2+3\varepsilon_3)=-0.645. \tag{4}$$

The key idea of Ou-Yang *et al.*'s approach is that the surface of SWCNT can be perfectly embedded by six-member carbon rings [31], the carbon-carbon bond of a SWCNT is curved on the surface of SWCNT, which is a part of a geodesic on the surface. In the continuum limit, the bond vector can be expressed as,

$$\begin{aligned} \vec{r}^{\,0}(M) &= \vec{r}_{ij}^{\,0}=\left[1-r_0^2\kappa^2(M)/6\right]r_0\vec{t}(M) \\ &+ \left[r_0\kappa(M)/2+r_0^2\kappa_s(M)/6\right]r_0\vec{N}(M) \\ &+ \left[\kappa(M)\tau(M)r_0^2/6\right]r_0\vec{b}(M),\ M=1,2,3 \end{aligned} \tag{5}$$

Here, $r_0 = 1.42\AA$ is carbon-carbon bond length without strains, the same as that used in Eq. (1), $M = 1,2,3$ denote three sp^2–bonded curves from atom i to atoms j on the surface of a SWCNT, $\{\vec{t}, \vec{N}, \vec{b}\}$ is Frenet frame at the location of atom i on the surface [31], and κ, τ and s are curvature, torsion, and arc parameter of a bond curve, respectively, $\kappa_s \equiv d\kappa/ds$. [31].

Then by differential geometric approach, up to $O(r_0/R)$, the bending energy can be expressed by geometric quantities. In Eq. (2), there are mean curvature H, Gaussian curvature K of a surface, and dA is area element.

In 2002, Tu and Ou-Yang obtained the total free energy [32] of a strained SWCNT with in-plane strain $\varepsilon_i = \begin{pmatrix} \varepsilon_1 & \varepsilon_6/2 \\ \varepsilon_6/2 & \varepsilon_2 \end{pmatrix}$ at the i-atom site, where ε_1, ε_2, and ε_6 are the axial, circumferential, and shear strains, respectively [33]. The total free energy contains two parts: one is the curvature energy expressed as Eq. (2); the other is the deformation energy [32]

$$E_d = \int \left[\frac{1}{2}k_d(2J)^2 + \bar{k}_2 Q\right] dA, \tag{6}$$

where $2J = \varepsilon_1 + \varepsilon_2$ and $Q = \varepsilon_1\varepsilon_2 - \varepsilon_6^2/4$, are respectively named "mean" and "Gaussian" strains, and

$$k_d = 9\left(\varepsilon_0 r_0^2 + \varepsilon_1\right)/(16\Omega) = 24.88\text{eV}/\text{Å}^2, \tag{7}$$

$$\bar{k}_2 = -3\left(\varepsilon_0 r_0^2 + 3\varepsilon_1\right)/(8\Omega) = -0.678k_d. \tag{8}$$

Tu and Ou-Yang, found that the total elastic energy (bending energy and strain energy)

$$E_d^{(s)} = \int \left[\frac{1}{2}k_c(2H)^2 + \bar{k}_1 K\right] dA + \int \left[\frac{1}{2}k_d(2J)^2 + \bar{k}_2 Q\right] dA \tag{9}$$

are comparable with the elastic energy of classic shell theory [34]:

$$\begin{aligned} E_c &= \frac{1}{2}\int D\left[(2H)^2 - 2(1-\nu)K\right] dA \\ &+ \frac{1}{2}\int \frac{C}{1-\nu^2}\left[(2J)^2 - 2(1-\nu)Q\right] dA, \end{aligned} \tag{10}$$

where $D = (1/12)Yh^3/(1-\nu^2)$ and $C = Yh$ are bending rigidity and in-plane stiffness of a shell. The ν is the Poisson ratio and h is the thickness of shell. Comparing Eq.(9) with Eq.(10), Tu and Ou-Yang obtained the Poisson ratio, effective wall thickness, and Young's modulus of SWCNT are $\nu = 0.34$, $h = 0.75\text{Å}$ and, $Y = 4.70\text{TPa}$, which are close to those given by Yakobson *et al.* [16].

In 2009, we improved the approach of Ou-Yang and Tu, studying the chirality-dependent elasticity of SWCNT with small diameter. Chiral correction of SWCNT's elasticity is also a curvature effect. In a limit $R \to \infty$, the elastic properties of SWCNT should be the same as those of a planar graphene sheet.

In general, the 2D continuum limit of strain energy per unit area of SWCNT can be written as,

$$\mathcal{E}_{elsticity} = \frac{1}{2}\sum_{ij} c_{ij}\varepsilon_i\varepsilon_j \tag{11}$$

where c_{ij} are in-plane elastic constants, $i, j, k = 1, 2, 6$.

Keep $O(r_0/R)^2$ terms in Eq. (5), based on Lenosky's type of interatomic interaction energy Eq. (1), we have chirality-dependent elastic constants of c_{11}, c_{12}, c_{22}, and c_{66},

$$\begin{aligned}
c_{11} &= \frac{9(\varepsilon_0 r_0^2 + 3\varepsilon_1)}{16\Omega_0} + \alpha^2 \cdot \frac{-4\varepsilon_0 r_0^2 + 9\varepsilon_1}{128\Omega_0} + \alpha^2 \cdot \frac{-4\varepsilon_0 r_0^2 + 9\varepsilon_1}{128\Omega_0} \cos(6\theta), \\
c_{12} &= \frac{3(\varepsilon_0 r_0^2 + 9\varepsilon_1)}{32\Omega_0} + \alpha^2 \cdot \frac{-8\varepsilon_0 r_0^2 + 75\varepsilon_1}{256\Omega_0} + \alpha^2 \cdot \frac{4\varepsilon_0 r_0^2 - 15\varepsilon_1}{128\Omega_0} \cos(6\theta), \\
c_{22} &= \frac{9(\varepsilon_0 r_0^2 + 3\varepsilon_1)}{16\Omega_0} + \alpha^2 \cdot \frac{-2\varepsilon_0 r_0^2 - 21\varepsilon_1}{32\Omega_0} + \alpha^2 \cdot \frac{2\varepsilon_0 r_0^2 + 3\varepsilon_1}{64\Omega_0} \cos(6\theta), \\
c_{66} &= \frac{3(\varepsilon_0 r_0^2 + 9\varepsilon_1)}{16\Omega_0} + \alpha^2 \cdot \frac{-4\varepsilon_0 r_0^2 - 57\varepsilon_1}{128\Omega_0} + \alpha^2 \cdot \frac{2\varepsilon_0 r_0^2 - 3\varepsilon_1}{128\Omega_0} \cos(6\theta),
\end{aligned}$$

Here, $\theta = \pi/6 - \theta_c$, and $\alpha \equiv (r_0/R)^2$. Substitute the values of ε_0, ε_1, r_0, and Ω_0 to Eq. (12), these in-plane elastic constants of SWCNT are given in unit of eV/$\AA^2$,

$$\begin{aligned}
c_{11} &= 23.0 - \alpha^2 + \alpha^2 \cos(6\theta_c), \\
c_{12} &= 9.6 - 0.6\,\alpha^2 - 0.9\,\alpha^2 \cos(6\theta_c), \\
c_{22} &= 23.0 - 3.4\,\alpha^2 - 1.2\,\alpha^2 \cos(6\theta_c), \\
c_{66} &= 9.6 - 1.9\,\alpha^2 - 0.5\,\alpha^2 \cos(6\theta_c).
\end{aligned}$$

In particular, we also get c_{16} (c_{26}), the harmonic elastic constant for coupling between axial strain (circumferential) strain and torsional twist, which are proportional to $(a_0/R)^2 \sin(6\theta_c)$, manifesting these coupling are different for left- and right-handed SWCNTs. We also calculated the Young's modulus and Possion's ratio based on these elastic constants, the curvature-chirality correction terms are even functions of chiral angle θ_c, implying we can not separate left- and right-handed SWCNTs by measuring the Young's modulus and Possion's ratio.

Axial-Strain Induced Torsion (a-SIT) of SWCNTs

The nonzero elastic constant $c_{16} \propto (r_0/R)^2 \sin(6\theta_c)$ implies the coupling between axial strain and torsional strain, which clearly shows a-SIT response is one of the curvature and chirality effects, only occurs in chiral SWCNT. From the expression of c_{16}, it is easy to understand that linear a-SIT response is distinct at $\theta_c = \pi/12$ and significant for SWCNT with small diameter, which is in accord with previous theoretical and simulation works [22–24].

To study the chirality-dependent elasticity of SWCNTs, we theoretically reproduced a recent MD simulations on SWCNTs' asymmetric a-SIT response reported by Liang *et al* [23]. The main character of asymmetric a-SIT is that a-SIT responses for tension or compression at large strain are much different. Torsion angle per unit length increases when the strain increases in tension. However, with increasing strain under compression, the torsion angle firstly increases, then decreases to zero, and increases again after changing the direction of twist.

This phenomenon, as a type of curvature-chirality effect, cannot be explained by the usual isotropic elastic shell theory of SWCNT. Moreover, asymmetric a-SIT behaviors for axial tensile and compressing strains suggests the anharmonic elasticity of SWCNT also plays an important role in asymmetric a-SIT responses.

We used second generation reactive empirical bond order (REBO-II) potential instead of Lenosky's energy [28], which is a classical many-body potential for solid carbon and hydrocarbons [29]. The advantages of the REBO-II potential are double-fold: it has an analytical form of carbon-carbon pair potentials with the bond length and bond angle being variables of energy functions; all the parameters of the REBO-II potential were fitted from a large data set of experiments and *abinitio* calculations, which make sure it can accurately reproduce the elastic properties of diamond and graphite. Liang *et al.*'s molecular dynamic simulation was also based on the REBO-II potential [23].

With the inclusion of a cubic anharmonic elastic terms, the elastic energy has a form [27]

$$\mathcal{E}_{elsticity} = \frac{1}{2}\sum_{ij} c_{ij}\varepsilon_i\varepsilon_j + \sum_{i\leq j\leq k} c_{ijk}\varepsilon_i\varepsilon_j\varepsilon_k, \tag{12}$$

all the elastic constants based on REBO-II potential have been obtained analytically [27]. Elastic constants $c_{11}, c_{12}, \ldots, c_{666}$ presented in Eq. (12) can be described by,

$$\mathbf{cc} = \mathbf{Mb},$$

where, **cc** is a column vector with the components cc_1 to cc_{16} being the sixteen elastic constants of SWCNT, i.e., c_{11} to c_{666}, respectively, **M** is a 16×5 matrix, **b** is a column vector with components [27],

$$b_1 = \left(\frac{\partial^2 V}{\partial r_{ij}^2}\right)_0 \cdot \frac{r_0^2}{\Omega_0},\ b_2 = \left(\frac{\partial^2 V}{\partial\left(\cos\theta_{ijk}\right)^2}\right)_0 \cdot \frac{1}{\Omega_0},$$

$$b_3 = \left(\frac{\partial^2 V}{\partial r_{ij}\partial\cos\theta_{ijk}}\right)_0 \cdot \frac{r_0}{\Omega_0}, b_4 = \left(\frac{\partial^2 V}{\partial\cos\theta_{ijk}\partial\cos\theta_{ijl}}\right)_0 \cdot \frac{1}{\Omega_0},$$

$$b_5 = \left(\frac{\partial^3 V}{\partial r_{ij}^3}\right)_0 \cdot \frac{r_0^3}{\Omega_0}.$$

Here, $r_0 = 1.42\,\text{Å}$ is carbon-carbon bond length without strains, and $\Omega_0 = 2.6\,\text{Å}^2$ is the area occupied by one carbon atom at the surface of SWCNTs. The $V\left(r_{ij}, \theta_{ijk}\right)$ is REBO-II potential, and r_{ij} is bond length between atom-i and atom-j, θ_{ijk} is the bond angle between adjacent bonds i-j and i-k. The derivatives can be obtained from analytical form of REBO-II potential [27],

$$\left(\frac{\partial^2 V}{\partial r_{ij}^2}\right)_0 \approx 43.67\text{eV}\cdot\text{Å}^{-2},\ \left(\frac{\partial^2 V}{\partial r_{ij}\partial\cos\theta_{ijk}}\right)_0 \approx -5.924\text{eV}\cdot\text{Å}^{-1},$$

$$\left(\frac{\partial^2 V}{\partial\left(\cos\theta_{ijk}\right)^2}\right)_0 \approx 3.187\text{eV},\ \left(\frac{\partial^2 V}{\partial\cos\theta_{ijk}\partial\cos\theta_{ijl}}\right)_0 \approx -0.367\text{eV},$$

and

$$\left(\frac{\partial^3 V}{\partial r_{ij}^3}\right)_0 \approx -333.4\mathrm{eV}\cdot\text{Å}^{-3}.$$

All 34 non-zero elements of matrix **M** are analytically written as [27]

$$\begin{aligned}
M_{1,1} &= \frac{9}{16}+\left(\frac{-45}{1024}\right)\alpha^2,\\
M_{2,1} &= \frac{3}{16}+\left(\frac{-43}{1024}\right)\alpha^2+\left(\frac{-11}{256}\right)\alpha^2\cos(6\theta),\\
M_{3,1} &= \frac{9}{16}+\left(\frac{-205}{1024}\right)\alpha^2+\left(\frac{11}{128}\right)\alpha^2\cos(6\theta),\\
M_{4,1} &= \left(\frac{11}{512}\right)\alpha^2\sin(6\theta),\\
M_{5,1} &= \left(\frac{-33}{512}\right)\alpha^2\sin(6\theta),\\
M_{6,1} &= \frac{3}{16}+\left(\frac{-43}{1024}\right)\alpha^2+\left(\frac{-11}{256}\right)\alpha^2\cos(6\theta),\\
M_{1,2} &= \frac{27}{16}+\left(\frac{-27}{64}\right)\alpha^2+\left(\frac{-189}{512}\right)\alpha^2\cos(6\theta),\\
M_{2,2} &= \frac{-27}{16}+\left(\frac{27}{128}\right)\alpha^2+\left(\frac{27}{64}\right)\alpha^2\cos(6\theta),\\
M_{3,2} &= \frac{27}{16}+\left(\frac{-243}{256}\right)\alpha^2\cos(6\theta),\\
M_{4,2} &= \left(\frac{-405}{1024}\right)\alpha^2\sin(6\theta),\\
M_{5,2} &= \left(\frac{459}{1024}\right)\alpha^2\sin(6\theta),\\
M_{6,2} &= \frac{27}{16}+\left(\frac{-27}{256}\right)\alpha^2+\left(\frac{27}{64}\right)\alpha^2\cos(6\theta),\\
M_{1,3} &= \frac{-9}{8}+\left(\frac{-285}{1024}\right)\alpha^2+\left(\frac{-3}{64}\right)\alpha^2\cos(6\theta),\\
M_{2,3} &= \frac{9}{8}+\left(\frac{-627}{1024}\right)\alpha^2+\left(\frac{-3}{256}\right)\alpha^2\cos(6\theta),\\
M_{3,3} &= \frac{-9}{8}+\left(\frac{-189}{1024}\right)\alpha^2+\left(\frac{9}{128}\right)\alpha^2\cos(6\theta),\\
M_{4,3} &= \left(\frac{-9}{512}\right)\alpha^2\sin(6\theta),\\
M_{5,3} &= \left(\frac{-21}{512}\right)\alpha^2\sin(6\theta),
\end{aligned}$$

$$M_{6,3} = \frac{-9}{8} + \left(\frac{165}{1024}\right)\alpha^2 + \left(\frac{-3}{256}\right)\alpha^2\cos(6\theta),$$

$$M_{1,4} = \frac{-27}{32} + \left(\frac{27}{128}\right)\alpha^2 + \left(\frac{189}{1024}\right)\alpha^2\cos(6\theta),$$

$$M_{2,4} = \frac{27}{32} + \left(\frac{-27}{256}\right)\alpha^2 + \left(\frac{-27}{128}\right)\alpha^2\cos(6\theta),$$

$$M_{3,4} = \frac{-27}{32} + \left(\frac{243}{1024}\right)\alpha^2\cos(6\theta),$$

$$M_{4,4} = \left(\frac{-459}{2048}\right)\alpha^2\sin(6\theta),$$

$$M_{5,4} = \left(\frac{-459}{2048}\right)\alpha^2\sin(6\theta),$$

$$M_{6,4} = \frac{-27}{32} + \left(\frac{27}{512}\right)\alpha^2 + \left(\frac{-27}{128}\right)\alpha^2\cos(6\theta),$$

$$M_{7,5} = \frac{5}{64} + \left(\frac{1}{128}\right)\cos(6\theta),$$

$$M_{8,5} = \frac{3}{64} + \left(\frac{-3}{128}\right)\cos(6\theta),$$

$$M_{9,5} = \left(\frac{3}{128}\right)\sin(6\theta),$$

$$M_{10,5} = \frac{3}{64} + \left(\frac{3}{128}\right)\cos(6\theta),$$

$$M_{11,5} = \left(\frac{-3}{64}\right)\sin(6\theta),$$

$$M_{12,5} = \frac{3}{64} + \left(\frac{-3}{128}\right)\cos(6\theta),$$

$$M_{13,5} = \left(\frac{5}{64}\right) + \left(\frac{-1}{128}\right)\cos(6\theta),$$

$$M_{14,5} = \left(\frac{3}{128}\right)\sin(6\theta),$$

$$M_{15,5} = \left(\frac{3}{64}\right) + \left(\frac{3}{128}\right)\cos(6\theta),$$

$$M_{16,5} = \left(\frac{-1}{128}\right)\sin(6\theta).$$

Here, $\alpha \equiv r_0/R$ and $\theta = \pi/6 - \theta_c$.

For tubes with large diameters and small strains, anharmonic elastic energy can be ignored along with c_{16} and c_{26} terms, thus the isotropic thin shell model for SWCNT is recovered, and the calculated in-plane Young's modulus and Possion's ratio are similar to the results in Ref. [32].

We reproduced the Liang *et al.*'s MD simulation accurately, as shown in Fig. 2.

Our continuum elastic theory has some advantages, compared to previous simulations,

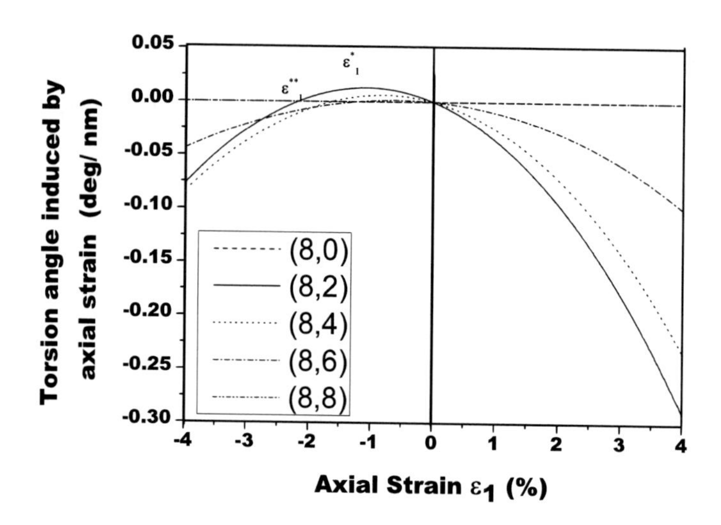

Figure 2. Torsion angle-axial strain relations for a series of $(8,m), m = 0,2,4,6,8$ SWCNTs, which shows the chirality dependence of the a-SIT response. Only chiral SWCNTs $(8,2)$, $(8,4)$ and $(8,6)$ have an a-SIT response, as shown.

since it is suitable to study SWCNTs with actual sizes, and all the elastic constants in the theory are obtained analytically.

Conclusion

In this chapter, we focus on chirality dependent elasticity of SWCNTs, shown the theoretical frame of continuum elasticity approaches developed by us. To illustrate the power of our approach, we discuss asymmetric a-SIT response, a phenomenon of SWCNTs induced by their chiral elasticity in detail. Analytical expressions of anharmonic elastic energy, as well as curvature-chirality induced anisotropic elasticity (chiral elasticity) are obtained based on the REBO-II potential. The calculated results are in reasonable agreement with recent molecular dynamic simulations. Our method can be used to calculate the chiral elastic properties and anharmonic elastic properties of SWCNTs with arbitrary radii and chiralities analytically.

References

[1] Iijima, S.; Ichihashi, T. *Nature* 1993, *363*, 603.

[2] Saito, R.; Dresselhaus, M. S.; Dresselhaus, G. *Physical Properties of Carbon Nanotubes*; Imperial College Press: London, 1998; pp 35-48.

[3] Mintmire, J. W.; Dunlap, B. I.; White, C. T. *Phys. Rev. Lett.* 1992, *68*, 631.

[4] Hamada, N.; Sawada, S. I.; Oshiyama, A. *Phys. Rev. Lett.* 1992, *68*, 1579.

[5] Saito, R.; Fujita, M.; Dresselhaus, G.; Dresselhaus, M. S. *Appl. Phys. Lett.* 1992, *60*, 2204.

[6] Tans, S. J. *et al. Nature* 1997, *386*, 474.

[7] Baughman, R. H.; Zakhidov, A. A.; and de Heer, W. A., *Science* 2002, *297*, 787.

[8] Craighead, H. G., Science, **290**, 1532 (2000).

[9] Sapmaz, S.; Blanter, Y. M.; Gurevich, L.; and van der Zant, H. S. J., *Phys. Rev. B* 2003, *67*, 235414.

[10] Baughman, R. H., *et al.*, *Science* 1999, *284*, 1340.

[11] Kim, P.; and Lieber, C. M., *Science* 1999, *286*, 2148.

[12] Kinaret, J. M.; Nord, T.; and Viefers, S., *Appl. Phys. Lett.* 2003, *82*, 1287.

[13] Lee, H. W. *et al*, *Nano Lett.* 2004, *4*, 2027.

[14] Jang, J. E. *et al*, *Appl. Phys. Lett.* 2005, *87*, 163114.

[15] Jang H. W. *et al.*, *Appl. Phys. Lett.* 2008, *93*, 113105.

[16] Yakobson, B. I.; Brabec, C. J.; Bernholc, J. *Phys. Rev. Lett.* 1996, *76*, 2511.

[17] Zhou, X.; Zhou, J. J.; Ou-Yang, Z. C. *Phys. Rev. B* 2000, *62*, 13692.

[18] Hernández, E. *et al. Phys. Rev. Lett.* 1998, *80*, 4502.

[19] Lu, J. P. *Phys. Rev. Lett.* 1997, *79*, 1297.

[20] Krishnan, A. *et al. Phys. Rev. B* 1998, *58*, 14013.

[21] Salvetat, J. -P., *et al.*, *Adv. Mat.* 1999, *11*, 161.

[22] Gartstein, Y. N.; Zakhidov, A. A.; Baughman, R. H., *Phys. Rev. B* 2003, *68*, 115415.

[23] Liang, H.; and Upmanyu, M., *Phys. Rev. Lett.* 2006, *96*, 165501.

[24] Geng, J.; and Chang, T., *Phys. Rev. B* 2006, *74*, 245428.

[25] Zhang, H. W.; Wang, L.; Wang, J. B.; Zhang, Z. Q.; and Zheng, Y. G., Phys. Lett. A 2008, *372*, 3488.

[26] Upmanyu, M.; Wang, H.L., Liang, H.; Mahajan. R, *J. R. S. interface* 2008, *5*, 303.

[27] Mu, W.; Li, M.; Wang, W.; Ou-Yang, Z-c., *New. J. Phys.* 2009, *11*, 113049.

[28] Lenosky, T. *et al. Nature* 1992 *355*, 333.

[29] Brenner, D. W.; Shenderova, O. A.; Harrison, J. A.; Stuart, S. J.; Ni, B.; Sinott, S. B., *J. Phys. Condens. Matter* 2002, *14*, 783.

[30] Zhou, X.; Zhou, J. J.; Ou-Yang, Z. C. *Physica B* 2001, *304*, 86.

[31] Ou-Yang, Z. C.; Su, Z. B.; Wang, C. L. *Phys. Rev. Lett.* 1997, *78*, 4055.

[32] Tu, Z. C.; Ou-Yang, Z. C. *Phys. Rev. B* 2002, *65*, 233407.

[33] Nye, J. F. *Physical Properties of Crystals*; Clarendon Press: Oxford, 1985; pp 33-148.

[34] Landau, L. D.; Lifshiz, E. M. *Elasticity Theory*; Pergamon: Oxford, 1986.

In: Nanowires: Properties, Synthesis and Applications
Editor: Vincent Lefevre
ISBN: 978-1-61470-129-3

Chapter 7

ROADMAP TO NANOSCALE HETEROSTRUCTURES: FROM SINGLE-COMPONENT NANOWIRES TO AXIALLY HETEROSTRUCTURED NANOWIRES

Nitin Chopra *

Department of Metallurgical and Materials Engineering, Center for Materials for Information Technology (MINT), The University of Alabama, Tuscaloosa, AL, U. S.

ABSTRACT

Growth of multicomponent nanostructures (or "nanoscale heterostructures") is critical for the development of complex nanodevices and multifunctional platforms. However, to fully understand the field of nanoscale heterostructures, it is important for nanotechnologists and nanoscientists to fundamentally understand single-component nanostructures. This review article provides an outline on critical concepts in single-component nanowires that could be further utilized to fabricate multicomponent axially heterostructured nanowires. This article comprises of an introduction focused on nanostructures and their applications, followed by a classification of nanostructures with a particular desire to develop and study nanowires, growth and characterization studies of single-component nanowires (semiconducting, metallic, etc.), using this knowledge to design and fabricate multicomponent axially heterostructured nanowires, characterization advances for understanding such nanowires, and applications of axially heterostructured nanowires. This article attempts to introduce a complete roadmap for the development of multicomponent axially heterostructured nanowires from single-component nanowires and approaches to integrate the former in device architectures. Such nanowire systems hold immense promise for applications in chemical and biological sensors, imaging tools, nanomagnetics, optoelectronics, and nanoelectronics.

* Department of Metallurgical and Materials Engineering, Center for Materials for Information Technology (MINT), 301 Seventh Avenue, 116 Houser Hall, The University of Alabama, Tuscaloosa, AL 35401, E mail: nchopra@eng.ua.edu, Tel: 205-348-4153, Fax: 205-348-2164.

INTRODUCTION

Nanomaterials exhibit distinct chemical, electrical, optical, thermal, and mechanical properties making them exciting for a range of applications from nanomedicine to nano-opto-electronics and sensors to energy devices [1-18]. This has led to comprehensive studies on understanding nanostructured carbides, oxides, nitrides, metals, carbons, alloys, polymers, semiconductors, and composites [1, 2, 19-78]. Furthermore, combination of different materials at nanoscale is of significant importance due to their enhanced functionality, self-stabilizing character, tunable interfaces, and novel properties. All these are possible due to the combined effect of structure, morphology, and size-scale of such multicomponent systems. Towards this end, heterostructures based on one-dimensional (1-D) nanostructures such as nanowires are very critical for the future development, miniaturization, and enhanced performance of devices and analytical platforms. Of particular importance are radially or axially heterostructured nanowires that can have composition modulation in radial or axial directions with abrupt or non-abrupt interfaces. These multifunctional heterostructures hold promise in and broad range of current and future technologies such as nanoelectronics, optoelectronics, sensors, and energy. The fundamental approach in developing such heterostructured nanowires is by mixing and matching different materials or physically dissimilar nanostructures to result in compositional variations along radial or axial direction of the nanowire. Both these heterostructure configurations necessitate understanding of the nanowire growth processes, interface development, and compatible growth methods that do not allow for disintegration of such a multicomponent system during multi-step processing schemes. Of particular interest are axially heterostructured nanowires due to their high aspect ratio and yet incorporating different material segments along the nanowire length. This geometrical aspect of these heterostructures also challenges the ability of state-of-the-art lithography techniques and ofcourse, multicomponent character imparts them unique surface chemistry that surpasses the limitations of a single-component nanowire system. Thus, the focus of this chapter is to bring forward the the most recent and pioneering developments in the area of axially heterostructured nanowires and basic underlying concepts necessary to grow or fabricate them. The latter necessitates our thorough understanding of single-component nanowire systems and their growth, characterization, and properties. Furthermore, the link between this knowledge and realization of axially heterostructured nanowire is established with an emphasis on device architectures, interface development and characterization, cumulative properties of such a multicomponent nanosystems, and their applications. The chapter also outlines major challenges in this area of research and the technical motivations for researchers to pursue the development and implementation of these heterostructured nanowires is also elaborated. Finally, the chapter is concluded by outlining future directions and developments required to advance this field. For an overview of heterostructures based on nanoparticles or polymeric nanostructures as well as radially heterostructured nanowires, the reader is referred to other excellent reviews and literature. [79-83, 98].

SYNTHESIS OF SINGLE-COMPONENT NANOWIRES

Nanowires have been subject of intense scientific research due to their unique properties that make them suitable for developing advanced devices [1]. Well-controlled physical characteristics such as diameter, length, and crytal structure are critical for their electronic and optical properties. In addition, nanowire surface functionality could lead to novel photocatalysts, chemical and biological sensors, and scanning probe tips that can provide atomic scale resolution. However, understanding nanowire synthesis and developing newer and simpler methods to grow them is becoming of particular importance.

For systematically controlling the dimensions, chemical compositions, and properties of nanowires, it is critical to design well-controlled synthesis strategies with predicatiable outcomes. For example, in regard to thin films, processes such as physical vapor deposition (PVD) and more advanced methods including molecular beam epitaxy can allow for precise variations in structure and composition of the thin films. Similarly, to enable such a growth control for nanowires, it will be necessary to understand the growth from a fundamental point of view. Nanowires can be synthesized by chemical methods, electrodeposition, chemical and physical vapor deposition (CVD, PVD), plasma synthesis, laser-based synthesis, and mechanical and nano/microfabrication methods [1, 2, 13, 84-92]. These methods have also resulted in various other forms of nanostructures such as thin films, nanofibers, nanotubes, nanobelts, nanoparticles, nanocomposites, and nanoporous structures [93-100]. However, the major challenges that still exist in the growth of any type of nanostructures are: 1) manipulation of their physical and chemical characteristics during the growth, 2) controlling impurity levels in nanostructures at atomic scale, 3) producing clean nanostructures with minimal surface contamination, and 4) direct integration of nanostructures into devices. In addition, if it desired to achieve multifunctionality by heterostructuring of nanostructures then these challenges are magnified by several times and warrants fundamental understanding of growth at an ab-initio level. Especially, in case of axially heterostructured nanowires, material selection, surface energy compatibility, and thermodynamics of growth become very critical factors [1, 101]. There are two dominant routes for producing single-component nanowires – high temperature synthesis and chemical synthesis. In regard to high temperature synthesis, nanowires can be grown in a vapor-liquid-solid (VLS) or vapor-solid (VS) growth method [1, 102, 103, 104]. The former process was first proposed by Wagner [105] and requires a seed or a catalyst that forms a suitable alloy phase with the nanowire material at high temperature. Once the supersaturation of the nanowire material takes place in this alloy phase, the former results in the form of nanowires [1,6]. This approach is most utilized approach to grow nanowires and a good understanding of phase diagrams can allow for developing scalable nanowire growth process coupled with appropriate material selection. VLS growth processes can be further categorized under laser-assisted catalytic growth process and metal-catalyzed CVD [1]. The former process involves laser ablation of target comprised of nanowire material and catalyst. The process takes place in a heated quartz tube connected with pressure controlling pump. The latter helps manipulating the pressure inside the tube and affects the condensation of the ablated material. The growth temperature is an important parameter determining both, nucleation and growth of the nanowire, and is selected based on the phase diagram of the nanowire material and catalyst. Overall, laser-assisted catalytic growth allows for the growth of wide variety of semiconductor nanowires including various combinations of elements group III, IV, V, VI, and VI in the periodic table [1]. In order to achieve better control over the nanowire dimensions, chemical composition, and crystal structure, metal-catalyzed CVD process for nanowire growth is a good option. This approach also necessitate a thorough understanding of phase diagrams and works most easily for eutectic couples such as Si-Au system for the growth of Si nanowires. The process involves the use of high temperature quartz tube furnace equipped with pressure controlling pump, precursor chemical and gas feeds, and inert carier gas line. A substrate coated with catalyst nanoparticles or a thin film of the same could be utilized and inserted inside the tube furnace. Slowly the appropriate conditions for the growth of nanowires are achieved and this includes precursor or gas flow rates, location of the substrate inside the tube, growth temperature and pressure, and growth duration. It has to be kept in mind that this high temperature growth process will also result in significant surface migration of catalyst nanoparticles on the substrate and thus, making this approach challenging when it comes to vertical-alignment, uniform spatial density, and diameters of the grown nanowires. However, a very well-controlled CVD set-up may result in desired outcomes. Apart from the growth parameters mentioned above, CVD tube length, tube diameter, and temperature profile inside the furnace are very critical. Presence of any contimation could further lead to undesirable nanowire growth.

An interesting example is Si nanowires that can be grown in a laser–assisted catalytic growth process and Au catalyzed CVD growth method [1]. In either case, it is helpful to consider the binary phase diagram of Si with other metals such as Fe and Au. In case of Fe-Si phase diagram, it can be noted that region above 1200 °C is a Si rich region along with $FeSi_x$ in liquid form. If the laser ablation of $FeSi_x$ process is conducted in a growth environment then Fe and Si vapors are incorporated into Fe-Si liquid and Si supersaturation in this results in precipitation in the form of Si nanowires. Similarly, binary phase diagram of Si-Au system is a eutectic system and is the basis of formation of Si nanowires in a Au-catalyzed CVD process. A CVD furnace equipped with a quartz tube will be fed with Si precursor (silane gas), carier gas and dilutant (Ar), and a vacuum pump to control pressure. At temperatures between 700 and 1000 °C, Si-Au alloy droplet is formed and addition of Si to this further leads to Si nanowire precipitation. In such nanowire growth processes, it is interesting to note that the catalytic nanoparticle of Au migrates along the length of growing nanowires and could end up either at the tip or at the root of the nanowire. The diameter of the nanowire could be closely correlated to the diameter of the catalytic nanoparticle and has been utilized as one of the approaches to manipulate nanowire physical dimensions in several studies. However, if the growth conditions are manipulated then this catalytic nanoparticle, in case of ZnO nanowires CVD growth, was observed along the nanowire length and further allows for branched nanowire formation. An important aspect of grown nanowires in VLS growth is the growth direction of the nanowires as this could lead to distinct properties by imparting anisotropic character. In addition, this also leads to an unanswered question: How to control the nanowire growth direction? The solution to this problem lies in the choice of growth substrate, surface energies of the catalyst nanoparticle and the growing nanowire, and thermodynamic stability of the interfaces during the growth.

On the other hand, VS method is non-catalytic and high temperature approach for the growth of nanowires and was proposed by Brenner and Sears [104, 106]. The earlier studies were focused on the growth of a variety of whiskers comprised of mercury, zinc, cadmium, cadmium sulfide, and silver. This approach does not require a catalyst for the growth of nanowires but a suitable vapor pressure of the nanowire material near the substrate that further leads to thermodynamically favaroable growth of nanowires. For example, growth of CuO nanowires by heating copper substrate in the presence of oxygen [102, 107]. Similarly metal oxide powders were heated at high temperature under controlled pressure environemnet to result in metal oxide nanowires. This approach has allowed for synthesis of ZnO, SnO_2, In_2O_3, and CdO nanowires [108]. There has been several growth mechanisms proposed for VS growth process and mainly attributed to different growth rate of different facets in a crystalline system, 2) prescence of defects such as dislocations that function as a sink for the nanowire vapor material and result in preferential growth of nanowire. This theory is also proposed to explain the faster nanowire growth rates than species condensation rate, and 3) variations in surface energy and chemical potential in different crystal orientation that allows for anisotropic growth in one dimension only. This has inspired a significant amount of research work in characterizing the nanowires grown via VS method and various techniques such as electron microscopy used could not help showing the presence of dislocations as a driving force. However, twin boundaries and stacking faults were observed in such nanowires. Thus, more work is warrented in understanding VS method for nanowire growth.

There also exist numerous other methods to synthesize single-component nanowires in solution-based or nano/microfabrication-based methods [109-114]. The advantage of these methods is mainly due to the ambient temperature growth conditions, where problems associated with high temperature surface migration of catalytic nanoparticles could be easily avoided. Solution-based methods can result in large yields of defect-free nanowires but their controlled arrangement on a substrate is not possible during the growth. In addition, this approach necessitates cleaning and stabilization of nanowires after the growth. In solution-based nanowire growth, the nanowire species is generated from a metal salt or a chemical precursor upon reaction with the reducing agent or complex compound formation. Important ingredients of the reaction mixture contain reducing agent, solvent, and surfactants. Once the chemical precursor dissolves and reacts with the solution components, it results in the nanowires as a precipitate. In this regard, formation of selenium nanowires is a good example [115-116], where amorphous selenium particles are prepared by the reduction of selenious acid with excess hydrazine at 100 °C. Cooling this solution resulted in nanocrystalline trigonal selenium. Subsequent step of aging this solution at room temperature resulted in their anisotropic growth in [001] direction in the form of nanowires. Inspite of solution environemnet, nanowires could be grown on a seed nanoparticle or a substrate. However, success in this direction is very limited only applicable to a few material systems resulting in sub 500 nm diameters of the produced nanowires. For example, platinum nanoparticles seeds were utilized to grow silver nanowires in presence of reducing agent and polymer surfactant[117]. Another interesting work is on growth of ZnO nanorods on a glass substrate using zinc salts and stabilizer/surfactant [118]. Nanowires of metals can also be grown in a similar approach as that for nanoparticle synthesis such as decomposition of organometallic compounds in presence of other organic chemical species. A simple synthesis in this regard is the decomposition of barium titanium isopropoxide in a solution containing hydrogen peroxide, heptadecane, and oleic acid. The reaction is performed at temperatures above 200 $^{o-}$C and for more than 5 hours to result in perovskite $BaTiO_3$ nanowires [119, 120]. The major challenge in this kind of approach is control over the nanowire dimensions and structural characteristics. In order to overcome this problem, template directed synthesis is an interesting option. Typical examples of the templates include solid substrate with step edges, cyclindrical channels in a membrane geometry, mesoscale and self-assembled structures, macromolecules such as DNA, and even other one dimensional nanostructures synthesized in a different approach. These templates can be etched away after the nanowire growth and the latter could be grown in a solution process (e.g., electrodeposition), room temperature dry process (e.g., PVD), and high temperature process (e.g., CVD). The template could be assumed as a guiding track for a nanowire during this process. To further elaborate on the templates and the templating process, some of the methods are discussed further.

Selective deposition of materials in the form of nanowires on the step edges of a specific substrate such as highly oriented pyrolytic graphite or chemically grooved silicon or InP substrates have been demonstrated [121-128]. The approach is to utilize the step edge sites as thermodynamically favorable locations for the deposition in the form of nanowires. Metal nanowires are most commonly formed by metal electrodeposition, solution-based metal nucleation, and high temperature treatment in reducing gas environment when starting with metal oxide systems. This approach can result in aligned arrays of nanowires and even a three-dimensional architecture of nanowires is also possible.

Nanowires of metals, magnetic materials, and alloyed phased have been synthesized by using porous templates [129-131]. The material can be deposited inside the nanochannels of the porous membranes through vapor phase evaporation, solution phase precipitation, and electrochemical deposition. The nanowires formed as in the form of arrays and once the template is dissolved, it is possible to achieve free-standing vertically-aligned nanowires. Two kinds of templates are commonly used: 1) polycarbonate track-etched membranes and 2) porous alumina membranes. The latter has high density of cylindrical, straight, and uniform pores. On the other hand, class of mesoporous materials with pore size as small as 1.5 nm have also been used for synthesizing sub 5 nm-sized nanowires. Under this classification, mesoporous silica (e.g., MCM-41 and MCM-48) and SBA-15 family are commonly used. The method to deposit nanowires within these templates involves infusion of metal precursor and its subsequent precipitation in metal nanowire. The major problem associated with porous templating method is the identification of optimal nanowire deposition conditions and if this is not achieved then it could lead to non-uniform deposition, metal nanoparticle formation, and incomplete filling of the pores.

All the above mentioned templating methods are also termed as hard or physical templating methods and there exist another class of templates also referred as soft templating method. In this case, the template is comprised of orderly arrangement of surfactant or polymer molecules. The process involves nucleation of metal nuclei from precursor metal salt, immediate encapsulation in a soft template micelle, and then anisotropic growth resulting in metal nanorods or nanowires. Reports in synthesis of Au and Ag nanowires and nanorods have demonstrated the use of this approach, where metal salts such as $HAuCl_4$ or $AgNO_3$ are added to a solution of reducing agent (e.g., ascorbic acid, sodium borohydride) [132-136]. This step results in seeds of corresponding Au or Ag nanoparticles and are further mixed in a solution containing more metal salt, soft template surfactant (cetyltrimethylammonium bromide-CTAB, and tetraoctylammonium bromide-TOAB [137]), and reducing agent. This last step, if performed in a controlled manner, results in various shapes and sizes of the Au or Ag nanostructures and a precise manipulation of growth parameters will result in nanorods or nanowires. The reason being that soft template micelle has certain affinity for the crystal facet of the growing nanowire and favors anisotropic growth. Several parameters can be controlled to result in desired nanowire characteristics and this includes metal salt concentration, reducing agent strength, type of surfactant and its concentration, and duration of growth.

Large yields of nanowires and nanorods could be obtained in this synthesis method. However, this techinique is very sensitive to the change in solution environment and could lead to severe aggregation. In addition, sample heterogeneities are yet to be understood in such an approach. In order to bring greater stability to the system, control over the growth process, and more chemical functionality, polymer-based micelle is helpful [138]. In this regard, diblock polymers such as poly(ethylene oxide)-poly (methacrylic acid) [139] could be used to form the cylindrical soft template that can capture growing nanowire nuclei and allows for its anisotropic growth. Due to different chain lengths and chemistry of the blocks building these polymers, it is possible to have better control over the nanowire growth process but the resulting product is polycrystalline and aggregation is inevitable.

Till now we have described various nanowire growth methods and they are dominantly suitable for the growth of single-component nanowires and, in some cases, for compound nanowires. Each of the approach has its own pros and cons. For example, metal-catalyzed CVD growth of nanowires can result in uniform arrays of nanowires but it is impossible to control the growth direction of nanowires. The latter problem could be overcome by using micelle-based synthesis method but method involving porous templates may not result in single crystalline nanowires. Heterogenieties in produced nanowires in regard to their crystal structures, defects, diameter distributions, and compositions in any of the processes stated above are unpredictable. The use of templates results in arrays of uniform size of nanowires but the scaling of this approach is limited by the batch nature of the process. In addition, removal of templates may result in damage to the nanowires. Thus, significant efforts are still required to grow nanowires in desired physico-chemical characteristics and fine tuning of these characteristics. At the same time, it is necessary to achieve complex architectures of nanowires that also incorporate multifunctional character in them. Ultimately, the knowledge of all described methods for singly-component nanowires becomes extremely critical for the growth/fabrication of heterostructured 1-D nanostructures, a route to achieve both complex architectures and multifunctional nanowire systems. Pioneering research work from groups led by Lieber, Wang, Yang, and Samuelson have reported on the synthesis of such heterostructures starting with the synthesis of Si nanowire-based heterostructures [140, 183]. Such heterostructured nanowires can be synthesized by using PVD, CVD, wet-chemistry, and other nano/microfabrication techniques. If necessary, two or more of these approaches can be merged to achieve desired heterostructured configuration.

NANOSCALE HETEROSTRUCTURES

For many decades thin film heterostructures have been employed in semiconductor devices and physics [141-146]. The main focus in this area is on improving and modulating device characteristics (e.g., band gap energies and charge carrier mobilities) by composition variations in epitaxially grown thin films with close interfacial lattice match [147]. Such research endeavors have resulted in devices such as light emitting diodes, solar cells, and transistors [148-151]. However, demands for device miniaturization as well as a need for characteristics like high efficiency, multifunctionality, and high surface area are already pushing the limits of lithography and microfabrication techniques [152]. Additionaly, heterostructures are not only critical for electronics and magnetic devices but also for applications in chemical and biological sensors, new and efficient energy technologies, and functional substrates or analytical platforms [153]. Apart from thin films, single-component nanostructures are also of interest [154-158]. Furthermore, with advances in growth and characterization tools, nanotechnology research is eventually shifting from simple single-component nanostructures to complex multicomponent nanostructures or nanoscale heterostructures [159-163]. The latter can be comprised of a combination of nanostructures of different or same material(s). Nanoparticle on nanoparticle, alloyed nanoparticles, bimetallic nanoparticles, nanoparticles on nanotubes or nanowires, nanowires on nanowires, and nanotubes on nanowires or nanotubes are some examples of such nanoscale heterostructures. Presently, the number of materials that can be simultaneously incorporated into such heterostructures is limited to two. Heterostructured configurations such as nanoparticles on nanoparticles, bimetallic nanoparticles, or alloyed nanoparticles can be synthesized in a variety of methods and have been utilized as sensors, device components, energy materials, catalysts, and biomedical systems [164-172]. However, nanotubes and nanowire-based heterostructures (also called heterostructured 1-D nanostructures) are not well-understood heterostructured configurations [173-177]. Among these, nanowire-based heterostructures [159,178] are of great importance because of their unique properties and potential applications. However, controlled synthesis of these heterostructures demands understanding nucleation and unidirectional crystal growth processes, realizing bottom-up and/or top-down growth methods to form them, and the ability to manipulate their chemical composition along radial or axial direction. The last one relies on material choices and compatible growth methods that allow for an abrupt, yet a continual change in the growth conditions. Of particular interest are axially heterostructured nanowires that exhibit unique functionality due to material stacking in axial direction of the 1-D nanowire [159]. Towards this end, vapor deposition and solution methods have been utilized to grow such nanowires [159,179, 180]. Overall this area of research is fast growing and demands attention of the scientific community. The challenge is also to attain growth strategies that can result in heterostructured 1-D nanostructures with desired compositions, interfaces, sizes, morphologies, and properties. This will further enable fast developing devices and technologies based on multifunctional and multicomponent heterostructured nanowires.

Emergence Of Heterostructured 1-D Nanostructures

Before detailed review of axially heterostructured nanowires is presented here, it is critical to understand how such heterostrcutured configurations emerged and the basic challenges and motivations involved in this area of research.

1-D nanostructures such as nanowires and nanotubes have high aspect ratios and have reached significant levels of commercialization or prototype device testsing [181-205]. These efforts and a desire to achieve multifunctionality in single-component nanostructures have led to recent developments in heterostructured 1-D nanostructures that can have orderly arrangement of different materials and inherit advantages and properties of 1-D configuration. This will further result in high performance and long-life to the devices incorporating such heterostructures. However, reaching this stage of development will first require gaining fundamental knowledge of growth, properties, and structure-property relationships of heterostructures. Generally, heterostructured 1-D nanostructures can be classified according to the junctions or interfaces present (Fig. 1): 1) segmented (axial) heterostructures, 2) co-axial core/shell (radial) heterostructures, 3) and hierarchical heterostructures. This chapter will focus on only axially heterostructured 1-D nanostructures. Such heterostructures have resulted in multifunctionality not achievable by single-component nanowires and also allowed for the manipulation of electronic and optical properties of the nanostructures.

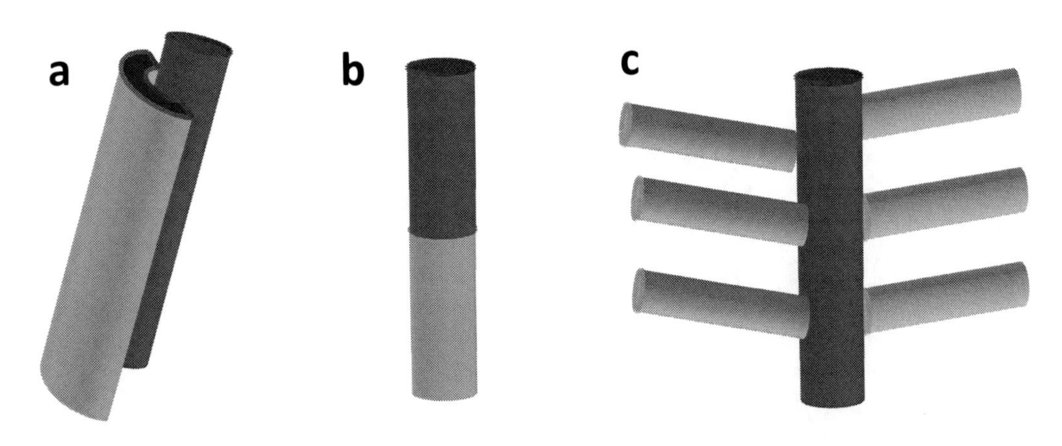

Figure 1. Different a) Radially heterostructured nanowires with a core and a shell of two different materials b) axially heterostructured nanowires with two segments of two different materials, and c) hierarchically heterostructured nanowires incorporating nanowires of two different materials.

Challenges and Motivations

As compared to compound nanowires or nanotubes [206, 207] heterostructured 1-D nanostructures take advantage of distinct functions and properties of different materials and precise interfaces. However, current fabrication approaches have limitations such as multi-step growth processes, harsh chemical conditions, and necessity to control several growth parameters simultaneously. This is further complicated by material incompatibility issues that hinder the heterostructure growth and allow for less understanding of surface properties preventing robust and reproducible self-assembly. Contamination of the nanostructures during wet synthetic routes or introduction of crystal defects during high temperature processing makes real-life device design far from reality. Finally, bundling due to capillary forces between synthesized heterostructures can prevent their assembly on a substrate and achieve desired architectures. All these challenges will atleast necessitate developing newer synthesis approaches combining bottom-up and top down methods as well as complete understanding of heterostructure growth mechanisms must be acquired. In regarding to combining growth approaches then one can envision coupling of CVD/PVD and self-assembly methods to achieve heterostructured 1-D nanostructures. Invention of such a hybrid fabrication methodology will lead to inherently parallel and often scalable processing that will also allow for direct integration of novel heterostructures into devices and useful platforms.

AXIALLY HETEROSTRUCTURED NANOWIRES

Axially heterostructured nanowires (also referred as superlattice nanowires [215] Fig. 1b), where composition is modulated in the axial direction or multiple segments of nanowires of different materials are joined end-to-end have potential to enable new functions and result in complex architectures that are not achievable by current microfabrication techniques. The importance of complex architectures has been observed in case of crossbar nanowire-based devices, where the junction point of two nanowires resulted in unique electronic properties [208, 209, 210]. The size, length, number of segments, materials choices, and physical characteristics of the junctions/heterointerfaces in such heterostructures determine their properties. This makes them very promising for applications in electronics, optoelectronics, piezoelectrics, thermoelectric, solar cell, and chemical and biological sensing [239, 247,251]. The growth strategies and materials selection for the axial heterostructures have been the major limitations. Thus, synthesizing heterostructures comprised of noble metals, a few transition metals, and elemental or compound semiconductors have been reported till now[211, 212, 213, 214]. The synthesis methods are dominated by CVD and wet-chemistry approaches [211,213]. The latter approach generally involves electrochemical deposition of different material such as noble or transition metals segments inside porous templates, the same growth strategy as discussed earlier for templating method in this chapter. Most of the reports in this direction have been focused on two material systems and more than two segments comprised of different materials are difficult to realize [211]. Similarly, metal-catalyzed CVD grown semiconducting axially heterostructured nanowires (Si-Ge, n-Si-p-Si, n-InP-p-InP, GaP-GaAs, InP–InAs, Si-NiSi, etc.) and nanowire-nanotube (ZnO nanowire-carbon nanotube, Si nanowire-CNT, Au nanowire-CNT, etc.) junctions have also been reported [215-223]. Lieber's group is well-known for studying these systems in laser ablation-based catalytic growth and CVD growth. The approach is based on sequential feed of precursor chemicals in the growth chamber to result in compositional modulation. The key to the successful growth of different segments is by selecting a catalyst and nanowire material systems compatible with each other inspite of chaning growth conditions. In this regard, Au nanoparticles as a catalyst have been of focus as it meets the criteria for the growth of group III-V and group IV materials in axially heterostructured configuration.

Axially heterostructured nanowires can also accommodate larger misfit strains due to their cylindrical geometry and interfaces [222,224]. If the growth process is not well-understood and the defects such as dislocations are formed then it can result in strained axially heterostructured nanowires [224]. This defect formation is also indirectly related to the critical dimensions of such heterostructures and also responsible for maintaining high level of structural, electrical, chemical, optical stability in devices based on them. Theoretical analysis and mathematical techniques such as finite element method (FEM) can help understanding misfit strains in the axially heterostructured nanowires on a thick and rigid substrate [224]. In this study, the axially heterostructured InGa1-xAs/GaAs nanowires were investigated and strains were introduced into them through a pair of orthogonal edge dislocations and lattice defects. Critical thickness of InGa1-xAs segment over the GaAs segment was calculated and showed an exponentially increasing trend as a function of decreasing radius of the latter as well as the lattice mismatch between the two segments [224]. Theoretically it was found that this thickness can reach infinity for a particular value of the critical radius and can result in no misfit strains. These results were consistent with the experimental results [224]. The critical radius was 8 nm for InAs/GaAs system with 6.7% misfit and a 3-D model was developed for fundamentally understanding axially heterostructured nanowires with controlled characteristics [224].

The composition and interfacial quality are two important parameters determining the properties of axially heterostructured nanowires. Both these parameters are dependant on each other because if the interface is abrupt then composition of each segment same as the segment material while for a diffused interface, there could be a decaying composition gradient of one segment material into another or vice versa. Control over these parameters is important for devices based on axially heterostructured semiconducting nanowires such as $Si/Si_{1-x}Ge_x$ [219]. Growth approaches such as CVD and CVD-pulsed laser deposition (PLD) methods can meet the demands of controlling the growth of these heterostructures [225]. In CVD process, silane and germane are used as precursors while the CVD-PLD employs silane as Si precursor with laser ablated Ge [219,225]. However, the interfaces created in these methods were not abrupt and led to interdiffusion of nanowire segments at the junction point. Other methods such as metal organic vapor phase epitaxy (MOVPE), chemical beam epitaxy (CBE), or MBE method have greater control and resulted in the growth of GaAsP, GaP, GaAs, and InAs based axially heterostructured nanowires resulted in an abrupt interfaces [219, 226, 227, 228, 229, 230]. The role of solute segregation in the formation of a particular interface (abrupt or non-abrupt) is yet to be understood. Another study on the growth of $Si/Si_{1-x}Ge_x$ explains this aspect and also the effect of multicomponent nanowire diameter on the Ge composition profiles, growth rates, as well as the interfacial characteristics [219]. These axial heterostructures were grown in a Au catalyzed VLS growth process in a low pressure CVD (LPCVD) method. These heterostructures were characterized for their structure, morphology, and composition using high-angle annular dark-field (HAADF) detector, high resolution TEM, and EDS. It was observed that the composition profile was asymmetric and interfaces were non-planar in shape due to the lattice strains between the two segments of the nanowire [219]. A common problem associated with CVD growth methods is the presence of catalyst nanoparticle at the tip of the grown nanowire and diffusion of nanowire material into this catalyst. Thus, Ge was also found in the catalytic Au nanoparticle present at the tip of the nanowire and this may have some implication in the growth of such axial nanowire heterostructures and the formation of diffused interfaces [219, 231, 232]. In another study on axially heterostructured Si/SiGe nanowires [233], a Gaussian composition profile was observed for Ge and the electrical properties of individual nanowire (155 nm diameter and 200 nm length) were also studied in an in-situ SEM-based micromanipulator as shown in the Fig. 2. The heterostructure showed Ohmic behavior at currents within 1 μA at 500 mV, resistance of 470 kΩ, and resistivity of 4 Ω-cm [233]. However, the effect of native oxide and gold nanoparticles on the electrical properties is an important and needs to be accounted for.

Laser-assisted catalytic growth has also been utilized in successfully developing axially heterostrostructured nanowires comprised of GaAs/GaP segments [1]. This superlattice architecture was easily controlled by manipulating the growth time and resulted in defect-free

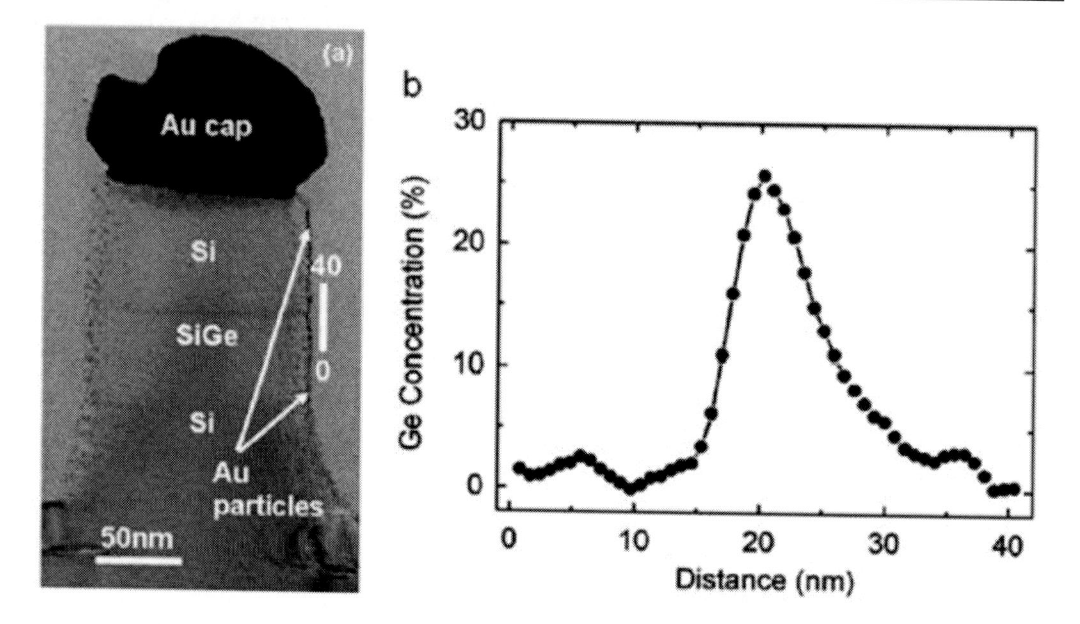

Figure 2. Axially heterostructured Si/SiGe nanowires. (a) TEM image of the heterostructured nanowire with Au nanoparticle at the tip,.(b) Concentration profile of Ge as a function of nanowire length, (c) The approach for electrically characterizing heterostructures, (d) SEM image of a contacted NW with electrochemically prepared Pt/Ir tip attached to the micromanipulator. Reprinted from Microelectronics Journal, 40, 3, P. D. Kanungo, A. Wolfsteller, N. D. Zakharov, P. Werner and U. Gösele, Enhanced electrical properties of nominally undoped Si/SiGe heterostructure nanowires grown by molecular beam epitaxy, 452-455, Copyright (2009), with permission from Elsevier.

heterostructures with diffused interfaces. The species that was diffused at the junctions was phosphorous and the length of diffused area was around ~ 15-20 nm. The diffused interface was attributed to the migration of the Au catalyst nanoparticle along the growing nanowire tip. Moreover, composition variation at interface also helps in minimizing the lattice strains leading to defect-free nanowires. However, abrupt interface is possible by reducing the diameters of the nanowires or by using of different catalyst nanoparticle or chemical precursor.

Recently, a report on ferromagnetic system comprised of Ge/$Mn_{11}Ge_8$ axially heterostructured nanowires were synthesized [218]. In these nanowires $Mn_{11}Ge_8$ phase shows a transition from antiferromagnetic to ferromagnetic above 150 K and retains the latter state till its Curie temperature (274 K) [234, 235]. The growth approach utilized a Mn catalyzed CVD process [218]. The catalytic seeds were grown on Si_3N_4/Si wafers by decomposing tricarbonyl-(methylcyclopentadienyl) manganese (TCMn) in a controlled pressure and temperature environment. Subsequently, feed precursors were introduced in the CVD chamber and were switched between germane and TCMn to result in axially heterostructured nanowires. Interestingly, the $Mn_{11}Ge_8$ segment was further oxidized into native oxide and thus, the axial heterostructure also had core/shell characteristics [218]. Surface diffusion and subsequent reaction of Ge and Mn species at the interface (Fig. 3) explained the multi-step growth mechanism for these nanowires [218].

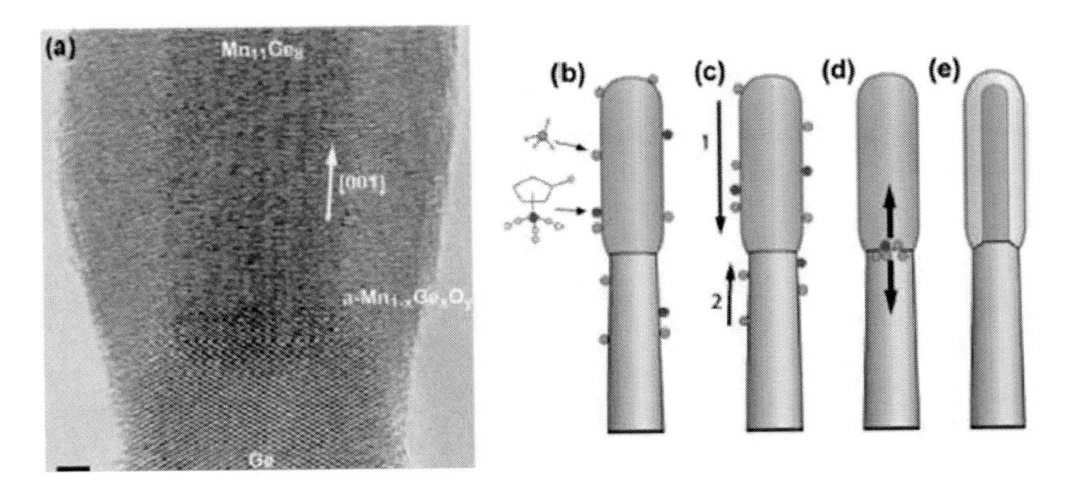

Figure 3. (a) TEM image of the crystalline germanium nanowire segment junctioned with crystalline $Mn_{11}Ge_8$, and amorphous $Mn_{1-x}Ge_xO_y$ phases. The scale bar is 2 nm. (b-d) heterostructured nanowire growth mechanism: (b) adsorption and decomposition of precursor molecules, (c) surface (1 and 2) diffusion of Ge (green) and Mn (purple) adatoms, and (d) incorporation of Ge and Mn atoms into the crystalline phases. (e) Postgrowth exposure to air leads to the formation of an amorphous Mn1-xGexOy shell. Reprinted with permission from (218). Copyright 2008 American Chemical Society.

Lieber and co-workers published several studies on axially heterostructured nanowires [215,217]. Radially heterostructured Si/Ni nanowires [217] were fabricated and high temperature annealing of these nanowires resulted in NiSi nanowires. By cleverly coupling this approach with lithography techniques and selective masking of nanowire segments, it was possible to realize axially heterostructured Si/NiSi nanowires (Fig. 4) [217]. These ohmic heterostructures and acted as a novel interconnects for the Si segment. These heterostructures were further demonstrated for Schottky barrier field effect transistors [236].

In order to enhance chemical functionality and unique electronic properties, nanowire-CNT axially heterostructured geometry of significance [216, 233, 237]. CNTs with unique properties and characteristics such as rich surface chemistry, thermal conductivity, electronic structure, and multi- or single-walled geomtry make them most ideal for heterostructured configurations. Lieber and co-workers reported [216] for the first time the fabrication of axially heterostructured Si nanowire-CNT heterostructures in a template-less CVD route. The approach utilized Au-FE catalyst, which unique catalyzed Si nanowires as well as CNTs when the chemical precursor feeds were changed sequentially [216]. The growth process was controlled to allow for this catalyst nanoparticle to be at the tip of the growing nanowire segments. This growth mechanism allows for Fe content in the catalyst to be utilized for the growth of multi-walled CNTs in an ethylene-based CVD process [216]. The process resulted in junction between multi-walled CNTs with Si nanowires [216]. Two kinds of junctions were observed between CNTs and Si nanowires: 1) a clear junction where the catalytic nanoparticle was excluded from the heterostructure and trapped on the outer side and 2) a junction that was interrupted by a catalyst nanoparticle and was trapped inside. These heterostructures were then grown on the scanning tunneling microscope (STM) tip and then the other end of the heterostructure was slowly dipped into liquid Ga-In to measure the electrical properties of heterostructured nanowire [216]. The measurement showed a diode like behavior for the heterostructures and was observed to be a function of the length of the heterostructure dipped inside the liquid [216]. To make the growth process for CNT-nanowire heterostructures simple, Ajayan and co-workers used porous alumina (AAO) as guiding templates [223]. It is well know that CNTs can be grown in AAO without a transition metal catalyst and their diameter is controlled by the pore size of AAO [238, 239]. Three segmented Au nanowire-CNT-Au nanowire and Au nanowire-Ni nanowire-CNT axially heterostructured 1-D nanostructures were prepared by coupling electrochemical deposition and CVD growth [223]. Even though this approach is less complex but it requires diameter of the AAO to be manipulated in order to control diameter of grown heterostructures.

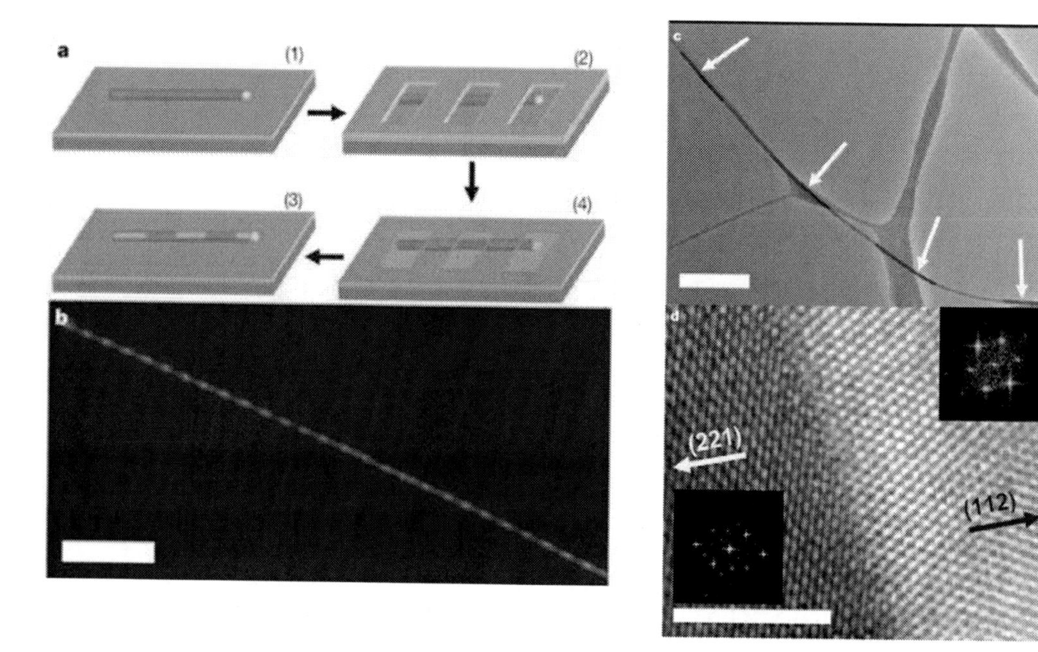

Figure 4. a) Fabrication of axially heterostructured NiSi/Si nanowire. (1) silicon nanowires (blue) are (2) coated with photoresist (grey) and lithographically masked, (3) selectively coated with nickel metal (green) with shell thickness similar to the core nanowire diameter, and (4) processed at 550 °C to form NiSi nanowire segments. b, Dark-field optical image of a axially heterostructured NiSi/Si nanowire. The bright green segments correspond to silicon and the dark segments to NiSi. Scale bar is 10 mm. c) TEM image of a NiSi/Si heterostructured nanowire. Arrows indicate different segments in the heterostructured nanowire (Si: bright segments and NiSi: the dark segments) Scale bar is 1 mm. d, High-resolution TEM image of the interface between NiSi and Si segments showing an atomically abrupt interface. Insets, two-dimensional Fourier transforms of the image depicting the [110] and [111] zone axes of NiSi and Si, respectively, where the arrows highlight the growth fronts of the NiSi (221) and Si (112). Scale bar is 5 nm. Reprinted by permission from Macmillan Publishers Ltd: Nature (217), copyright (2004).

Apart from nanoelectronics, these heterostructures have potential for applications where chemistry and surfaces are critical such as advanced sensors and biomedical devices. For example, Au nanowire-CNT axially heterostructured system incorporate hydrophillic metal segment and hydrophobic CNTs [223]. This kind of multifunctionality is useful for developing analytical devices or self-assembled architectures. These heterostructures were further dispersed in immiscible liquids mixture such as water and dichloromethane. The latter is hydrophobic and forms a globule in the water and the interface of this globule attracts hydrophobic CNT segment of the heterostructures resulting in their controlled assembly in the form of macroscopic spheres [223]. These spheres had Au nanowires facing the water side due to their hydrophilic nature. Furthermore, heterostructured nanowires were utilized as a proof-of-concept electrochemical supercapacitor device and showed specific capacitance of 72 F/g and a high power density of 48 kW/kg [240]. This high power density for CNT-based electrodes was attributed to a precise interfacing between the CNT and the Au nanowire. In similar direction of research, there is a need for exploring novel inorganic nanotubes-based axial heterostructures. AAO template was used to deposit Cu in the form of nanotubes followed by the formation of electrodeposited Bi nanowire. AAO template removal resulted in Cu nanotube-Bi nanowire axially heterostructured 1-D nanostructures [241].

Axially heterostructured configuration can allow for segment-specific surface chemistry and thus is promising for the chemical and biological applications. With the material selection of magnetic and noble metals, realization of these applications is possible. Towards this end, studies performed by Searson and co-workers are a significant achievement [211, 242, 244, 245]. In addition, their research in magnetically aligning axially heterostructured Au/Ni nanowires on a pre-patterned substrate is a step forward for developing functional platforms in an simple approach. Electrochemical deposition approach was utilized to synthesize these heterostructured nanowires in an AAO template. The fundamental concept behind the sensing mechanism was to utilize complementary biological pairs and surface chemistry of substrate and axially heterostructured nanowires. In regard to complementary biological pairs, des-thio biotin-streptavidin, biotin-streptavidin, and biotin-avidin are well known for their high association constants [243]. Thiol (-SH) terminated groups have high affinity for Au and histidine-tagged molecules or species such as palmitic acid have high affinity to Ni [211]. This kind of chemistry has been commonly utilized for demonstrating proof-of-concept chemical and biological sensors. First avidin molecules on a silver coated substrate were selectively patterned and then biotinalyted Au segment in Au/Ni nanowires was covalently linked onto the patterned substrate [211]. The magnetic field was applied to this substrate and the attached nanowires were aligned by utilizing magnetic nature of the Ni segment. Such an assembly of heterostructured nanowires is critical for complex device architectures and enables their direct integration into analytical platforms. Further efforts in this direction included selective binding chemistry to develop nanogaps, embedded catalyst systems, protein immobilization substrates, and multifunctional sensors [244, 245, 256, 247, 248, 249]. Besides metal-based axially heterostructured nanowires, integration of a conducting polymer segment is also an exciting option to introduce flexibility and greater multifunctionality to such axially heterostructured nanowires [250]. Thus, axially heterostructured Au/polypyrrole/Ni/Au nanowires were fabricated and assembled in device architecture. The Au segment facilitates self-assembly of a suspended nanowire and also acts as a conductor, Ni segment is magnetic in nature, and polypyrrole shows an excellent interaction with gas molecules. These nanowires showed a diode-like behavior and a constant photoresponse under photoexcitation [250].

More complex axially heterostructured CdTe/Au/CdTe nanowires have been fabricated for energy and nanoelectronics applications [251]. With high absorption coefficient, CdTe is a direct band gap semiconductor (Eg = 1.44 eV) [252] while Au-thiol (-SH) chemistry provides selective sites. This heterostructured configuration has strong potential for chemical field effect transistor (FET)-based biosensor (Fig. 5), where the device was fabricated using e-beam lithography [251]. Au segment was functionalized with thiolated single strand (ss) DNA (I) that can easily and selectively bind to complementary ss DNA (II) and this was done prior to fabricating the device [251]. The binding event was identified by the conductance change of the nanowire in the FET device, which is a common approach for any FET-based chemical sensors. The sensor could detect 1 μM solution of ss-DNA (II) and the change in the conductance was also dependent on the nanowire surface charges. These multifunctional heterostructures can lead the path towards multiplexed nanowire sensors that can detect multiple species from the mixture with minimal interferences. A few recent reports show great promise in this direction [251, 253].

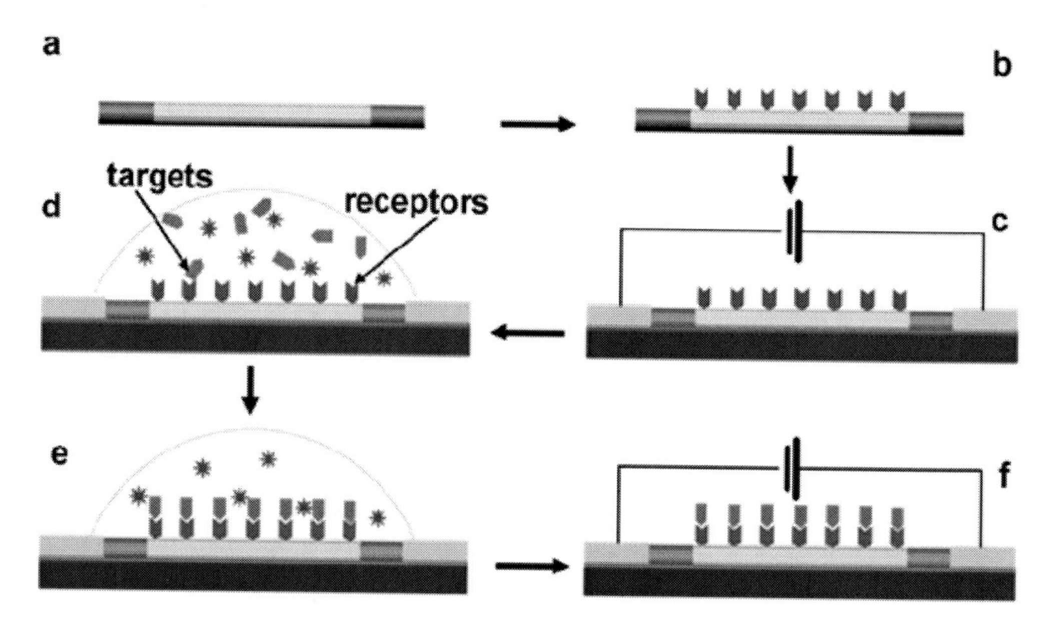

Figure 5. Schematic of CdTe-Au-CdTe striped nanowires for sensing specific biomolecules. (a) Electrochemically fabricated CdTe-Au-CdTe multisegment nanowire. (b) Au segment surface functionalization with thiol-terminated molecules. (c) Fabrication of nanowire FET device using lithography. (d) Immersion of nanowires into biomolecule solution,. (e) Specific biomolecules having affinity with the receptors at the nanowire permission from (251). Copyright 2008 American Chemical Society. Surface and attracted towards the device. And (f) Conductance responses from the device are measured by by I-V curves after targets are detected and bound. Reprinted with

Axially heterostructured nanowires uniquely provide an opportunity for merging molecular biology with nanotechnology. This includes studies on bionanomechamical systems such as the utilization of motor proteins and fundamental understanding of their actuating mechanisms [254, 255, 256, 257]. These motor proteins could be myosin, kinesin, flagellar, and adenosine triphosphate synthase (ATPase) [255, 258, 259]. Myosin and kinesin are linear motors while the last two are rotation-based proteins and the hydrolysis of ATP results in a mechanical motion [260]. In one example [261], ATPase was selectively bound to Ni segment of axially heterostructured Ni/Au/Ni nanowires and then ATP was introduced inside the reaction cell. The hydrolysis of ATP led to the rotation of nanowire in the solution. This has direct implictaions in developing bio-nano robots, forced bioactuator, smart biosensors, and delivery systems.

The biochemical activity associated with biomolecules and surface chemistry of axially heterostructured nanowires will definitely lead us to unique biocatalytic platforms. In this direction of research, magnetoswitchable bioelectrocatalysis was initially demonstrated for relay modified magnetic spheres [262]. Axially heterostructured Au/Ni nanowires have been shown to perform biocatalytic oxidation of glucose [212]. The magnetic properties of Ni segment led to change in the orientation of the nanowire that affected the catalytic process for glucose oxidation. The enzyme used for oxidation was glucose oxidase (GOx) that converts glucose to gluconic acid [263] in the presence of ferrocene as electron-transfer mediator [212]. This enzyme was coated onto Au segment of the axially heterostructured Au/Ni nanowires while ferrocene and nanowires were functionalized onto the electrode surface. Electron transfer process could be manipulated by the adjusting the orientation of nanowire on the substrate, and that further affects the catalytic process [212]. This is also referred as nanowire-based electrochemical [212] magnetoswitchable bioelectrocatalytic processes. The magnetic field application allowed for the heterostructured nanowire in horizontal and vertical direction as well as actuate away from the mediator. This activated, hindered, and blocked interaction of the nanowire with surface functionalized ferrocene mediator molecules. Cyclic voltammetry is used to observe the changes and the overall approach shows the importance this mechanism for the development of bioreactors, biofuel cells, and biosensors [264].

CONCLUSION

In this chapter, we have learnt about an important research direction in nanostructure growth and applications by way of developing multicomponent axially heterostructured nanowires with discrete domains of physically dissimilar materials. It is to be noted that the fundamental knowledge of single-component nanowire growth methods is extremely critical for developing such heterostructures. The comprehensive review of axially heterostructured nanowires in this chapter shows that heterostructuring of 1-D nanostructures aims to develop multifunctionality on the smallest possible size scale and beyond the limits of convental techniques such as lithography-based nanostructure fabrication. The use of high temperature and solution growth strategies to fabricate or synthesize these heterostructures can result in well-defined architectures of such heterostructures and eventually, lead towards their direct integration into novel multifunctional devices. The case studies described in this chapter explain fundamental understanding of heterostructure formation and chemistry, manipulation of their properties, structure-property relationships, and also discusses new nanofabrication techniques and assembly of novel devices. These studies also indicate that heterostructured nanostructures have potential for revolutionizing nanoelectronics, optoelectronics, sensing, catalysis, biomedical devices, and energy technologies.

In conclusion, fundamental studies related to axially heterostructured nanowires are necessary for advancing this field of research and at the same time will be a rewarding and unique research endeavor. This is also a fast emerging area of research with rapid advancements at every step of development and thus, opens numerous opportunities to scientists and engineers. Obviously, there is still a need for simple and scalable growth methods that can result in controlled composition modulations, interfacial characteristics, size, and morphologies of the heterostructures. In this regard, it is also important to solve the problem of material compatibility with the growth methods employed, which does not hinder the development of variety of other materials systems in heterostructured form.

Characterizing these axially heterostructured nanowires for their properties will allow for understanding the novel physical and chemical phenomena they exhibit. All these efforts will have a powerful impact on civilian safety and health, homeland and border security, and fulfill customer demands of increasingly miniaturized electronics and optoelectronics.

ACKNOWLEDGMENTS

The author thanks the University of Alabama and National Science Foundation (NSF award #: 0925445) for supporting his nanoscale heterostructure research program and Dr. Shweta Kapoor for proof-reading, reference formatting, and editing this manuscript.

REFERENCES

[1] Wang, Z. L.; *Nanowires and nanobelts,* Springer: New York, NY, 2006, 1.
[2] Meyyappan, M.; *Carbon nanotubes: Science and applications*, CRC Press LLC: Boca Raton, FL, 2005,.
[3] Cui, Y.; Lauhon, L. J.; Gudiksen, M. S.; Wang, J.; Lieber, C. M.; Diameter-controlled synthesis of single-crystal silicon nanowires, *App. Phys. Lett.* 2001, 78, 2214.
[4] Tian, B.; Kempa, T. J.; and Lieber, C. M.; Single nanowire photovoltaics, *Chem. Soc. Rev.* 2009, 38, 16.
[5] Lieber, C. M.; Wang, Z. L.; Functional nanowires, *Mater. Res. Bull.*, 2007, 32, 99.
[6] Lauhon, L. J.;, Gudiksen, M. S.; Lieber, C. M.; Semiconductor nanowire heterostructure, *Phil. Trans. R. Soc. Lond. A*, 2004, 362, 1247.
[7] S. Iijima; Helical microtubules of graphitic carbon, *Nature,* 1991, 354, 56.
[8] Daniel, M. C.; Astruc, D.; Gold nanoparticles: Assembly, supramolecular chemistry, quantum-size-related properties, and applications toward biology, catalysis, and nanotechnology, *Chem. Rev.*, 2004, 104, 293.
[9] Esfanda, R.; Tomalia, D. A.; Poly(amidoamine) (PAMAM) dendrimers: from biomimicry to drug delivery and biomedical applications, *Drug Dis. Today*, 2001, 6, 427.
[10] Lu, W.; Lieber, C. M; Nanoelectronics from bottom up, *Nature Mater.*, 2007, 6, 841.
[11] Chopra, N.; Gavalas, V. G.; Hinds, B. J.; Bachas, L. G.; Functional one-dimensional nanomaterials: Applications in nanoscale biosensors, *Anal. Lett.*, 2007, 40, 2067.
[12] Arico, A. S.; Bruce, P.; Scrosati, B,; Tarascon, J.-M.; Schalkwijk, W. V.; Nanostructured materials for advanced energy conversion and storage devices, *Nature Mater.*, 2005, 4, 366.
[13] G. Timp; Nanotechnology, 1999, Springer: New York, NY,.
[14] Oesterling, E.; Chopra, N.; Gavalas, V.; Arzuaga, X.; Lim, E. J.; Sultana, R.; Butterfield, D. A.; Bachas, L. G.; Hennig, B.; Alumina nanoparticles induce expression of endothelial cell adhesion molecules, *Toxicol. Lett.*, 2008, 178, 160.
[15] Kolmakov, A.; Moskovits, M.; Chemical sensing and catalysis by one-dimensional metal oxide nanostructures, *Ann. Rev. Mater. Res.*, 2004, 34, 151.

[16] Rosi, N. L.; and Mirkin, C. A.; Nanostructures in Biodiagnostics, *Chem. Rev.*, 2005, 105, 1547.

[17] Xia, Y.; Yang, P.; Sun, Y.; Wu, Y.; Mayers, B.; Gates, B.; Yin, Y.; Kim, F.; Yan, H.; One-dimensional nanostructures: Synthesis, characterization, and applications, *Adv. Mater.*, 2003, 15, 353.

[18] Acharya, S.; Sarma, D. D.; Golan; Y.; Sengupta, S.; Ariga, K.; Shape-dependent confinement in ultrasmall zero-, one-, and two-dimensional PbS nanostructures, *J. Am. Chem. Soc.*, 2009, 131, 11282.

[19] Kalpakjian, S.; Schmid, S.; 'Manufacturing engineering and technology', 2008, New Jersey, Prentice Hall.

[20] Patrick, A. D.; Dong, X.; Allison, T. C.; Blaisten-Barojas, E.; Silicon carbide nanostructures: A tight binding approach, *J. Chem. Phys.*, 2009, 130, 244704.

[21] Shim, H. W.; Kuppers, J. D.; Huang, H.; Strong friction of silicon carbide nanowire films, *Nanotechnology*, 2009, 20, 025704.

[22] Hyeon, T.; Fang, M.; Suslick, K. S.; Nanostructured molybdenum carbide: Sonochemical synthesis and catalytic properties, *J. Am. Chem. Soc.*, 1996, 118, 5492.

[23] Wong, E. W.; ,Maynor, B. W.; Burns, L. D.; Lieber, C. M.; Growth of metal carbide Nanotubes and nanorods, *Chem. Mater.*, 1996, 8, 2041.

[24] Arie, T.; Akita, S.; Nakayama, Y.; Growth of tungsten carbide nano-needle and its application as a scanning tunnelling microscope tip, *J. Phys. D: Appl. Phys.*, 1998, 31, L49.

[25] Lee, S. -K.; Zetterling, C. –M.; Östling, M.; Åberg, I.; Magnusson, M. H.; Deppert, K.; Wernersson, L. –E.; Samuelson, L.; Litwin, A.; Reduction of the Schottky barrier height on silicon carbide using Au nano-particles, *Solid State Elect.*, 2002, 46, 1433.

[26] Shah, M. A.; Al-Shahry, M. S.; Asiri, A. M.; Biomedical applications of iron oxide nanostructures, *Int. J. Nano Biomater.*, 2009, 2, 164.

[27] Savua, R.; Joanni, E.; Low-temperature, self-nucleated growth of indium–tin oxide nanostructures by pulsed laser deposition on amorphous substrates, *Scripta Mater.*, 2006, 55, 979.

[28] Cheng, F.; Shen, J.; Ji, W.; Tao. Z.; Chen, J.; Selective synthesis of manganese oxide nanostructures for electrocatalytic oxygen reduction, *ACS Appl. Mater. Interfaces*, 2009, 1, 460.

[29] Chopra, N.; Claypoole, L.; Bachas, L. G.; Formation of Ni/NiO core/shell nanostructures and their attachment on carbon nanotubes, *NSTI Nanotech 2009 Proceedings*, 2009, 1, 187.

[30] Kasuga, T.; Hiramatsu, M.; Hoson, A.; Sekino, T.; Niihara, K.; Formation of titanium oxide nanotube, *Langmuir*, 1998, 14, 3160.

[31] Waldauf, C.; Morana, M.; Denk, P.; Schilinsky, P.; Coakley, K.; Choulis, S. A.; Brabec, C. J.; Highly efficient inverted organic photovoltaics using solution based titanium oxide as electron selective contact, *Appl. Phys. Lett.*, 2006, 89, 233517.

[32] Wang, B.; Bates, J. B.; Hart, F. X.; Sales, B. C.; Zuhr, R. A.; Robertson, J. D.; Characterization of thin-film rechargeable lithium batteries with lithium cobalt oxide cathodes, *J. Electrochem. Soc.* 1996, 143, 3203.

[33] Negro, L. D.; Yi, J. H.; Kimerling, L. C.; Hamel, S.; Williamson, A.; Galli, G.; Light emission from silicon-rich nitride nanostructures, *Appl. Phys. Lett.* 2006, 88, 183103.

[34] Zhou, X. T.; Sham, T. K.; Shan, Y. Y.; X. F. Duan, X. F.; S. T. Lee, S. T.; Rosenberg, R. A.; One-dimensional zigzag gallium nitride nanostructures, *J. Appl. Phys.* 2005, 97, 104315.

[35] Johnson, J. C.; Choi, H. –J.; Knutson, K. P.; Schaller, R. D.; Yang, P.; Saykally, R. J.; Single gallium nitride nanowire lasers, *Nature*. 2002, 1, 106.

[36] Chopra, N. G.; Luyken, R. J.; Cherrey, K.; Crespi, V. H.; Cohen, M. L.; Louie, S. G.; Zettl, A.; Boron nitride nanotubes, *Science*. 1995, 269, 966.

[37] Balasubramanian, C.; Godbole, V. P.; Rohatgi, V. K.; Das, A. K.; Bhoraskar, S. V.; Synthesis of nanowires and nanoparticles of cubic aluminium nitride, *Nanotechnology*. 2004, 15, 370.

[38] Srivastava, D.; Menon, M.; Cho, K.; Anisotropic nanomechanics of boron nitride nanotubes: Nanostructured "skin" effect, *Phys. Rev. B*. 2001, 63, 195413.

[39] Vaddiraju, S.; Mohite, A.; Chin, A.; Meyyappan, M.; Sumanasekera, G.; Alphenaar, B. W.; Sunkara, M. K.; Mechanisms of 1D crystal growth in reactive vapor transport: Indium nitride nanowires, *Nano Lett.* 2005, 5, 1625.

[40] Hua, Z.; Yang, S.; Huang, H.; Lv, L.; Lu, M.; Gu, B.; Du, Y.; Metal nanotubes prepared by a sol–gel method followed by a hydrogen reduction procedure, *Nanotechnology*. 2006, 17, 5106.

[41] Walter, E. C.; Zach, M. P.; Favier, F.; Murray, B. J.; Inazu, K.; Hemminger, J. C.; Penner, R. M.; Metal nanowire arrays by electrodeposition, *ChemPhysChem*. 2003, 4, 131.

[42] Walter, E. C.; Penner, R. M.; Liu, H.; Ng, K. H.; Zach, M. P.; Favier, F.; Sensors from electrodeposited metal nanowires, *Surf. Interface Anal.* 2004, 34, 409.

[43] Venkatachalam, K.; Arzuaga, X.; Chopra, N.; Gavalas, V.; Xu, J.; Bhattacharya, D.; Hennig, B.; L. G.; Reductive dechlorination of polychlorinated biphenyl (PCB) using palladium nanoparticles and assessment of the toxic potency in vascular endothelial cells, J. *Hazardous Mater.* 2008, 159, 483.

[44] Zhang, J. Z.; Noguez, C.; Plasmonic Optical properties and applications of metal nanostructures, *Plasmonics*. 2008, 3, 127.

[45] Wolf, S. A.; Awschalom, D. D.; Buhrman, R. A.; Daughton, J. M.; von Molnár, S.; Roukes, M. L.; Chtchelkanova, A. Y.; Treger, D. M.; Spintronics: A spin-based electronics vision for the future, *Science*. 2001, 294, 1488.

[46] Chopra, N.; Bachas, L. G.; Knecht, M.; Fabrication and biofunctionalization of carbon-encapsulated Au nanoparticles, *Chem. Mater.* 2009, 21, 1176.

[47] Nednoor, P.; Gavalas, V. G.; Chopra, N.; Hinds, B. J.; Bachas, L. G.; Carbon nanotube based biomimetic membranes: Mimicking protein channels regulated by phosphorylation: *J. Mater. Chem.* 2007, 17, 1755.

[48] Matranga, C.; Bockrath, B.; Chopra, N.; Hinds, B. J.; Andrews, R.; Raman spectroscopic investigation of gas interactions with an aligned multiwalled carbon nanotube membrane, *Langmuir*. 2006, 22, 1235.

[49] Chopra, N.; Majumder, M.; Hinds, B. J.; Bi-functional carbon nanotubes by sidewall protection, *Adv. Funct. Mater.* 2005, 15, 858.

[50] Majumder, M.; Chopra, N.; Hinds, B. J.; Effect of tip functionalization on transport through vertically oriented carbon nanotube membranes, *Journal of the American Chemical Society*. 2005, 127, 9062.

[51] Nednoor, P.; Chopra, N.; Gavalas, V. G.; Bachas, L. G.; Hinds, B. J.; Reversible biochemical switching of ionic transport through aligned carbon nanotube membranes, *Chem.Mater.* 2005, 17, 3595.

[52] Majumder, M.; Chopra, N.; Hinds, B. J.; Enhanced flow in carbon nanotube, *Nature.* 2005, 438, 44.

[53] Chopra, N.; Xu, W.; Delong, L. E.; Hinds, B. J.; Incident angle dependence of nanogap size in suspended carbon nanotube shadow lithography, *Nanotechnology.* 2005, 16, 133.

[54] Hinds, B. J.; Chopra, N.; Rantell, T.; Andrews, R.; Gavalas, V.; Bachas, L. G.; Aligned multiwalled carbon nanotube membranes, *Science.* 303, 2004, 62.

[55] Chopra, N.; Hinds, B. J.; Catalytic size control of multiwalled carbon nanotube diameter in xylene chemical vapor deposition process, *Inorg. Chim. Acta.* 2004, 357 3920.

[56] Chopra, N.; Kichambare, P. D.; Andrews, R.; Hinds, B. J.; Control of multiwalled carbon nanotube diameter by selective growth on the exposed edge of a thin film multilayer structure, *Nano Lett.* 2002, 2, 1177.

[57] Kroto, H. W.; Heath, J. R.; O'Brien, S. C.; Curl, R. F.; ,Smalley, R. E.; C60: Buckminsterfullerene, *Nature.* 1985, 318, 162.

[58] van Ruitenbeek, J.; Nanomaterials: Live-action alloy nanowires, *Nature Nanotech.* 2006, 1, 164.

[59] Xu, C. –L.; Li, H.; Xue, T.; Li, H. –L.; Fabrication of CoPd alloy nanowire arrays on an anodic aluminum oxide/Ti/Si substrate and their enhanced magnetic properties, *Scripta Mater.* 2006, 54, 1605.

[60] Xiao, Y.; Yua, G.; Yuana, J.; Wang, J.; Chen, Z.; Fabrication of Pd–Ni alloy nanowire arrays on HOPG surface by electrodeposition, *Electrochim. Acta.* 2006, 51, 4218.

[61] Senapati, S.; Ahmad, A.; Khan, M. I.; Sastry, M.; Kumar, R.; Extracellular biosynthesis of bimetallic Au-Ag alloy nanoparticles, *Small.* 2005, 1, 517.

[62] Yasuda, K.; Schmuki, P.; Formation of self-organized zirconium titanate nanotube layers by alloy anodization, *Adv. Mater.* 2007, 19, 1757.

[63] Xue, S.; Cao, C.; Wang, D.; Zhu, H.; Synthesis and magnetic properties of Fe0.32Ni0.68 alloy nanotubes, *Nanotechnology.* 2005, 16, 1495.

[64] Varadan, V. K.; Chen, L.; Xie, J.; Magnetic nanotubes and their biomedical applications, *Nanomedicine.* 2008,

[65] O'Brien, G. A.; Quinn, A. J.; Tanner, D. A.; Redmond, G.; A Single Polymer Nanowire Photodetector, *Adv. Mater.* 2006, 18, (18), 2379.

[66] Kemp, N. T.; McGrouther, D.; Cochrane, J. W.; Newbury, R.; Bridging the Gap: Polymer Nanowire Devices, *Adv. Mater.* 2007, 19, (18), 2634.

[67] Dan, Y.; Cao, Y.; Mallouk, T. E.; Johnson, A. T.; Evoy, S.; Dielectrophoretically assembled polymer nanowires for gas sensing, *Sensor Actuat. B-Chem.* 2007, 125, (1), 55.

[68] Ravichandran, J.; Manoj, A. G.; Liu, J.; Manna, I.; Carroll, D. L.; A novel polymer nanotube composite for photovoltaic packaging applications, *Nanotechnology.* 2008, 19, 085712.

[69] Ahir, S. V.; Terentjev, E. M.; Photomechanical actuation in polymer-nanotube composites, *Nature Mater.* 2005, 4, 491.

[70] McCarthy, B.; Coleman, J. N.; Curran, S. A.; Dalton, A. B.; Davey, A. P.; Konya, Z.; Fonseca, A.; Nagy, J. B.; Blau, W. J.; Observation of site selective binding in a polymer nanotube composite, *J. Mater. Sci. Lett.* 2000, 19, 1573.

[71] Koombua, K.; Pidaparti, R. M.; Tepper, G. C.; A Drug Delivery System Based on Polymer Nanotubes, *2nd IEEE International Conference on Nano/Micro Engineered and Molecular Systems*. 2007, 785.

[72] Martin, C. R.; Template synthesis of electronically conductive polymer nanostructures, *Acc. Chem. Res.* 1995, 28, 61.

[73] Maynor, B. W.; Filocamo, S. F.; Grinstaff, M. W.; Liu, J.; Direct-writing of polymer nanostructures: Poly(thiophene) nanowires on semiconducting and insulating surfaces, *J. Am. Chem. Soc.* 2002, 124, 522.

[74] Cammas, S.; Suzuki, K.; Sone, C.; Sakurai, Y.; Kataoka, K.; Okano, T.; Thermo-responsive polymer nanoparticles with a core-shell micelle structure as site-specific drug carriers, *J. Control. Release*. 1997, 48, 157.

[75] Huang, Z. –M.; Zhang, Y. –Z.; Kotaki, M.; Ramakrishna, S.; A review on polymer nanofibers by electrospinning and their applications in nanocomposites, *Compos. Sci. Technol.* 2003, 63, 2223.

[76] Majumdar, A.: Thermoelectricity in Semiconductor Nanostructures, *Science*. 2004, 303, 777.

[77] T. Steiner: Semiconductor nanostructures for optoelectronics applications, Artech House Inc.: Norwood, MA, 2004

[78] Alexson, D.; Chen, H.; Cho, M.; Dutta, M.; Li, Y.; Shi, P.; Raichura, A.; Ramadurai, D.; Parikh, S.; Stroscio, M. A.; Vasudev, M.; Semiconductor nanostructures in biological applications, *J. Phys.: Condens. Matter*. 2005, 17, R637.

[79] Mohanraj, V. J.; Chen, Y.;: Nanoparticles – A review, *Trop. J. Pharm. Res.* 2006, 5, 561.

[80] Martin, C. R.; Template synthesis of electronically conductive polymer nanostructures, *Acc. Chem. Res.* 1995, 28, 61.

[81] Ramanathan, K.; Bangar, M. A.; Yun, M.; Chen, W.; Myung, N. V.; Mulchandani, A.; Bioaffinity sensing using biologically functionalized conducting-polymer nanowire, *J. Am. Chem. Soc.* 2005, 127, 496.

[82] Friedrich, K.; Fakirov, S.; Zhang, Z.; Polymer composites: From nano- to macro- scale; Springer: New York, NY, 2005.

[83] Chopra, N.; Multifunctional and multicomponent heterostructured one-dimensional nanostructures: Advances in growth, characterization, and applications, *Mater. Technol.* 2010, 25, 212-230.

[84] Saupe, G. B.; Waraksa, C. C.; Kim, H. –N.; Han, Y. J.; Kaschak, D. M.; Skinner, D. M.; Mallouk, T. E.; Nanoscale tubules formed by exfoliation of potassium hexaniobate, *Chem. Mater.* 2000, 12, 1556.

[85] Caswell, K. K.; Bender, C. M.; Murphy, C. J.; Seedless, surfactantless wet chemical synthesis of silver nanowires, *Nano Lett.* 2003, 3, 667.

[86] Cao, H.; Xu, Y.; Hong, J.; Liu, H.; Yin, G.; Li, B.; Tie, C.; Xu, Z.; Sol-gel template synthesis of an array of single crystal CdS nanowires on a porous alumina template, *Adv. Mater.* 2001, 13, 1393.

[87] Liu, K.; Chien, C. L.; Searson, P. C.; Yu-Zhang, K.; Structural and magneto-transport properties of electrodeposited bismuth nanowires, *Appl. Phys. Lett.*, 1998, 73, 1436.

[88] G.W. Yang, G. W.; Laser ablation in liquids: Applications in the synthesis of nanocrystals, *Prog. Mater. Sci.*, 2007, 52, 648.

[89] Choi, W. K.; Li, L.; Chew, H. G.; Zheng, F.; Synthesis and structural characterization of germanium nanowires from glancing angle deposition, *Nanotechnology*. 2007, 18, 385302.

[90] Suryanarayana, C.; Mechanical alloying and milling, *Prog. Mater.Sci.* 2001, 46, 1.

[91] Saleem, A. M.; Berg, J.; Desmaris, V.; Kabir, M. S.; Nanoimprint lithography using vertically aligned carbon nanostructures as stamps, *Nanotechnology*. 2009, 20, 375302.

[92] Mao, Y.; Park, T.-J. Zhang, F.; Zhou, H.; Wong, S.S. Environmentally Friendly Methodologies of Nanostructure Synthesis, *Small*. 3, 1122-1139.

[93] Grunes, J.; Zhu, J.; Somorjai, G. A.; Catalysis and nanoscience, *Chem. Comm.* 2003, 2257.

[94] M. Meyyappan, M.; Carbon nanotubes: Science and applications, CRC Press LLC: ,Boca Raton, FL, 2005..

[95] Cumberland, S. L.; Hanif, K. M.; Javier, A.; Khitrov, G. A.; Strouse, G. F.; Woessner, S. M.; Yun, C.S.; Inorganic clusters as single-source precursors for preparation of CdSe, ZnSe, and CdSe/ZnS nanomaterials, *Chem. Mater.* 2002, 14, 1576.

[96] Wang, Y. W.; Zhang, L. D.; Wang, G. Z.; Peng, X. S.; Chu, Z. Q.; Liang, C. H.; Catalytic growth of semiconducting zinc oxide nanowires and their photoluminescence properties, *Journal of Crystal Growth*. 2002, 234, 171–175.

[97] Slutsker, J.; Tan, Z.; Roytburd, A. L.; Levin, I. Thermodynamic aspects of epitaxial self-assembly and magnetoelectric response in multiferroic nanostructures, *Journal of Materials Research*, 2007, 22, 2087-2095.

[98] Rotello, V.; Lockwood, D. J.; Nanoparticles: Building Blocks for Nanotechnology; Springer: New York, NY, 2004.

[99] Lu, G. Q.; Zhao, X. S.; Nanoporous materials: Science and Engineering, Imperial College Press: Hackensak, NJ..

[100] Roco, M. S.; Willams, R. S.; Alivisatos, P.; 'Nanotechnology Research Directions', 2001, Norwell, MA, Kluwer Academic Publishers,.

[101] Rao, C. N. R.; Govindaraj, A.; Nanotubes and Nanowires'; RSC Publications: USA, 2005.

[102] Chopra, N.; Hu, B.; Hinds, B. J.; Selective growth and kinetic study of copper oxide nanowires from patterned thin film multilayer structures, *J. Mater. Res.* 2007, 22, 2691.

[103] Zhang, J.; Zhang, L.; Peng, X.; Wang, X.; Vapor–solid growth route to single-crystalline indium nitride nanowires, *J. Mater. Chem.* 2002, 12, 802.

[104] Brenner, S. S.; Sears, G. W.; Mechanism of whisker growth—III nature of growth sites, *Acta Metall.* 1956, 4, 268.

[105] Wagner, R. S.; Ellis, W. C.; Vapor–liquid–solid mechanism of single crystal growth. *Appl. Phys. Lett.* 1964, 4, 89.

[106] Sears, G. W.;, A mechanism of whisker growth, *Acta Metal.* 1955, 3, 367-369.

[107] Jiang, X.; Herricks, T.; Xia, Y.; CuO nanowires can be synthesized by heating copper substrates in air, *Nano Lett.* 2002, 2, 1333.

[108] Pan, Z. W.; Dai, Z. R.; Wang, Z. L.; Nanobelts of semiconducting oxides, *Science*. 2001, 291, 1947-1949.

[109] Gates, B.; Yin, Y.; Xia, Y.; A solution-phase approach to the synthesis of uniform nanowires of crystalline selenium with lateral dimensions in the range of 10–30 nm, *J. Am. Chem. Soc.* 2000, 122, 12582.

[110] Sun, Y.; Gates, B.; Mayers, B.; Xi, Y.; Crystalline silver nanowires by soft solution processing, *Nano Lett.* 2002, 2, 165.

[111] Yin, A. J.; Li, J.; Jian, W.; Bennett, A. J.; Xu, J. M.; Fabrication of highly ordered metallic nanowire arrays by electrodeposition, *Appl. Phys. Lett.* 2001, 79, (7), 1039.

[112] Gates, B. D.; Xu, Q.; Love, J. C.; Wolfe, D. B.; Whitesides, G. M.; Unconventional nanofabrication, *Ann. Rev. Chem. Res.* 2004, 34, 339.

[113] Yan, X. –M.; Kwon, S.; Contreras, A. M.; Bokor, J.; Somorjai, G. A.; Fabrication of Large Number Density Platinum Nanowire Arrays by Size Reduction Lithography and Nanoimprint Lithography, *Nano Lett*, 5, (4), 745.

[114] Namatsu, H.; Watanabe, Y.; Yamazaki, K.; Yamaguchi, T.; Nagase, M.; Ono, Y.; Fujiwara, A.; Horiguchi, S.; Fabrication of Si single-electron transistors with precise dimensions by electron-beam nanolithography, *J. Vac. Sci. Technol. B.* 2003, 21, 1.

[115] Gates, B.; Yin, Y.; Xia, Y.; A Solution-Phase Approach to the Synthesis of Uniform Nanowires of Crystalline Selenium with Lateral Dimensions in the Range of 10–30 nm, *J Am. Chem. Soc.* 2000, 122, 12582-12583.

[116] Wunderlich, B.; Shu, H. –C.; The crystallization and melting of selenium, *J. Cryst. Growth.* 1980, 48, 227-239.

[117] Sun, Y.; Gates, B.; Mayers, B.; Xia, Y.; Crystalline Silver Nanowires by Soft Solution Processing, *Nano Lett.* 2002, 2, 165-168.

[118] Govender, K.; Boyle, D. S.; O'Brien, P.; Brinks, D.; West, D.; Coleman, D.; Room-Temperature Lasing Observed from ZnO Nanocolumns Grown by Aqueous Solution Deposition, *Adv. Mater.* 2002, 14, 1221-1224.

[119] Urban, J. J.; Yun, W. S.; Gu, Q.; Park, H.; Synthesis of Single-Crystalline Perovskite Nanorods Composed of Barium Titanate and Strontium Titanate, *J. Am. Chem. Soc.* 2002, 124, 1186-1187.

[120] Urban, J. J.; Spanier, J. E.; Ouyang, L,; Yun, W. S.; Park, H.; Single-Crystalline Barium Titanate Nanowires, *Adv. Mater.* 2003, 15,423-426.

[121] Favier, F.; Walter, E. C.; Zach, M. P.; Bentor, T.; Penner, R. M.; Hydrogen Sensors and Switches from Electrodeposited Palladium Mesowire Arrays, *Science.* 2001, 293, 2227-2231.

[122] Walter, E. C.; Favier, F.; Penner, R. M.; Palladium Mesowire Arrays for Fast Hydrogen Sensors and Hydrogen-Actuated Switches, *Anal. Chem.* 2002, 74, 1546-1553.

[123] Himpsel, F. J.; Jung, T.; Ortega, J. E.; Nanowires on stepped metal surfaces, *Surf. Rev. Lett.* 1997, 4, 371-380.

[124] Petrovykh, D. Y.; Himpsel, F. J.; Jung, T.; Width distribution of nanowires grown by step decoration, *Surf. Sci.* 1998, 407, 189-199.

[125] Jorritsma, J.; Gijs, M. A. M.; Kerkhof, J. M.; Stienen, J. G. H.; General technique for fabricating large arrays of nanowires, *Nanotechnology.* 1996, 7, 263-265.

[126] Muller, T.; Hennig, K. –H.; Schimidt, B.; Formation of Ge nanowires in oxidized silicon V-grooves by ion beam synthesis, *Nucl. Intrum. Methods Phys. Res. B.* 2001,175-177, 468-473.

[127] Song, H. H.; Jones, K. M.; Baski, A. A.; Growth of Ag rows on Si(5 5 12), *J. Vac. Sci. Technol. A.* 1999, 17, 1696-1699.

[128] Sugawara, A.; Coyle, T.; Hembree, G. G.; Scheinfein, M. R.; Self-organized Fe nanowire arrays prepared by shadow deposition on NaCl(110) templates, *Appl. Phys. Lett.* 1997, 70, 1043-1045.

[129] Cepak, V. M.; Martin, C. R.; Preparation and stability of template-synthesized metal nanorod sols in organic solvents, *J. Phys. Chem. B*, 1998, 102, 9985-9990.

[130] Preston, C. K.; Moskovits, M.;, Optical characterization of anodic aluminum oxide films containing electrochemically deposited metal particles. 1. Gold in phosphoric acid anodic aluminum oxide films, *J. Phys. Chem.* 1993, 97, 8495–8503.

[131] Liu, K.; Chien, C. L.; Searson, P. C.; Finite-size effects in bismuth nanowires, *Phys. Rev. B*, 1998, 58, R14681-R14684.

[132] Murphy, C. J.; Jana, N. R.; Controlling the Aspect Ratio of Inorganic Nanorods and Nanowires, *Adv. Mater.* 2002, 14, 80.

[133] Johnson, C. J.; Dujardin, E.; Davis, S. A.; Murphy, C. J.; Mann, S.; *J.* Growth and Form of Gold Nanorods Prepared by Seed-Mediated, Surfactant-Directed Synthesis, *Mater. Chem.* 2002, 12, 1765.

[134] Jana, N. R.; Gearheart, L.; Murphy, C. J.; Seed-Mediated Growth Approach for Shape Controlled Synthesis of Spheroidal and Rodlike Gold Nanoparticles using a Surfactant Template, *Adv. Mater.* 2001, 13, 1389.

[135] Jana, N. R.; Gearheart, L.; Murphy, C. J.; Wet chemical synthesis of silver nanorods and nanowiresof controllable aspect ratio, *Chem. Commun.* 2001, 617.

[136] Jana, N. R.; Gearheart, L.; Murphy, C. J.; Wet Chemical Synthesis of High Aspect Ratio Cylindrical Gold Nanorods, *J. Phys. Chem. B.* 2001, 105, 4065-4067.

[137] Yu, Y. –Y.; Chang, S. S.; Lee, C. –L.; Wang, C. R. C.; Gold nanorods: Electrochemical synthesis and optical properties, *J. Phys. Chem. B*, 1997, 101, 6661-6664.

[138] Bates, F. S.; Fredrickson, G. H.; Block copolymer thermodynamics: Theory and experiment, *Ann. Rev. Phys. Chem.* 1990, 41, 525-557.

[139] Zhang, D.; Qi, L.; Ma, J.; Cheng, H.; Formation of Silver Nanowires in Aqueous Solutions of a Double-Hydrophilic Block Copolymer, *Chem. Mater.* 2001, 13, 2753-2755.

[140] Thelander, C.; Mårtensson, T.; Björk, M. T.; Ohlsson, B. J.; Larsson, M. W.; Wallenberg, L. R.; Samuelson, L.; Single-electron transistors in heterostructure nanowires, *Appl. Phys. Lett.* 2003, 83, 2052.

[141] Alferov, Z .I.; Classical heterostructures paved the way, *III-Vs Review*. 1998, 11, 26-31.

[142] Sze, S. M.; Physics of semiconductor devices, John Wiley and Sons: Canada, 1981.

[143] Ahn, C. H.; Rabe, K. M.; Triscone J.-M.; Ferroelectricity at the Nanoscale: Local Polarization in Oxide Thin Films and Heterostructures, *Science*. 2004, 303, 488-491.

[144] Alivov, Y. I.; Kalinina, E. V.; Cheenkov, A. E.; Look, D. C.; Ataev, B. M.; Omaev, A. K.; Chukichev, M. V.; Bagnall, D. M. Fabrication and characterization of n-ZnO/p-AlGaN heterojunction light-emitting diodes on 6H-SiC substrates, *Applied Physics Letters*. 2003, 83, 4719-4721.

[145] Sze, S. M.; Physics of semiconductor devices; John Wiley and Sons: Hoboken, NJ, 1981,.

[146] Wang, J.; Neaton, J. B.; Zheng, H.; Nagarajan, V.; Ogale, S. B.; Liu, B.; Viehland, D.; Vaithyanathan, V.; Schlom, D. G.; Waghmare, U. V.; Spaldin, N. A.; Rabe, K. M.; Wuttig, M.; Ramesh, R.; Epitaxial BiFeO3 multiferroic thin film heterostructures, *Science*. 2003, 299, 1719.

[147] Alferov, Z. I.; Classical heterostructures paved the way, *III-V Reviews*. 1998, 11, 26.

[148] Hwang, D. -K; Kang, S. –H.; Lim, J. –H.; Yang, E. –J.; Oh, J. –Y.; Yang, J. –H.; Park, S. –J. p-ZnO/n-GaN heterostructure ZnO light-emitting diodes, *Applied Physics Letters*. 2005, 86, 222101/1-222101/3.

[149] Heber, J.; Enter the oxides, *Nature*. 2009, 459, 28-30.

[150] Wang, C. K.; Chiou, Y. Z.; Chang, S. J.; Su, Y. K.; Huang, B. R.; Lin, T. K.; Chen, S. C. AlGaN/GaN metal-oxide semiconductor heterostructure field-effect transistor with photo-chemical-vapor deposition SiO2 gate oxide, *Journal of Electronics Materials*. 2007, 32, 407-410.

[151] Kasap, S.; Principles of Electronic Materials and Devices; McGraw Hill: New York, NY, McGraw Hill, 2000.

[152] Li, Y.; Qian, F.; Xiang, J.; Lieber, C. M.; Nanowire electronic and optoelectronic devices, *Mater. Today*. 2006, 9, 18.

[153] Birenbaum, N. S.; Lai, B. T.; Chen, C. S.; Reich, D. H.; Meyer, G. J.; Selective noncovalent adsorption of protein to bifunctional metallic nanowire surfaces, *Langmuir*. 2003, 19, 9580.

[154] Wu, Y.; Yan, H.; Huang, M.; Messer, B.; Song, J. H.; Yang, P.; Inorganic Semiconductor Nanowires: Rational Growth, Assembly, and Novel Properties, *Chemistry-A European Journal*. 2002, 8, 1260-1268.

[155] Wu, Y.; Yan, H.; Yang, P.; Semiconductor Nanowire Array: Potential Substrates for Photocatalysis and Photovoltaics, *Topics in Catalysis*. 2002, 19, 197-202.

[156] Patolsky, F.; Zheng, G.; Lieber, C. M.; Nanowire-based biosensors, *Analytical Chemistry*. 2006, 4260-4269.

[157] Chopra, N.; Gavalas, V.; Hinds, B.J.; Bachas, L.G.; Functional one-dimensional nanomaterials: Applications in nanoscale biosensors, *Analytical Letters*. 2007, 40, 2067-2096.

[158] Wolf, S.; Tauber, R. N.; Silicon processing for the VLSI Era, Vol. 1: Process technology, Taylor & Francis group: 1991,.

[159] Li, Y.; Qian, F.; Xiang, J.; Lieber, C. M.; Nanowire electronics and optoelectronics, *Materials Today*. 2006, 9, 18-27.

[160] Salem, A. K.; Chao, J.; Leong, K. W.; Searson, P. C.; Receptor-Mediated Self-Assembly of Multicomponent Magnetic Nanowires, *Advanced Materials*. 2004, 16, 268-271.

[161] Dick, K. A.; Kodambaka, S.; Reuter, M. C.; Deppert, K.; Samuelson, L.; Seifert, W.; Wallenberg, L. R.; Ross, F. M.; The Morphology of Axial and Branched Nanowire Heterostructures, *Nano Letters*. 2007, 7, 1817-1822.

[162] Chopra, N.; Bachas, L. G.; Knecht, M. Fabrication and Biofunctionalization of Carbon-Encapsulated Au Nanoparticles, *Chemistry of Materials*. 2009, 21, 1176-1178.

[163] Chopra, N.; Majumder, M.; Hinds, B.J.; Bi-functional carbon nanotubes by sidewall protection, *Advanced Functional Materials*. 2005, 15(5), 858-864.

[164] Caruso, R. A.; Susha, A.; Caruso, F.; Multilayered titania, silica, and laponite nanoparticle coatings on polystyrene colloidal templates and resulting inorganic hollow spheres, *Chem. Mater*. 2001, 13, 400.

[165] Vestal, C. R.; Zhang, Z. J.; Atom transfer radical polymerization synthesis and magnetic characterization of MnFe2O4/polystyrene core/shell nanoparticles, *J. Am. Chem. Soc*. 2002, 124, 14312.

[166] Liu, H. L.; Wu, J. H.; Min, J. H; Kim, Y. K.; Synthesis of monosized magnetic-optical AuFe alloy nanoparticles, *J. Appl. Phys.* 2008, 103, 07D529.

[167] Zhong, C. J.; Maye, M. M.; Core-shell assembled nanoparticles as catalysts, *Adv. Mater.* 2001, 13, 1507.

[168] Guchhait, A.; Rath, A. K.; Pal, A. J.; Hybrid core–shell nanoparticles: Photoinduced electron-transfer for charge separation and solar cell applications, *Chem. Mater.* 2009, 21, 5292.

[169] Hafiz, J.; Mukherjee, R.; Wang, X.; Cullinan, M.; Heberlein, J. V. R.; McMurry, P. H.; Girshick, S. L.; Nanoparticle-coated silicon nanowires, *J. Nanopart. Research,* 2006, 8, 995.

[170] Tak, Y.; Joon Hong, S.; Sung Lee, J.; Yong, K.; Solution-based synthesis of a CdS nanoparticle/ZnO nanowire heterostructure array, *Cryst. Growth Des.* 2009, 9, 2627.

[171] Yang, C. –M.; Kalwei, M.; Schüth, F.; Chao, K. –J.; Gold nanoparticles in SBA-15 showing catalytic activity in CO oxidation, *Appl. Catal. A.* 2003, 253, 289.

[172] Su, H.; Jing, L.; Shi, K.;Yao, C.; Fu, H.; Synthesis of large surface area LaFeO3 nanoparticles by SBA-16 template method as high active visible photocatalysts, *J. Nanopart. Res.* 2009, doi: 10.1007/s11051-009-9647-5.

[173] Jianfeng, Y. E.; Limin, Q. L.; Solution-phase Synthesis of One-dimensional Semiconductor Nanostructures, *J. Mater. Sci. Technol.* 2008, 24, 529.

[174] Thelander, C.; Agarwal, P.; Brongersma, S.; Eymery, J.; Feiner, L. F.; Forchel, A.; Scheffler, M.; Riess, W.; Ohlsson, B. J.; Gösele, U.; Samuelson, L.; Nanowire-based one-dimensional electronics, *Mater. Today*. 2006, 9, 28.

[175] Yan, X.; Tay, B. –K.; Miel, P.; Field emission from ordered carbon nanotube-ZnO heterojunction arrays, *Carbon*. 2008, 46, 753.

[176] Lu, M.; Li, M. K.; Li, H. L.; Guo, X. –Y.; Well-aligned heterojunctions of carbon nanotubes and silicon nanowires synthesized by chemical vapor deposition, *J. Mater. Sci. Lett.* 2003, 22, 1107.

[177] Liu, H.; Cheng, G. –A.; Liang, C.; Zheng, R.; Fabrication of silicon carbide nanowires/carbon nanotubes heterojunction arrays by high-flux Si ion implantation, *Nanotechnology*. 2008, 19, 245606.

[178] Shi, W.; Chopra N.; A simple and surfactant-free route for the synthesis of copper oxide (CuO) nanowire-cobalt oxide (Co3O4) nanoparticle heterostructures with unique optical characteristics, Submitted to a Journal.

[179] Shen, G.; Chen, D.; Bando, Y.; Goldberg, D.; One dimensional nanoscale heterostructures, J. *Mater. Sci. Technol.* 2008, 24, 541.

[180] Choi, J. R.; Oh, S. J.; Ju, H.; Cheon, J. Massive Fabrication of Free-Standing One-Dimensional Co/Pt Nanostructures and Modulation of Ferromagnetism via a Programmable Barcode Layer Effect, *Nano Letters*. 2005, 5, 2179-2183.

[181] Wu, Y.; Yang, P.; Direct observation of vapor–liquid–solid nanowire growth, *J. Am. Chem. Soc.* 2001, 123, 3165.

[182] Hu, J.; Odom, T. W.; Lieber, C. M.; Chemistry and physics in one dimension: Synthesis and properties of nanowires and nanotubes, *Acc. Chem. Res.* 1999, 32, 435.

[183] Law, M.; J. Goldburger, J.; Yang, P.; Semiconductor nanowires and nanotubes, *Annu. Rev. Mater. Res.*,2004, 34, 83.

[184] Kodambaka, S.; Tersoff, J.; Reuter, M. C.; Ross, F. M.; Germanium nanowire growth below the eutectic temperature, *Science*. 2007, 316, 729.

[185] Kodambaka, S.; Tersoff, J.; Reuter, M. C.; Ross, F. M.; Germanium nanowire growth below the eutectic temperature, *Science*, 2007, 316, 729.

[186] Shankar, K. S.; Raychaudhuri, A. K.; Fabrication of nanowires of multicomponent oxides: Review of recent advances, *Mater. Sci. Eng. C.* 2005, 25, 738.

[187] Prasher, R.; Thermal conductivity of hollow nanowires, *The Tenth Intersociety Conference on Thermal and Thermomechanical Phenomena in Electronics Systems, ITHERM '06*, 2006, 1170.

[188] Mazzola, L.; Commercializing nanotechnology, *Nature Biotechnol.* 2003, 21, 1137.

[189] Thostenson, E. T.; Ren, Z.; Chou, T. –W.; Advances in the science and technology of carbon nanotubes and their composites: a review, *Compos. Sci. Technol.* 2001, 61, 1899.

[190] Xu, W. –C.; Takahashi, K.; Matsuo, Y.; Hattori, Y.; Kumagai, M.; Ishiyama, S.; Kaneko, K.; Iijima, S.; Investigation of hydrogen storage capacity of various carbon materials. *Int. J. Hydrogen Energ.* 2007, 32, 2504.

[191] Ajayan, P. M.; Tour, J.; Nanotube composites, *Nature.* 2007, 447, 1066.

[192] Vajtai, R.; Wei, B. Q.; George, T. F.; Ajayan, P. M.; Chemical vapor deposition of organized architectures of carbon nanotubes for applications, *Top. Appl. Phys.* 2007, 109, 188.

[193] Tenne, R.; Margulis, L.; Genut, M.; Hodes, G.; Polyhedral and cylindrical structures of tungsten disulphide, *Nature.* 1992, 360, 444.

[194] Tenne, R.; Inorganic nanotubes and fullerene-like nanoparticles. *Nature.* 2006, 1, 103-111.

[195] Tenne, R.; Advances in the synthesis of inorganic nanotubes and fullerene-Like nanoparticles, *Angew. Chem. Int. Ed.* 2003, 42, 5124.

[196] Jiang, Y.; Wu, Y.; Xie, B.; Zhang, S.; Qian, Y.; Room temperature preparation of novel Cu2−xSe nanotubes in organic solvent, *Nanotechnology.* 2004, 15, 283.

[197] Fan, R.; Karnik, R.; Yue, M.; Li, D.; Majumdar, A.; Yang, P.; DNA translocation in inorganic nanotubes, *Nano Lett.* 2005, 5, 1633.

[198] Tenne, R.; Rao, C. N. R.; Inorganic nanotubes, *Philos. T. Roy Soc Lond.* 2004, 362, 2099.

[199] Rao, C. N. R.; Nath, M.; Inorganic nanotubes, *Dalton T.*, 2003, 1.

[200] Morales, A. M.; Lieber, C. M.; A laser ablation method for the synthesis of crystalline semiconductor nanowires, *Science.* 1998, 279, 208.

[201] Duan, X.; Lieber, C. M.; General synthesis of compound semiconductor nanowires, *Adv. Mater.* 2000, 12, 298.

[202] Thelander, C.; Nilsson, H. A.; Jensen, L. E.; Samuelson, L.; Nanowire single-electron memory, *Nano Lett.* 2005, 5, 635.

[203] Kanungo, P. D.; Zakharov, N.; Bauer, J.; Breitenstein, O.; Werner, P.; Goesele, U.; Controlled in situ boron doping of short silicon nanowires grown by molecular beam epitaxy, *Appl.Phys. Lett.* 2008, 92, 263107.

[204] Huang, M. H.; Mao, S.; Feick, H.; Yan, H.; Wu, Y.; Kind, H.; Weber, E.; Russo, R.; Yang, P.; Room-temperature ultraviolet nanowire nanolasers, *Science.* 2001, 292, 1897.

[205] Lu, X.; Yavuz, M. S.; H.-Y. Tuan, H. –Y.; Korgel, B. A.; Xia, Y.; Ultrathin gold nanowires can be obtained by reducing polymeric strands of oleylamine−AuCl complexes formed via aurophilic interaction, *J. Am. Chem. Soc.* 2008, 130, 8900.

[206] Li, G. –R.; Tong, Y. –X.; Kay, L. –G.; Liu, G. –K.; Electrodeposition of BixFe1-x intermetallic compound nanowire arrays and their magnetic properties, *J. Phys. Chem. B*. 2006, 110, 8965.

[207] Wang, X.; Sun, X. –M.; Yu, D.; Zou, B. –S.; Li, Y.; Rare earth compound nanotubes, *Adv. Mater.* 2003, 15, 1442.

[208] Cui, Y.; Lieber, C. M.; Functional nanoscale electronic devices assembled using silicon nanowire building blocks , *Science*. 2001, 291, 851.

[209] Huang, Y.; Duan, X.; Cui, Y.; Lieber, C. M.; Gallium nitride nanowire nanodevices, *Nano Lett.* 2002, 2, 101.

[210] Duan, X.; Huang, Y.; Lieber, C. M.; Nonvolatile memory and programmable logic from molecule-gated nanowires, *Nano Lett.* 2002, 2, 487.

[211] Salem, A. K.; Chao, J.; Leong, K. W.; Searson, P. C.; Receptor mediated self assembly of multicomponent magnetic nanowires, *Adv. Mater.* 2004, 16, 268.

[212] Loaiza, Q. A.; Laocharoensuk, R.; Burdick, J.; Rodriguez, M. C.; Pingarron, J. M.; Pedrero, M.; Wang, J.; Adaptive orientation of multifunctional nanowires for magnetic control of bioelectrocatalytic processes, *Angew. Chem. Int. Ed.* 2007, 46, 1508.

[213] Wu, Y.; Xiang, J.; Yang, C.; Lu, W.; Lieber, C. M.; Single-crystal metallic nanowires and metal/semiconductor nanowire heterostructures, *Nature*. 2004, 430, 61.

[214] Björk, M. T.; Ohlsson, B. J.; Sass, T.; Persson, A. I.; Thelander, C.; Magnusson, M. H.; Deppert, K.; Wallenberg, L. R.; Samuelson, L.; One-dimensional Steeplechase for Electrons Realized, *Nano Lett.* 2002, 2, 87.

[215] Gudiksen, M. S.; Lauhon, L. J.; Wang, J.; Smith, D. C.; Lieber, C. M.; Growth of nanowire superlattice structures for nanoscale photonics and electronics, *Nature*. 2002, 415, 617.

[216] Hu, J.; Ouyang, M.; Yang, P. Lieber, C. M.; Controlled growth and electrical properties of heterojunctions of carbon nanotubes and silicon nanowires, *Nature*, 1999, 399, 48.

[217] Wu, Y.; Xiang, J.; Yang, C.; Lu, W.; Lieber, C. M.; Single-crystal metallic nanowires and metal/semiconductor nanowire heterostructures, *Nature*. 2004, 430, 61.

[218] Lensch-Falk, J. L.; Hemesath, E. R.; Lauhon, L. J.; Syntaxial growth of Ge/Mn-germanide nanowire heterostructure, *Nano Lett.* 2008, 8, 2669.

[219] Clark, T. E.; Nimmatoori, P.; Lew, K. –K.; Pan, L.; Redwing, J. M.; Dickey, E. C.; Diameter dependent growth rate and interfacial abruptness in vapor-liquid-solid Si/Si1-xGex heterostructure nanowires, *Nano Lett.* 2008, 8, 1246.

[220] Dick, K. A.; Kodambaka, S.; Reuter, M. C.; Deppert, K.; Samuelson, L.; Seifert, W.; Wallenberg, R.; Ross, F. M.; The morphology of axial and branched nanowire heterostructures, *Nano Lett.* 2007, 7, 1817.

[221] Zakharov, N. D.; Werner, P.; Gerth, G.; Schubert, L.; Sokolov, L.; Gosele, U.; Growth phenomena of Si and Si/Ge nanowires on Si (1 1 1) by molecular beam epitaxy, *J. Cryst. Growth*. 2006, 290, 6.

[222] Swadener, J. G.; Picraux, S. T.; Strain distribution and electronic property modification in Si/Ge axial nanowire heterostructures, *J. Appl. Phys.* 2009, 105, 044310.

[223] Ou, F. S.; Shaijumon, M. M.; Ajayan, P. M.; Controlled manipulation of giant hybrid inorganic nanowire assemblies, *Nano Lett.* 2008, 8, 1853.

[224] Ye, H.; Lu, P.; Yu, Z.; Song, Y.; Wang, D.; Wang, S.; Critical thickness and radius for axial heterostructure nanowires using finite-element method, *Nano Lett.* 2009, 9, 1921.

[225] Wu, Y.; Fan, R.; Yang, P.; Block-by-block growth of single-crystalline Si/SiGe superlattice nanowires, *Nano Lett.* 2002, 2, 83.
[226] Bauer, J.; Gottschalch, V.; Paetzelt, H.; Wagner, G.; VLS growth of GaAs/InGaAs/GaAs axial double heterostructure nanowire by MOVPE, *J. Crystal Growth*. 2008, 310, 5106.
[227] Krogstrup, P.; Yamasaki, J.; Sorensen, C. B.; Johnson, E.; Wagner, J. B.; Pennington, R.; Aagesen, M.; Tanaka, N.; Nygard, J.; Junctions in axial III-V heterostructure nanowires obtained visa an interchange of group III elements, *Nano Lett.* 2009, 9, 3689.
[228] . Dheeraj, D. L.; Patriarche, G.; Zhou, H.; Harmand, J. C.; Weman, H.; Fimland, B. O.; Growth and structural characterization of GaAs/GaAsSb axial heterostructured nanowires, *J. Crystal Growth*. 2009, 311, 1847.
[229] Heib, M.; Gustafsson, A.; Conesa-Boj, S.; Peiro, F.; Morante, J. R.; Abstreiter, G.; Arbiol, J.; Samuelson, L.; Morral, A. F.; Catalyst-free nanowires with axial InxGa1-xAs/GaAs heterostructures, *Nanotechnology*. 2009, 20, 075603.
[230] Mose, A. F.; Hoang, T. B.; Dheeraj, D. L.; Zhou, H. L.; van Helvoort, A. T. J.; Fimland, B. O.; Weman, H.; Micro-photoluminescence study of single GaAsSb/GaAs radial and axial heterostructured core-shell nanowires, *IOP Conf. Series: Mater. Sci. Eng.* 2009, 6, 012001.
[231] Paladugu, M.; Zou, J.; Guo, Y. –N.; Zhang, X.; Kim, Y.; Joyce, H. J.; Gao, Q.; Tan, H. H.; and Jagadish, C.; Nature of heterointerfaces in GaAs/InAs and InAs/GaAs axial nanowire heterostructures, *Appl. Phys. Lett.* 2008, 93, 101911.
[232] Paladugu, M.; Zou, J.; Guo, Y. –N.; Zhang, X.; Joyce, H. J.; Gao, Q.; Tan, H. H.; Jagadish, C.; Kim, Y.; Crystallographically driven Au catalyst movement during growth of InAs/GaAs axial nanowire heterostructures, *J. Appl. Phys.* 2009, 105, 073503.
[233] Kanungo, P. D.; Wolfsteller, A.; Zakharov, N. D.; Werner, P.; Gösele, U.; Enhanced electrical properties of nominally undoped Si/SiGe heterostructure nanowires grown by molecular beam epitaxy, *Micorelect. J.* 2009, 40, 452.
[234] Kaibe, H.; Nemoto, T.; Sato, T.; Sakata, M.; Nishida, I.; Composition and physical properties of the most germanium-rich germanide of manganese, *J. Less Com. Metals*. 1988, 138, 303.
[235] Yamada, N.; Atomic Magnetic moment and exchange interaction between Mn atoms in intermetallic compounds in Mn-Ge system, *J. Phys. Soc. Jpn.*,1990, 59, 273.
[236] Weber, W. M.; Geelhaar, L.; Unger, E.; Cheze, C.; Kreupl, F.; Riechert, H.; Lugli, P.; Silicon to nickel-silicide axial nanowire heterostructures for high performance electronics, *Phys. Stat. Sol. (b)*. 2007, 244, 4170.
[237] Liu, J.; Li, X.; Dai, L.; Water-assisted growth of aligned carbon nanotube-ZnO heterojunction arrays, *Adv. Mater.* 2006, 18, 1740.
[238] Hwang, S. –K.; Lee, J.; Jeong, S. –H.; Lee, P. –S.; Lee, K. –H.; Fabrication of carbon nanotube emitters in an anodic aluminium oxide nanotemplate on a Si wafer by multi-step anodization, *Nanotechnology*. 2005, 16, 850.
[239] Meng, G.; Jung, Y. J.; Cao, A.; Vajtai, R.; Ajayan, P. M.; Controlled fabrication of hierarchically branched nanopores, nanotubes, and nanowires, *Proc. National Acad. Sci. USA*. 2005, 102, 7074.

[240] Shaijumon, M. M.; Ou, F. S.; Ci, L.; Ajayan, P. M.; Synthesis of hybrid nanowire arrays and their application as high power supercapacitor electrodes, *Chem. Comm.* 2008, 2373.

[241] Yang, D.; Meng, G.; Zhang, S.; Hao, Y.; Au, X.; Wei, Q.; M. Ye, M.; and Zhang, L.; Electrochemical synthesis of metal and semimetal nanotube-nanowire heterojunctions and their electronic properties, *Chem. Comm.* 2007, 1733.

[242] Ji, C.; Searson, P. C.; Fabrication of nanoporous gold nanowires, *Appl. Phys. Lett.* 2002, 81, 4437.

[243] Hermanson, G. T.; Bioconjugate techniques, Elsevier Science: San Diego, CA, 1996.

[244] Wildt, B.; Mali, P.; Searson, P. C.; Electrochemical template synthesis of multisegment nanowires: Fabrication and protein functionalization, *Langmuir*. 2006, 22, 10528.

[245] Chen, M.; Chien, C. –L.; Searson, P. C.; Potential modulated multilayer deposition of multisegment Cu/Ni nanowires with tunable magnetic properties, *Chem. Mater.* 2006, 18, 1595.

[246] Wang, A. A.; Lee, J.; Jenikova, G.; Mulchandani, A.; Myung, N. V.; W.; Controlled assembly of multi-segment nanowires by histidine-tagged peptides, *Nanotechnology*. 2006, 17, 3375.

[247] Fond, A. M.; Birenbaum, N. S.; Felton, E. J.; Reich, D. H.; Meyer, G. J.; Preferential noncovalent immunoglobulin G adsorption onto hydrophobic segments of multifunctional metallic nanowires, *J. Photochem. Photobio. A: Chem.* 2007, 186, 57.

[248] Stoermer, R. L.; Keating, C. D.; Distance-dependent emission from dye-labeled oligonucleotides on striped Au/Ag nanowires: Effect of secondary structures and hybridization efficiency, *J. Amer. Chem. Soc.* 2006, 128, 13243.

[249] Reich, D. H.; Tanase, M.; Hultgren, A.; Bauer, L. A.; Chen, C. S.; Meyer, G. J.; Biological applications of multifunctional magnetic nanowires, *J. Appl. Phys.* 2003, 93, 7275.

[250] Bangar, M. A.; Hangarter, C. M.; Yoo, B.; Rheem, Y.; Chen, W.; Mulchandani, A.; Myung, N. V.; Magnetically assembled multisegmented nanowires and their applications, *Electroanalysis*. 2009, 21, 61.

[251] Wang, X.; Ozkan, C. Z.; Multisegment nanowire sensors for the detection of DNA molecules, *Nano Lett.* 2008, 8, 398.

[252] Zhao, A. W.; Meng, G. W.; Zhang, L. D.; Gao, T.; Sun, S. H.; Pang, Y. T.; Electrochemical synthesis of ordered CdTe nanowire arrays, *Appl. Phys. A: Mater. Sci. Processing*. 2003, 76, 537.

[253] Burdick, J.; Alonas, E.; Huang, H. –C.; Rege, K.; Wang, J.; High-throughput template multisegment synthesis of gold nanowires and nanorods, *Nanotechnology*. 2009, 20, 065306.

[254] Howard, J.; Mechanics of motor proteins and the cytoskeleton, *Appl. Mech. Rev.* 2002, 55, B39.

[255] Hirokawa, N.; Kinesin and dynein superfamily proteins and the mechanism of organelle transport, *Science*. 1998, 279, 519.

[256] Vale, R. D.; Milligan, R. A.; The Way Things Move: Looking Under the Hood of Molecular Motor Proteins, *Science*. 2000, 288, 88.

[257] Hess, H.; Clemmens, J.; Qin, D.; Howard, J.; Vogel, V.; Light-controlled molecular shuttles made from motor proteins carrying cargo on engineered surfaces, *Nano Lett.* 2001, 1, 235.

[258] Veigel, C.; Coluccio, L. M.; Jontes, J. D.; Sparrow, J. C.; Milligan, R. A.; Molloy, J. E.; Themotor proteinmyosin-I produces itsworking stroke in two steps, *Nature*. 1999, 398, 530.

[259] Rice, S.; Lin, A. W.; Safer, D.; Hart, C. L.; Naberk, N.; Carragher, B. O.; Cain, S. M.; Pechatnikova, E.; Wilson-Kubalek, E. M.; Whittaker, M.; Patel, E.; Cookek, R.; Taylor, E. W.; Milligan, R. A.; Vale, R. D.; A structural change in the kinesin motor protein that drives motility, *Nature*. 1999, 402, 778.

[260] Dittrich, M.; Hayashi, S.; Schulten, K.; On the mechanism of ATP hydrolysis in F1-ATPase, *Biophys. J.* 2003, 85, 2253.

[261] Ren, Q.; Zhao, Y. –P.; Yue, J. C.; Cui, Y. B.; Biological application of multicomponent nanowires in hybrid devices powered by F1-ATPase motors, *Biomed. Microdevices*. 2006, 8, 201.

[262] Hirsch, R.; Katz, E.; Willner, I.; Magneto-switchable bioelectrocatalysis, *J. Am. Chem. Soc.*,2000, 122, 12053.

[263] White, B. J.; Harmon, H. J.; Novel optical solid-state glucose sensor using immobilized glucose oxidase, Biochem. *Biophys. Res. Comm.* 2002, 296, 1069.

[264] Wang, J.; Adaptive nanowires for on-demand control of electrochemical microsystems, *Electroanalysis*. 2008, 20, 611.

INDEX

A

B

C

D

E

F

G

H

I

J

K

L

M

N

O

P

Q

R

S

T

U

V

W

X

Y

Z